100 PROBLEMS

in Celestial Navigation

Self-Contained—with Answers

Leonard Gray

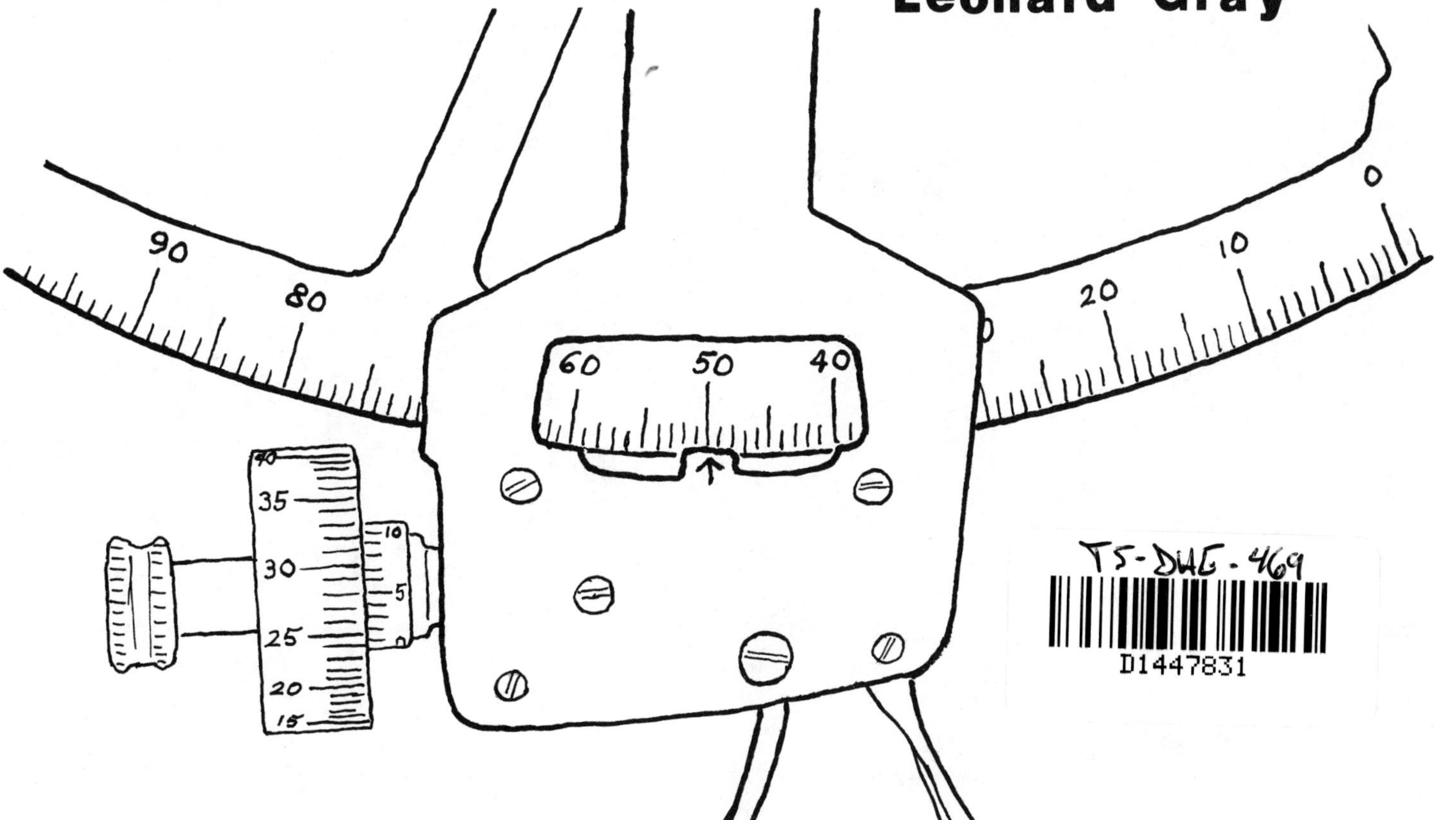

100 Problems In Celestial Navigation

Leonard Gray

PUBLISHED JOINTLY BY:

Paradise Cay Yacht Sales
Post Office Box 1351
Middletown, California 95461
Tel: 800-736-4509

Celestaire, Inc
416 S. Pershing
Wichita, Kansas 67218
Tel: 800-727-9785

ISBN
0-939837-14-5

Library of Congress Catalog Card Number
29-62315

Contents

Introduction

Navigation can be thought of as an attempt to reduce location uncertainties and evaluate the remaining ones. Since dealing with errors is a big part of the navigator's job, sighting inaccuracies and an occasional blunder have been included in these problems to make them authentic. Things go wrong, as in real life. When this happens, the reader should try reworking the problem using a different method—say, H.O. 249 Vol. 2, instead of Vol. 1—and should not rule out time or sighting mistakes.

Good preparation is important in navigation, and is included here, as in a real situation. Readers are urged to figure out twilight times and what bodies to shoot, as well as working the sights.

A navigator with a computer and a navigation program can do similar twilight planning for a cruise, at home before leaving, using two guesses—one for a boat speed slower than expected and the other for a slightly faster one—and put the printouts in a loose-leaf notebook for use at sea.

These problems can be worked with either the *Nautical Almanac* (or the yachting equivalent) or the *Air Almanac*. For sight reduction, the easiest method is H.O. 249 Vol. 1, and Vol. 2 or 3. (Vols. 2 and 3 overlap at 40° latitude.) H.O. 229 Vol. 3 can be used, if preferred—or any other method, such as H.O. 208 or H.O. 211—or a computer program or navigation calculator. (H.O. 249 and H.O. 229 have been renamed Pub. No. 249 and Pub. No. 229, but most navigators still use the old designations.) *Nautical Almanac* and H.O. 249 excerpts are included in the appendices.

The problems can be done in any order, but 1-1 has a few hints that are not repeated for the others. The exercises are realistic from a navigation standpoint, but any political reasons for avoiding certain destinations are ignored.

The usual conventions are followed—for example, if neither limb is stated for a sun sight, the lower limb is assumed; but LL or UL is always specified for the moon. Times, unless otherwise stated, are GMT.

Celestial navigation is not difficult, but it has lots of routines and conventions to remember—and many opportunities to go wrong. It helps to use as many bodies as possible during a round of sights—at least six—and, for sun and moon shots, three or more of each body.

In rounding off, some people (including the author) deal with .5 by dropping the .5 if it's preceded by an even number, but raising if the .5 is preceded by an odd number. So 34.5 would round to 34, and 35.5 would round to 36. This improves the chances of rounding errors canceling each other.

The invention of the marine chronometer must have seemed like a miracle to eighteenth-century British navigators. Each one cost a substantial sum, and was treated like a jewel, with rituals for winding and protecting it. Now we have an even greater marvel, the digital watch, and it may be that only navigators appreciate it properly. (The one the author wears cost $15, and loses about a second a month.) Three of them should be taken on a cruise, each tagged with its weekly error (having been checked for a month or so in advance). One should be set to boat time, and changed as time zones are crossed, and the others set to GMT. They should be compared every day or so to the constant time signals from WWV (on 20, 15, 10, or 5 MHz) or CHU (14.667, 7.335, or 3.333 MHz), or the hourly ones from the BBC or another station.

The problems in this book have been validated with the following computer programs:

1. PC Navigator (Pacific Marine Corp., 213-433-7815)—simple to use, superior menu; performs most navigation functions.
2. J.Henry (Somerset Publishing Co., 315-466-0614)—comprehensive; works all types of navigation problems.
3. Floppy Almanac (U. S. Naval Observatory; sold by U.S. Government Printing Office, Washington, D.C. 20402)—low cost, with many sight-reduction and astronomical routines for navigational and other bodies (including 200 stars). Computes Hc, Zn, GHA, and altitude corrections. Does not get fixes, DR positions, etc.

The sight-reduction form in Appendix F may be copied for personal, non-commercial use.

Chapter 1

New York to Flores (I)

Problem 1-1
We are sailing from New York to the Azores in a 37-foot ketch. It is early in the morning of Thursday, May 13, 1993, and we are up before daylight to prepare a plan for morning twilight. We are on a course of 090°, making 5.5 knots in a light following wind. Stars are visible, and an occasional thin cloud crosses the moon. Our DR position at 0730 (GMT) will be 40° 10′N, 50° 15′W.

Planning for a round of twilight sights makes things much easier at sighting time. Then, for each body, the navigator presets the sextant scale to the predicted altitude, turns to the predicted azimuth (after converting true to magnetic), and, if all goes well, spots the body without much searching. Identification errors are avoided, and the sighting goes a lot faster.

From the right-hand daily page of the *Nautical Almanac*, we find the LHA of nautical twilight, civil twilight, and sunrise. On the first yellow page in the almanac, we note that 50° 15′ longitude equals 3h 21m time, which we add to each LMT to get GMT. (We do all of our navigating and plotting with GMT.)

	LHA	GHA
Nautical twilight	0337	0658
Civil twilight	0414	0735
Sunrise	0446	0807

This table gives us the time to be ready for sights (0658), the middle of the sighting period (0735), and the time when stars will no longer be visible (0807).

Next we need the LHA γ at 0414 LMT. The easiest way to get this is to look on the left-hand white page of the almanac, under *GMT* (not LMT), and take out the *GHA* γ (not LHA γ). Adding to this the 14-minute increment from the yellow pages (3° 31′) gives 294° (rounded to the nearest degree).

From H.O. 249 Vol. 1, for latitude 40°N, on the line for LHA γ 294°, we list for our sighting plan the predicted altitude (rounded) and azimuth of each of the seven stars. From the *Nautical Almanac* separator page we get the magnitudes of these stars.

Now we need to add the moon and planets to the plan. For each one, we get RA body = GHA γ - GHA body (with RA stated in degrees). Then we use the red template to mark their positions on the base of the star finder, 2102D. (It helps to add the sun, to be sure that it shows just below the horizon at sighting time.) We replace the red template with the blue one for 35°N or 45°N, turn it to our LHA γ at sighting time, 294°, and record the expected altitudes and azimuths for the moon and planets. We check one or two stars to see that their 2102D altitudes and azimuths approximate those found in H.O. 249 Vol. 1.

Now we make up our twilight plan, listing the bodies in sighting order, from dimmest to brightest:

MAG.	BODY	ALT.	AZ.
2.5	Enif	49°	128°
2.2	Alpheratz	34°	079°
2.2	Kochab	43°	339°
2.1	Rasalhague	52°	233°
1.9	Alkaid	31°	311°
0.9	Altair	59°	173°
0.2	Arcturus	20°	279°
-4.6	Venus	14°	097°
-	Moon	34°	148°

We'd prefer to shoot Alpheratz, Kochab, Rasalhague, Alkaid, Altair, and Arcturus, for good azimuth spacing, and because all these sights can be worked with H.O. 249 Vol. 1.

We are on deck at 0700 GMT with the sextant, our digital watch set to GMT, and a small hand-bearing compass to check azimuths. When the horizon becomes visible, Enif and Alpheratz have faded, so we start with Kochab. We get sights as follows:

BODY	GMT	Hs
Kochab	07-33-45	43°23ʹ.8
Rasalhague	07-35-16	51°05ʹ.2
Alkaid	07-37-15	30°15ʹ.9
Altair	07-39-02	58°38ʹ.0
Venus	07-41-24	15°15ʹ.3
Moon (LL)	07-44-08	34°05ʹ.6

Our height of eye was 7 feet, and the sextant I.C. was -1ʹ.2. We work these sights with H.O. 249 Vol. 1 and Vol. 3 (or with H.O. 229, or an electronic method). The plot looks good (although not perfect), and we pick the center of the figure. What was our May 13 0744 fix? How reliable do we consider it?

Problem 1-2

Later Thursday morning, May 13, 1993, our course is still 090°, speed 5.5. Our last fix, at 0744 this morning, was at 40°14ʹN, 49°58ʹW. We note, on the right daily page of the *Nautical Almanac*, that the moon is near last quarter—so we can get a sun-moon fix between morning and early afternoon. The height of eye is again 7 feet, but the I.C. this time is -1ʹ.0. The increased waves are giving us some pitching, but we get sights as follows:

BODY	GMT	Hs
Sun	12-20-04	46° 42′.2
Sun	12-21-31	47° 02′.6
Sun	12-23-06	47° 21′.8
Moon (UL)	12-26-14	24° 54′.7
Moon (UL)	12-27-49	24° 42′.4
Moon (UL)	12-29-16	24° 33′.4

What is our May 13 1229 fix? How reliable is it?

Problem 1-3
Our last fix, at 1229 May 13, 1993, was 40° 13′N, 49° 19′W. The boat is pitching badly, but we'd like an afternoon fix. Our course is still 090°, speed 5.5. Since the moon is down, we will try for more sun sights to cross with the 12-21-31 sun sight (whose LOP was the middle of the three). The height of eye is still 7 feet, but the I.C. is again -1′.2. Results are:

BODY	GMT	Hs
Sun	12-21-31	47° 02′.6
Sun	15-25-17	67° 53′.5
Sun	15-26-36	67° 57′.6
Sun	15-28-02	67° 50′.2

What is our May 13 1222-1528 running fix? How reliable is it?

Problem 1-4
Our last reliable fix, at 1229 May 13, 1993, was 40° 13′N, 49° 19′W. There is still some pitching as we approach evening twilight, and some clouds are moving in. We decide to plan for a round of sights anyway. Course and speed are still 090°/5.5, the I.C. is now -1′.4, and the height of eye still 7 feet. Our plan is similar to the one for morning twilight.

	LHA	GHA
Sunset	1908	2220
Civil twilight	1939	2251
Nautical twilight	2017	2329

LHA γ at 1939 LMT = 166°. As before, we get the predicted altitudes and azimuths for selected stars from H.O. 249 Vol. 1, and for the other bodies from the Star-Finder 2102D. For evening twilight we list them from brightest to dimmest, the order in which we will sight them:

MAG.	BODY	ALT.	AZ.
-1.9	Jupiter	46°	152°
0.2	Capella	29°	307°
0.2	Arcturus	44°	103°
0.5	Procyon	32°	247°
1.2	Pollux	48°	270°
1.2	Spica	29°	140°
1.3	Regulus	59°	208°
2.1	Polaris	40°	359°
2.2	Kochab	47°	020°

Our preferred bodies from this list are Capella, Arcturus, Procyon, Spica, Regulus, and Polaris.

At sunset, the sea is still rough, and the sky is about half covered with cirrocumulus, but we decide to do our best. We spot Jupiter in the southeast and get a fair shot. When Capella and Arcturus become visible, we shoot them quickly, but we can't find Procyon, Pollux, or any of the others on the plan. We need something in the northeast or southwest—Kochab or Procyon—but they aren't visible, and the horizon is fading. We find Polaris, hastily preset the sextant scale to 40° and shoot, but that's the last star we see. Results are:

BODY	GMT	Hs
Jupiter	22-35-30	44° 07.'8
Capella	22-48-11	29° 42.'0
Arcturus	22-50-10	44° 23.'5
Polaris	22-51-47	39° 42.'9

What is our May 13 2252 fix? Is it usable?

Problem 1-5

From our May 13, 1993, 2252 fix, at 40° 09'N, 48° 02'W, we continue on course 090° at 5.5 knots. Our I.C. is -1.'2, height of eye still 7 feet. Early in the morning, May 14, 1993, the visibility has improved, and we prepare for a round of sights at A.M. twilight.

	LMT	GMT
Nautical twilight	0338	0646
Civil twilight	0415	0723
Sunrise	0446	0754

The LHA γ at 0414 LMT is 296°. Using the star-finder and H.O. 249 Vol. 1, we make up the sighting plan:

MAG.	BODY	ALT.	AZ.	
2.5	Enif	50°	130	X
2.2	Alpheratz	36°	080°	
2.2	Kochab	43°	339°	
2.1	Rasalhague	50°	235°	
1.9	Alkaid	30°	311°	
0.9	Altair	59°	177°	
0.2	Arcturus	18°	280°	
-4.6	Venus	14°	096°	X
-	Moon	32°	135°	X

We have more bodies than we need on this list. We will probably skip Enif (whose azimuth is close to Alkaid's reciprocal), Venus (close to Arcturus's reciprocal), and the moon (close to Alkaid's reciprocal).

By the time the horizon appears, some of the dimmer stars have faded, and the first one we shoot is Rasalhague. We spot Altair, then can see only Venus and the moon. Results are:

BODY	GMT	Hs
Rasalhague	07-18-45	51°09'.4
Altair	07-20-01	58°40'.1
Altair	07-21-12	58°44'.9
Venus	07-23-02	14°36'.9
Venus	07-23-55	14°47'.3
Moon (LL)	07-26-37	32°39'.8

What is our May 14 0727 fix? How reliable is it?

Problem 1-6

From our May 14, 1993, 0727 fix, at 40°09'N, 46°46'W, we continue on 090°/5.5, with I.C. -1.2, and height of eye 7 feet. We decide to get a sun-moon fix.

Results are:

BODY	GMT	Hs
Sun	12-15-30	48°26'.4
Sun	12-18-00	48°55'.0
Sun	12-19-41	49°15'.2
Moon (UL)	12-22-06	33°11'.4
Moon (UL)	12-23-31	32°54'.9
Moon (UL)	12-25-03	32°42'.3

What is our May 14 1225 fix?

Problem 1-7

There is a discussion among the crew. A and B say we should get a noonsight, but C says the noonsight is archaic—from the days before accurate time was available at sea—and H.O. 249 is so easy to use that the simple noonsight computations are no real advantage. Also, a running fix on the sun would give more information. B argues that the noonsight can give longitude as well as latitude. The decision is made to get a noonsight and see how it works out.

The navigator has compared our DR positions with our fixes for the last two days, and decided that we should have been allowing for an eastward-setting current of about 1 knot. We will use 6.5 knots, instead of 5.5, as our estimated speed since the last fix (40° 09′N, 46° 04′W, at 1225 May 14, 1993). Our I.C. is still −1′.2, height of eye 7 feet.

To plan for the noonsight, we need to estimate the GMT of LAN at our DR position at the time of LAN.

LMT of LAN	1200	(by definition)
Eq. of time	−0004	(from daily page; add if meridian passage is after 1200; subtract if before)
Longitude (W)	+0302	(time equivalent of 45° 30′—a guess; need not be precise)
GMT of LAN	1458	

The noonsight formulas are:

d contrary to L: L = 90° − Ho − d
d same name, <L: L = 90° − Ho + d
d same name, >L: L = Ho + d − 90°

The middle formula is the one we want. Rearranging and solving for Ho (or, in this case, the predicted altitude), we get:

$$90° - 40° 09' + 18° 43' = 68° 34'$$

Ten or fifteen minutes before 1500 GMT, we set our sextant to about 69° and go on deck. The boat is steady, moving nicely on a reach, and the sky is clear. We shoot every minute, and when the altitude begins to level off we shoot and read the scale as rapidly as we can. Results are:

GMT	Hs
14-53-04	68° 19′.3
14-54-10	68° 20′.6
14-54-32	68° 21′.9
14-55-41	68° 22′.5
14-56-40	68° 23′.0
14-57-24	68° 23′.7
14-58-15	68° 23′.7
14-59-08	68° 23′.6
14-59-52	68° 23′.8
15-00-38	68° 23′.1
15-02-37	68° 22′.1
15-03-55	68° 21′.4

The best way to find the greatest Hs is to plot GMT on the x axis and Hs on the y axis of a piece of graph paper, then draw a curve through the points, smoothing out the sighting errors. Our curve shows Hs to be about 68°23!8, at 14-59-07.

An Hs of 68°23!8 results in an Ho of 68°35!6. With the same formula we used for planning, we solve for L this time, and get 90° - 68°35!6 + 18°43!6 = 40°08!0 (north latitude).

Finding the longitude is simple. The GMT of LAN (the time at the highest point of our curve) was 14-59-07. At that time, the GHA of the sun was its longitude—therefore *our* longitude. (In east longitude, we would use 360° - GHA of the sun.) The almanac gives this figure as 45°42!4.

What is our May 14 noonsight-determined position? How reliable is it?

Problem 1-8

Our last good fix, May 14, 1993, at 1225, was 40°09′N, 46°04′W. Our course is still 090°, speed 6.5, I.C. -1!2, height of eye 7 feet. Well before sunset, we plot our expected DR position at the time of yesterday's GMT of P.M. civil twilight as an approximation to tonight's, and use this position to calculate tonight's predicted times:

	LMT	GMT
Sunset	1907	2158
Civil twilight	1939	2230
Nautical twilight	2017	2308

The LHA γ at 1939 LMT is 167°. We make up our twilight plan as before, using H.O. 249 Vol. 1 for stars and the marked-up star-finder for other bodies.

MAG.	BODY	ALT.	AZ.
-1.9	Jupiter	46°	154°
0.2	Capella	29°	308°
0.2	Arcturus	45°	103°
0.5	Procyon	32°	248°
1.2	Pollux	47°	270°
1.2	Spica	30°	141°
1.3	Regulus	59°	210°
2.1	Polaris	40°	359°
2.2	Kochab	47°	019°

We skipped Mars because it's within 1° of Procyon and not as bright, and added Polaris to include a body near 000° azimuth.

Two crew members will be shooting tonight, at the same time, and they agree to divide the list. No. 1 chooses Pollux, Regulus, and Polaris, and No. 2 picks Jupiter, Capella, Procyon, and Kochab. The results are:

Navigator No. 1

BODY	GMT	Hs
Pollux	22-36-06	46°54′.5
Regulus	22-37-48	58°49′.1
Polaris	22-39-39	39°40′.6
Regulus	22-43-02	58°15′.7

Navigator No. 2

BODY	GMT	Hs
Jupiter	22-30-04	45°21′.3
Capella	22-35-12	29°00′.2
Procyon	22-37-50	31°25′.9
Kochab	22-39-44	47°54′.6

What is Navigator No. 1's May 14 2243 fix? How reliable is it?

What is Navigator No. 2's May 14 2240 fix? How reliable is it?

Problem 1-9

Since our last fix, 40°07′N, 44°37°W, at 2243 May 14, we continue on course 090°, speed 6.5. Our I.C. is still -1′.2, and height of eye 7 feet. We figure the A.M. twilight times:

	LMT	GMT
Nautical twilight	0337	0631
Civil twilight	0414	0708
Sunrise	0446	0740

We find that the LHA γ at 0414 LMT is 296°. Our twilight plan is:

MAG.	BODY	ALT.	AZ.	
2.2	Alpheratz	36°	080°	
2.2	Kochab	43°	339°	
2.1	Rasalhague	50°	235°	
1.9	Alkaid	30°	311°	
0.9	Altair	59°	177°	
0.6	Saturn	28°	140°	
0.2	Arcturus	18°	280°	
-4.6	Venus	14°	096°	X
-	Moon	30°	122°	

The azimuth of Venus is close to the reciprocal of Arcturus, so we will probably skip it. We go on deck at 0630, and find that the dimmer stars are obscured by haze, although we can see the moon and planets and Arcturus. Luckily, the horizon becomes visible, and we get two shots of Saturn. But we can't find Arcturus, so we get two each of Venus and the Moon.

BODY	GMT	Hs
Saturn	07-16-31	28° 42'.6
Saturn	07-19-10	29° 03'.1
Venus	07-21-08	17° 08'.8
Venus	07-22-41	17° 20'.6
Moon (LL)	07-26-03	32° 26'.2
Moon (LL)	07-27-53	32° 46'.1

What is our May 15 0728 fix? How reliable is it?

Problem 1-10

From our May 15, 1993, 0728 position, 40° 08'N, 43° 24'W, with our course and speed unchanged at 090° and 6.5, I.C. -1'.2, and height of eye 7 feet, we will plan a morning sun-and-moon fix. We get the following:

BODY	GMT	Hs
Sun	12-29-30	53° 29'.7
Sun	12-30-59	53° 47'.1
Sun	12-32-25	54° 01'.8
Moon (UL)	12-35-17	38° 51'.2
Moon (UL)	12-36-51	38° 37'.1
Moon (UL)	12-38-16	38° 25'.8

What is our May 15 1238 fix? How reliable is it?

Problem 1-11

After our sun-moon fix, 40° 07'N, 42° 41'W, at 1238 May 15, 1993, the wind shifted and dropped. We are now on course 080°, speed 4.0. Our I.C. is still -1'.2, and height of eye 7 feet. The crew member who argued for the noonsight yesterday wants to take another one today, and has pointed out that it isn't necessary to have the exact time. Using the same formula we used yesterday, we can plan for the expected altitude at LAN: Hp = 90° - L + d. But we know that the altitude won't be much different today, so we assume that the Hs will be about 68° 24'. (Our latitude will be greater—therefore the sun's altitude will be lower—but not by much.)

We go on deck with the sextant and watch the sun. As its altitude begins to level off, we start shooting. An assistant gives us a "Mark" at one-minute intervals, and when the sun approaches its peak, we record each one—just the sextant reading, without the time.

SUN'S Hs
68° 37.'4
68° 37.'7
68° 38.'3
68° 38.'6
68° 38.'5
68° 38.'5
68° 38.'0
68.'37.'3

What is our latitude at LAN May 15? How reliable is it? Was the crew member correct in saying that we could get a noonsight without knowing the time?

Problem 1-12

Our DR plot from 40° 07'N, 42° 41'W, at 1238 May 15, 1993, is 080°, 4.0 knots. Our I.C. is -1.'0, and height of eye 7 feet. P.M. twilight times, from the almanac, are:

	LMT	GMT
Sunset	1908	2155
Civil twilight	1939	2226
Nautical twilight	2017	2304

The LHA γ at 1939 LMT is 169°. Our twilight plan is:

MAG.	BODY	ALT.	AZ.
-1.9	Jupiter	46°	155°
0.2	Capella	28°	308°
0.2	Arcturus	46°	105°
0.5	Procyon	30°	249°
1.2	Pollux	45°	271°
1.2	Spica	31°	143°
1.3	Regulus	58°	213°
2.1	Polaris	40°	359°
2.2	Kochab	48°	019°

We skipped Mars, because it has about the same azimuth as Procyon, and isn't as bright. We added Polaris because we can use a body near 000° azimuth. Here's what we get:

BODY	GMT	Hs
Jupiter	22-30-35	46° 39.'6
Capella	22-32-01	27° 17.'3
Arcturus	22-33-44	47° 24.'1
Procyon	22-35-16	29° 03.'4
Regulus	22-36-45	57° 15.'4
Polaris	22-38-18	39° 46.'0

What is our May 15 2238 fix? How reliable is it?

Chapter 2

Wellington to Valparaiso

Problem 2-1

We left Wellington, New Zealand, on Friday, and it is now December 18, 1993. From our fix at 1600 December 17, 40°55′S, 176°05′E, we changed course to 070°, making 7.5 knots. Our height of eye is 9 feet, and our sextant I.C. is + 0′.5. We decide to get a late-afternoon sun-moon fix.

BODY	GMT	Hs
Sun	05-31-17	20°40′.1
Sun	05-32-47	20°23′.0
Sun	05-34-30	20°01′.7
Moon (LL)	05-37-00	50°43′.0
Moon (LL)	05-38-41	50°32′.3
Moon (LL)	05-40-42	50°23′.9

What is our December 18 0541 fix? How reliable is it?

Problem 2-2

From our December 18, 1993, 0541 fix at 40°35′S, 178°18′E, we continue on 070° at 7.5 knots, with height of eye 9 feet and I.C. + 0′.5, and make plans for a round of sights at P.M. twilight. We find the LMT of civil twilight for 40°S from the almanac, 2001, and make a rough estimate of what our longitude will be around then: 179°E. The time equivalent of this is 11 hours 56 minutes, which (since we are in east longitude) we subtract from the LMT, to get 0805 GMT. Then, to get more accurate figures for the twilight times, we plot our DR position at 0805 (40°29′S, 178°40′E), and enter the almanac again. Now we get:

	LMT	GMT
Sunset	1930	0735
Civil twilight	2003	0808
Nautical twilight	2045	0850

We find from the almanac that the LHA γ at 2003 LMT is 28°, and with this we enter H.O. 249 Vol. 1 and the star finder, and make up a twilight plan.

MAG.	BODY	ALT.	AZ.	
-	Moon	27°	285°	
-1.6	Sirius	23°	093°	
-0.9	Canopus	43°	129°	
0.3	Rigel	35°	069°	
0.7	Saturn	33°	280°	X
1.1	Aldebaran	22°	043°	
1.3	Fomalhaut	53°	272°	X
2.1	Peacock	37°	223°	X
2.2	Hamal	26°	004°	
2.2	Alpheratz	17°	336°	

We added Sirius to the H.O. 249 Vol. 1 list, because it's so bright. We will probably not use Fomalhaut or Saturn, whose azimuths are close to the reciprocal of Sirius's, or Peacock, whose azimuth is close to the reciprocal of Aldebaran's.

The sky is clear at sunset, and we get a moon shot, then the stars, as they appear.

BODY	GMT	Hs
Moon (LL)	07-49-13	30° 02!7
Sirius	08-01-35	21° 58!0
Canopus	08-03-12	42° 44.4
Rigel	08-04-52	34° 00!8
Aldebaran	08-06-27	21° 12!3
Hamal	08-08-04	26° 00!2
Alpheratz	08-09-39	16° 29!3

What is our December 18 0810 fix? How reliable is it?

Problem 2-3

From our December 18, 1993, 0810 fix at 40° 30'S, 178° 41° E, we are making 6.8 knots on course 070°, with the wind on our starboard quarter. After a few hours' sleep, we plan for a round of A.M. twilight sights. Our sextant I.C. is still + 0!5, and our height of eye 9 feet.

The crew argued through supper about what will happen when we cross the international date line. The line runs down the 180th meridian, then jogs to the east, so that it's at 172° 30'W as it passes New Zealand. Some of the crew are not sure what day it will be when we cross the line. There are jokes about the A Team skipping a day on watch while the B Team serves two days in a row.

The navigator asserts that crossing the line will have no effect on the time or date in GMT (which we are using for navigation). But the LMT and LZT—and therefore the boat time—will change. In a few hours, when we cross from east longitude to west longitude, the local hour will not change, but the local day will suddenly go from Sunday to Saturday. The navigator will relabel Sunday's watch schedule as Saturday, because, in local time, while yesterday was Saturday, today will also be Saturday.

At twilight, we will still be in east longitude (but barely), and the times will be:

	LMT	GMT
Nautical twilight	0311 (12/19)	1512 (12/18)
Civil twilight	0353 (12/19)	1554 (12/18)
Sunrise	0426 (12/19)	1627 (12/18)

Note that if we had used a DR position a few minutes away, in west longitude, to compute these times, the result in the GMT column would have been the same as we got above.

The LHA γ at 0353 (12/19) LMT is 146°. We use H.O. 249 Vol. 1 and the star finder 2102D to get predicted altitudes and azimuths of the best bodies to sight, as usual.

MAG.	BODY	ALT.	AZ.	
1.3	Regulus	38°	007°	
1.2	Spica	34°	075°	X
1.1	Acrux	57°	147°	X
0.5	Procyon	36°	320°	
0.5	Betelgeuse	19°	298°	X
0.3	Rigel	22°	278°	X
-0.9	Canopus	54°	232°	
-1.3	Jupiter	24°	087°	
-1.6	Sirius	45°	288°	

We will probably not use Rigel or Betelgeuse, whose azimuths are close to Sirius; Acrux, close to the reciprocal of Procyon; or Spica, close to Jupiter; but we will leave them in the plan. (Note that we might have elected to sight only H.O. 249 Vol. 1 stars, for easy sight-reduction, skipping Jupiter and Sirius. The choice depends partly on the visibility.)

A little after 1500 GMT, we go on deck and prepare to shoot Regulus as soon as the horizon appears. We get the following sights:

BODY	GMT	Hs
Regulus	15-44-01	37° 15'.6
Procyon	15-45-39	37° 30'.1
Canopus	15-48-02	55° 20'.3
Jupiter	15-49-50	22° 54'.2
Sirius	15-51-35	45° 35'.3

What is our December 18 1552 fix? How reliable is it?

Problem 2-4

During the morning, still on course 070°, speed 6.8, we decide to get a running fix on the sun. Our I.C. remains at +0'.5 and height of eye 9 feet. We start a new DR from our 1552 fix, at 40° 10′S, 179° 53′E, remembering that last night when we converted the almanac's LMT twilight times to GMT, the date changed from Sunday, December 19, to Saturday, December 18. Another way of saying this is that when we crossed the international date line this morning from east longitude to west, the GMT date didn't change, but the LMT date (and zone-time date) changed from Sunday, December 19, to Saturday, December 18.

We shoot the sun before LAN, then get more sights when it has passed the meridian and is about 90° in azimuth from where we shot it earlier. Our results are:

	GMT	Hs
Sun	22-05-08 (12/18)	61° 35.′9
Sun	22-06-42 (12/18)	61° 52.′8
Sun	22-08-20 (12/18)	62° 09.′3
Sun	00-30-13 (12/19)	71° 30.′3
Sun	00-31-47 (12/19)	71° 23.′9
Sun	00-33-05 (12/19)	71° 14.′3

What is our 2208 (12/18)-0033 (12/19) running fix? How reliable is it?

Problem 2-5

From our 0033 December 19, 1993, fix, at 39° 48′S, 178° 49′W, we continue on course 070°, speed 6.8. Our I.C. is still +0.′5, and height of eye 9 feet. We decide to get an afternoon sun-moon fix. The results are:

BODY	GMT	Hs
Sun	02-12-43	55° 47.′7
Sun	02-14-16	55° 29.′2
Sun	02-15-50	55° 15.′5
Moon (UL)	02-18-40	43° 21.′8
Moon (UL)	02-20-08	43° 30.′9
Moon (UL)	02-21-38	43° 42.′7

What is our December 19 0222 fix? How reliable is it?

Problem 2-6

After the December 19, 1993, 0222 sun-moon fix, at 39° 44′S, 178° 31′W, we changed course to 095°, and our speed dropped to 5.5 knots. Our height of eye is still 9 feet and our I.C. still +0.′5. Our P.M. twilight-time plan is:

	LMT	GMT
Sunset	1927 (12/18)	0719 (12/19)
Civil twilight	1959 (12/18)	0751 (12/19)
Nautical twilight	2040 (12/18)	0832 (12/19)

Our LHA γ at 1959 LMT is 27°, and our twilight plan, made up using H.O. 249 Vol. 1 and the star finder, is:

MAG.	BODY	ALT.	AZ.
-	Moon	32°	297°
-1.6	Sirius	23°	093°
-0.9	Canopus	43°	130°
0.3	Rigel	35°	070°
1.1	Aldebaran	22°	043°
2.1	Peacock	37°	222°
2.2	Hamal	27°	004°
2.2	Alpheratz	18°	337°

Note that we skipped Fomalhaut (listed in H.O. 249 Vol. 1 for this LHA and location) and substituted Sirius. We might have added Saturn, but it has a similar azimuth and is not as bright as Sirius.

We go on deck ready to shoot just after sunset, and are disappointed to find that cirrus clouds have moved in. We get an early shot of the moon, then search for anything else as the sky darkens. We are able to shoot Sirius and Peacock, but no other bodies are visible. We repeat the moon shot, but now we can no longer see Sirius.

BODY	GMT	Hs
Moon (LL)	07-35-16	34°31.′3
Sirius	07-47-18	22°38.′5
Peacock	07-50-31	36°43.′7
Moon (LL)	07-52-03	31°39.′5

What is our December 19 0752 fix? How reliable is it?

Problem 2-7

At the time of our 0752 December 19, 1993, fix, at 39°46′S, 177°45′W, we changed course and speed to 085°/7.0. Our height of eye is still 9 feet, but our I.C. is now 0. We plan our A.M. twilight sights:

	LMT	GMT
Nautical twilight	0314	1500
Civil twilight	0355	1541
Sunrise	0428	1614

The LHA γ at 0355 LMT is 146°. Our sighting plan is:

MAG.	BODY	ALT.	AZ.
1.3	Regulus	38°	007°
1.2	Spica	34°	075°
1.1	Acrux	57°	147°
0.5	Procyon	36°	320°
0.5	Betelgeuse	19°	298°
0.3	Rigel	22°	278°
-0.9	Canopus	54°	232°
-1.3	Jupiter	25°	087°
-1.6	Sirius	44°	286°

When we go on deck a little after 1500, the sky is completely clear, and the boat is gliding along smoothly on course. We have an assistant to record for us, and as soon as the horizon appears, we start shooting rapidly. We are pleased to be able to include every body on our list.

BODY	GMT	Hs
Regulus	15-36-03	38° 06:8
Spica	15-37-21	33° 49:1
Acrux	15-38-58	56° 44:3
Procyon	15-40-33	36° 16:6
Betelgeuse	15-42-15	18° 29:6
Rigel	15-43-48	21° 17:6
Canopus	15-45-29	52° 34:3
Jupiter	15-47-01	25° 48:3
Sirius	15-48-40	42° 51:5

What is our December 19 1549 fix? How reliable is it?

Problem 2-8

We run a new DR from 39° 38′S, 176° 27′W, at 1549 December 19, 1993. Our course is 090°, speed 6.0, height of eye still 9 feet, and I.C. 0. We get sights for an afternoon sun-moon fix as follows:

BODY	GMT	Hs
Sun	01-15-17	63° 30:5
Sun	01-16-50	63° 13:8
Moon (UL)	01-20-04	26° 25:3
Moon (UL)	01-21-44	26° 40:1

What is our December 20 0122 fix? How reliable is it?

Problem 2-9

From 39° 45′S, 175° 02′W, at 0122 December 20, 1993, we change course to 105°, and our speed becomes 4.0 knots. Our height of eye is 9 feet, and I.C. 0. We plan for a round of P.M. twilight sights.

	LMT	GMT
Sunset	1928 (12/19)	0706 (12/20)
Civil twilight	2001 (12/19)	0739 (12/20)
Nautical twilight	2042 (12/19)	0820 (12/20)

The LHA γ at 2001 LMT is 29°, and our twilight plan is:

MAG.	BODY	ALT.	AZ.
-	Moon	35°	310°
-1.6	Sirius	24°	092°
-0.9	Canopus	44°	129°
0.3	Rigel	36°	068°
1.1	Aldebaran	22°	042°
2.1	Peacock	36°	222°
2.2	Hamal	27°	003°
2.2	Alpheratz	17°	336°

A few clouds are moving in, so we shoot the moon early, and Sirius as soon as it's visible. We get the rest of the sights through breaks in the clouds.

BODY	GMT	Hs
Moon (LL)	07-15-20	38° 07:7
Sirius	07-21-04	20° 50:5
Canopus	07-38-17	43° 50:9
Rigel	07-45-13	18° 50:2
Peacock	07-46-49	35° 06:0
Hamal	07-48-42	26° 47:0

What is our December 20 0749 fix? How reliable is it?

Problem 2-10
We continue from our last fix, 39°53'S, 174°28'W, at 0749 December 20, 1993, with course and speed still 105° and 4.0, I.C. 0, and height of eye 9 feet. From the radio forecast, we expect cirrus clouds to build up, changing to cirrostratus and altostratus, but we plan for A.M. twilight sights, just in case. We take yesterday's A.M. civil twilight time (1541) as a rough guide, extend our DR, and find more exact times as follows:

	LMT	GMT
Nautical twilight	0312	1447
Civil twilight	0354	1529
Sunrise	0427	1602

The LHA γ at 0354 LMT is 147°. With this we enter H.O. 249 Vol. 1, and we set our star finder (having already pencilled in the sun, moon, and planet positions on the base), for the plan:

MAG.	BODY	ALT.	AZ.
1.3	Regulus	38°	006°
1.2	Spica	34°	074°
1.1	Acrux	57°	148°
0.5	Procyon	36°	319°
0.5	Betelgeuse	18°	297°
0.3	Rigel	22°	278°
-0.9	Canopus	53°	232°
-1.4	Jupiter	25°	086°
-1.6	Sirius	44°	286°

At the time of nautical twilight, the dimmer stars are not visible, but we see some of the brighter ones, so we stay on deck with the sextant as the horizon becomes visible. We shoot Betelgeuse, then three of the brightest bodies, and, with nothing else available repeat our sights of Sirius and Jupiter.

BODY	GMT	Hs
Betelgeuse	15-35-01	16° 54.′9
Jupiter	15-36-35	26° 36.′8
Canopus	15-38-07	51° 38.′3
Sirius	15-39-54	41° 38.′3
Sirius	15-41-00	41° 24.′5
Jupiter	15-42-26	27° 40.′2

What is our December 20 1542 fix? How reliable is it?

Chapter 3

Anchorage to Hilo

Problem 3-1
We are moving along smoothly on course 185° in the late afternoon of May 14, 1993, on a glassy sea. Our speed has dropped to 3.6 knots. The forecast is for cirrostratus clouds to move in, but we decide to plan for a round of P.M. twilight sights, anyway. Our DR position at 0100 May 14 was 40°41'N, 153°12'W. Our sextant I.C. is -1!0, and height of eye 8 feet.

Using a rough guess of 40°N and 153°W, we enter the almanac and take out 1938 for the LMT of civil twilight. Adding the time equivalent of 153° (1012) from the first yellow page, we get 0550 for the GMT of civil twilight. We get a refined DR position, at 0550, of 40°24'N, 153°14'W, and for this position get a complete set of more accurate twilight times:

	LMT	GMT
Sunset	1908 (5/13)	0521 (5/14)
Civil twilight	1940 (5/13)	0553 (5/14)
Nautical twilight	2018 (5/13)	0631 (5/14)

The LHA γ at 1940 LMT is 167°. Using the star finder and H.O. 249 Vol. 1, we make up a sighting plan:

MAG.	BODY	ALT.	AZ.	
-1.9	Jupiter	46°	153°	
0.2	Capella	29°	308°	
0.2	Arcturus	45°	103°	
0.5	Procyon	32°	248°	
1.2	Pollux	47°	270°	
1.2	Spica	30°	141°	X
1.3	Regulus	59°	210°	
2.2	Kochab	47°	019°	X

We mark, as not very desirable, Spica (close to Jupiter's azimuth and Capella's reciprocal) and Kochab (close to the reciprocal of Regulus), but leave them on the list in case the more favorable bodies are obscured. We omit Mars, which has the same azimuth as Procyon, and is not as bright.

At sighting time, the visibility is, in fact, poor, but we manage to shoot a few of the brighter bodies:

BODY	GMT	Hs	
Jupiter	05-55-04	45°48!0	
Pollux?	05-57-12	30°47!5	(about 250° az.)
Capella	05-58-49	28°15!9	
Arcturus	05-59-13	46°06!0	
Jupiter	06-00-31	46°15!1	
Pollux?	06-02-15	29°51!5	(about 250° az.)

What was the unknown body we shot twice? What is our May 14 0602 fix? How reliable is it?

Problem 3-2

We have someone wake us early in the morning of May 14 so we can make up a plan for A.M. twilight, hoping the sky will be clear by then. Our May 14, 1993, 0602 fix was at 40°31′N, 153°08′W, and our course and speed are still 185°/3.6, I.C. -1!0, and height of eye 8 feet. Our planned times (using a first guess, then refining it, as before) are:

	LMT	GMT
Nautical twilight	0338	1351
Civil twilight	0415	1428
Sunrise	0446	1459

The LHA γ at 0415 LMT is 296°. Our sighting plan is:

MAG.	BODY	ALT.	AZ.	
2.5	Enif	50°	130°	X
2.2	Alpheratz	36°	080°	
2.2	Kochab	43°	339°	
2.1	Rasalhague	50°	235°	
1.9	Alkaid	30°	311°	X
0.9	Altair	59°	177°	
0.6	Saturn	28°	140°	X
0.2	Arcturus	18°	280°	X
-4.6	Venus	15°	097°	
-	Moon	32°	131°	

We have more bodies than we need. We list them all, but mark those that we probably won't use because their azimuths are close to those of other bodies or their reciprocals: Enif (same azimuth as moon), Alkaid (reciprocal of moon), Saturn (close to moon), and Arcturus (close to reciprocal of Venus).

Around the time of civil twilight we sight the following:

BODY	GMT	Hs
Alpheratz	14-28-17	36°05!1
Kochab	14-29-52	42°40!1
Rasalhague	14-31-20	49°54!2
Altair	14-33-01	58°56!1
Venus	14-37-04	16°18!6
Moon (LL)	14-39-33	33°15!8

What is our May 14 1440 fix? How reliable is it?

Problem 3-3

From our May 14, 1993, 1440 fix, at 40°01'N, 153°09'W, we continue on course 185°, but now our speed indicator is not functioning. Crew members' estimates range from 2.5 to 5 knots—but we can do better than that. We will drop a balled-up tissue at the bow and time its passage to the stern. Our boat is known to be 33 feet on the water line, and we use 6080 feet in a nautical mile, and 3600 seconds in an hour, to write a formula for computing the boat's speed through the water:

$$\text{Knots} = (.592 \times \text{feet}) \div \text{seconds}$$

We time the passage of the object from bow to stern at 5 seconds; this give us: $(.592 \times 33) \div 5 = 3.9$ knots, our speed estimate. Our I.C. is still -1:0, and height of eye 8 feet. We get the following sights for a morning sun-moon fix:

BODY	GMT	Hs
Moon (UL)	20-01-13	32° 00:4
Moon (UL)	20-02-51	31° 42:7
Moon (UL)	20-04-40	31° 29:4
Sun	20-07-30	56° 19:0
Sun	20-08-56	56° 31:3
Sun	20-09-45	56° 43:1

What is our May 14 2010 fix? How reliable is it?

Chapter 4

New York to Flores (II)

Problem 4-1
It is early in the morning of July 6, 1993, and we are headed for the Azores. The sky has been overcast for a week, and we badly need a fix. Light winds have alternated with flat calms and hot, humid weather, straining the gear and making the crew queasy. But now a cold front has passed, and we are moving nicely at 6 knots on course 090°. Our 0700 DR position was 40°00′N, 51°20′W. The sextant I.C. is 0, and the height of eye 9 feet. We compute the times for A.M. twilight:

	LMT	GMT
Nautical twilight	0325	0650
Civil twilight	0406	0731
Sunrise	0438	0803

The LHA γ at 0406 LMT is 346°. Our sighting plan is:

MAG.	BODY	ALT.	AZ.
2.2	Kochab	30°	345°
2.2	Diphda	28°	153°
1.3	Fomalhaut	20°	182°
0.9	Altair	37°	248°
0.4	Saturn	36°	196°
0.2	Capella	25°	050°
0.1	Vega	40°	292°
-3.9	Venus	24°	087°
-	Moon	28°	217°

This is an excellent plan, with nine bodies to sight, nicely spaced in azimuth. We omitted Aldebaran from the H.O. 249 Vol. 1 recommendations (it's close to Venus in azimuth), and added the solar-system bodies from the star finder.

But when we go on deck with the sextant at nautical twilight, the boat's motion is unsteady, and spray is coming aboard now and then. We stay, anyway, to see what we can get. We manage to shoot one star and the brighter bodies:

BODY	GMT	Hs
Altair	07-41-13	34°52!1
Saturn	07-43-10	34°40!0
Venus	07-44-45	27°21!7
Moon (LL)	07-47-36	25°25!3
Moon (LL)	07-51-12	25°02!4
Moon (LL)	07-59-53	23°53!7

What is our July 6 0800 fix?

Problem 4-2
From our July 6, 1993, 0800 fix, at 40°11'N, 50°31'W, we continue on 090° at 6.0 knots, with the same I.C. (0) and height of eye (9 feet). A few hours after sunrise, the boat's motion is steadier, and we plan for a running fix on the sun. Then we have a better idea. We use the star finder to determine that Venus is far enough in azimuth from the sun to give us a fix, if we can spot it in daylight. We pre-set the sextant to the predicted altitude, and go on deck with a hand-bearing compass. We search a few degrees up and down, left and right, but don't see Venus. We try to relax our eyes to focus for distance, then suddenly we see the planet. We shoot it, then the sun:

BODY	GMT	Hs
Venus	13-58-17	58°28!3
Venus	13-59-51	58°15!6
Venus	14-01-42	57°59!2
Sun	14-04-17	65°41!2
Sun	14-05-47	65°54!7
Sun	14-07-21	66°09!6

What is our July 6 1407 fix?

Problem 4-3
From our July 6, 1993, 1407 fix, at 40°10'N, 49°27'W, we continue on course 090° at 6.0 knots. Our I.C. is still 0, and our height of eye 9 feet. We make plans for P.M. twilight:

	LMT	GMT
Sunset	1932	2245
Civil twilight	2005	2318
Nautical twilight	2046	2349

The LHA γ at 2005 LMT is 226°. Our sighting plan is:

MAG.	BODY	ALT.	AZ.	
-1.5	Jupiter	35°	231°	
0.1	Vega	50°	074°	
0.2	Arcturus	67°	210°	
0.9	Altair	20°	095°	
1.2	Spica	34°	210°	X
1.2	Antares	21°	160°	
1.3	Deneb	31°	055°	X
2.0	Dubhe	48°	322°	
2.1	Polaris	39°	000°	
2.1	Rasalhague	47°	119°	

Note that we included Spica and Deneb in the list, but will probably not shoot them, because of duplications in azimuth. We skipped Mars, whose azimuth is close to Vega's reciprocal, and we added Polaris, because we need a body near 000° or 180°. This is a good list, with bodies whose azimuths look like spokes in a wheel.

At sighting, time, alas, we get sights of the three brightest bodies, but as the horizon fades, we can't find any of the dimmer objects. We can still see Jupiter, so we shoot it again.

BODY	GMT	Hs
Jupiter	23-14-17	35° 16.′6
Vega	23-15-52	49° 39.′7
Arcturus	23-17-40	66° 28.′1
Jupiter	23-19-03	34° 37.′5

This is a disappointing result from such an ambitious plan. Can we get a usable fix from these few shots?

Problem 4-4

From our July 6, 1993, 2319 fix, 40° 08′N, 47° 50′W, we proceed on course 090° at 6.0 knots, with I.C. still 0 and height of eye 9 feet. We make up a plan for A.M. twilight. The times are:

	LMT	GMT
Nautical twilight	0324	0631
Civil twilight	0405	0712
Sunrise	0437	0744

The LHA γ at 0405 is 346°. The sighting plan is:

MAG.	BODY	ALT.	AZ.
2.2	Kochab	30°	345°
2.2	Diphda	28°	153°
1.3	Fomalhaut	20°	182°
0.9	Altair	37°	248°
0.2	Capella	26°	050°
0.1	Vega	40°	292°
-3.9	Venus	24°	086°
-	Moon	36°	208°

We skipped Saturn and Aldebaran, because of duplications in azimuth. Note that, except for the different twilight times and the slightly different predicted altitude and azimuth for the moon, we could have used yesterday's plan instead of making up a new one for today.

Just before civil twilight, the horizon becomes visible, and sighting conditions are perfect. We shoot every body we had in the plan.

BODY	GMT	Hs
Kochab	07-09-13	30°25:0
Diphda	07-10-54	27°42:0
Fomalhaut	07-12-30	20°19:0
Altair	07-14-10	36°13:3
Capella	07-15-45	26°48:6
Vega	07-17-42	38°09:9
Venus	07-20-03	25°38:5
Moon (LL)	07-22-29	35°07:1

What is our July 7 0722 fix?

Problem 4-5

From our 0722 July 7, 1993, fix, 40°04′N, 46°36′W, we continue on course 090° at 6.0 knots, with I.C. 0 and height of eye 9 feet. Using the almanac and the star-finder, we note that we have a rare opportunity to get a three-body fix during the day—sun, moon, and Venus—if the visibility is good. We find predicted altitudes and azimuths, and in a clear sky we spot and shoot Venus, and proceed. Our results are:

BODY	GMT	Hs
Venus	09-40-16	52°04:8
Venus	09-43-12	52°36:1
Moon (UL)	09-45-53	16°07:3
Moon (UL)	09-47-31	15°52:1
Sun	09-50-11	21°36:4
Sun	09-51-44	21°55:8

What is our July 7 0952 fix?

Problem 4-6

From our 0952 July 7, 1993, fix, at 40°03′N, 46°12′W, we continue on course 090° at 6.0 knots. Our I.C. is still 0, and height of eye 9 feet. For P.M. twilight, we compute the following times:

	LMT	GMT
Sunset	1931	2229
Civil twilight	2004	2302
Nautical twilight	2045	2343

The LHA γ at 2004 LMT is 227°. We could use these times and yesterday's sighting plan (list of predicted altitudes and azimuths), since the changes are slight, but here is the way the new plan would look:

MAG.	BODY	ALT.	AZ.	
-1.5	Jupiter	35°	232°	
0.1	Vega	50°	074°	
0.9	Altair	21°	096°	
1.2	Spica	34°	211°	
1.2	Antares	21°	161°	
1.3	Deneb	31°	055°	X
2.0	Dubhe	48°	322°	
2.1	Polaris	39°	000°	
2.1	Rasalhague	48°	120°	
2.2	Denebola	40°	254°	X

We added Vega (because of its brightness) and Polaris (because of its azimuth). We listed Deneb and Denebola, but will probably not sight them (because of azimuth duplications), unless we have plenty of time.

At sighting time, conditions are as good as they were for A.M. twilight, and we again shoot all the bodies in the plan.

BODY	GMT	Hs
Jupiter	22-49-13	36° 20:8
Vega	22-50-48	48° 25:1
Altair	22-52-29	18° 54:6
Spica	22-54-02	34° 23:1
Antares	22-56-01	20° 43:0
Deneb	22-57-45	30° 43:4
Dubhe	22-59-12	48° 07:9
Polaris	23-00-50	39° 19:2
Rasalhague	23-02-25	47° 57:5
Denebola	23-04-09	39° 18:5

What is our July 7 2304 fix?

Problem 4-7

From our July 7, 1993, 2304 fix, at 39°59′N, 44°12′W, we continue on course 090° at 6.0 knots. Our I.C. is still 0, and height of eye 9 feet. We plan for A.M. twilight sights.

	LMT	GMT
Nautical twilight	0325	0618
Civil twilight	0406	0659
Sunrise	0438	0731

The LHA γ at 0406 LMT is 348°. Our plan is similar to yesterday's:

MAG.	BODY	ALT.	AZ.
2.2	Kochab	30°	345°
2.2	Diphda	28°	155°
1.3	Fomalhaut	20°	183°
0.9	Altair	36°	249°
0.2	Capella	27°	051°
0.1	Vega	38°	292°
-3.9	Venus	24°	086°
-	Moon	44°	197°

At twilight, the sky is clear, but the horizon is obscured, except toward the southwest. We shoot the few bodies available.

BODY	GMT	Hs
Fomalhaut	07-10-17	20° 10ʹ.1
Altair	07-12-03	33° 16ʹ.1
Saturn	07-13-48	34° 09ʹ.1
Moon (LL)	07-19-16	42° 00ʹ.5

What is our July 8 0719 fix?

Problem 4-8

From our 0719 July 8, 1993, fix, at 40°00′N, 42°58′W, we continue on course 090° at 6.0 knots. Our height of eye is still 9 feet, but we check the sextant and find that its I.C. is now -2ʹ.0. We'd like another fix when the haze clears. The moon and sun are nearly opposite in the sky, so we won't get a sun-moon fix. We will have to rely on a running fix of the sun.

When the sun rises, the horizon under it is still obscured, and stays that way all morning, although it is clear in the opposite direction. We decide to get a backsight—point the sextant toward the horizon opposite the sun, and measure the angle from there past the zenith to the sun. What we will be measuring is an angle that equals the altitude of the sun minus 180°. Therefore, the body we are backsighting must be high enough so that 180° - Hs is not beyond the range of our sextant (120° or so). We get three backsights of the sun, and three hours later, when the sun has moved to the southwest, get three regular lower-limb sights.

BODY	GMT	Hs
Sun (backsight)	14-10-05	109° 31ʹ.6
Sun (backsight)	14-11-46	109° 19ʹ.5
Sun (backsight)	14-13-23	109° 13ʹ.6
Sun	17-01-39	57° 07ʹ.3
Sun	17-03-20	56° 46ʹ.9
Sun	17-04-55	56° 30ʹ.0

There are two ways to work a backsight. We will use the easier, more logical one: apply the normal I.C. and dip corrections to Hs, subtract the result from 180°, then apply the

S.D. correction, but in reverse. If the sight appeared to be on the lower limb (as in this case), use the UL correction. If it looked like an upper-limb sight, use the LL correction. (For a star or a planet, apply the normal refraction correction in this third step.) What is our July 8 1413-1705 R. fix?

Problem 4-9
We'd like another fix, to confirm the backsights we took, so we plan for evening twilight. From our July 8, 1993, 1705 fix, at 39°59'N, 41°27'W, we continue on course 090°, speed 6.0 knots. Our height of eye is still 9 feet, and I.C. -2.′0. We find the twilight times:

	LMT	GMT
Sunset	1931	2214
Civil twilight	2004	2247
Nautical twilight	2045	2328

The LHA γ at 2004 LMT is 228°. We make up our sighting plan:

MAG.	BODY	ALT.	AZ.
-1.5	Jupiter	34°	232°
0.9	Altair	21°	096°
1.2	Spica	33°	212°
1.2	Antares	21°	162°
1.3	Deneb	32°	055°
2.0	Dubhe	47°	322°
2.1	Polaris	39°	000°
2.1	Rasalhague	48°	121°
2.2	Denebola	39°	255°

Our sights are:

BODY	GMT	Hs
Altair	22-40-03	20° 04.′1
Spica	22-41-49	33° 49.′9
Antares	22-43-40	21° 14.′0
Dubhe	22-45-21	47° 34.′7
Polaris	22-46-59	39° 13.′9
Rasalhague	22-48-55	48° 50.′5

What is our July 8 2249 fix?

Chapter 5

Furneaux Islands to Wanganui

Problem 5-1
It is early afternoon (local time) of December 18, 1993—a beautiful clear day in late spring—and we are making good time in a calm sea, a third of the way in our cruise from the Furneaux Group, off the north coast of Tasmania, to the North Island of New Zealand. Our 0300 December 18 DR position was at 40° 18′S, 157° 02′E. Our course is 090°, speed 7.5, sextant I.C. -1:0, and height of eye 8 feet. We decide to get an afternoon sun-and-moon fix. Results are:

BODY	GMT	Hs
Moon (LL)	05-15-44	56° 47:8
Moon (LL)	05-17-20	56° 51:8
Moon (LL)	05-18-41	56° 53:3
Sun	05-21-03	38° 17:5
Sun	05-22-37	37° 58:1
Sun	05-24-12	37° 39:0

What is our December 18 0524 fix?

Problem 5-2
From our 0524 December 18, 1993, fix, at 40° 18′S, 157° 26′E, we proceed on 090° at 7.5 knots, with the same I.C. (-1.0) and height of eye (8 feet). We plan for a round of sights at P.M. twilight, and find the following times:

	LMT	GMT
Sunset	1928	0856
Civil twilight	2002	0930
Nautical twilight	2043	1011

The LHA γ at 2002 LMT is 28°. Our sighting plan is:

MAG.	BODY	ALT.	AZ.	
-	Moon	28°	286°	
-1.6	Sirius	23°	093°	
-0.9	Canopus	43°	129°	
0.3	Rigel	35°	069°	
1.1	Aldebaran	22°	043°	
1.3	Fomalhaut	53°	272°	
2.1	Peacock	37°	223°	X
2.2	Hamal	26°	004°	
2.2	Alpheratz	17°	336°	

Peacock's azimuth is the reciprocal of Aldebaran's, so we probably won't use it. At sighting time, we get a few sights of the moon and Sirius, but conditions are so good that we shoot six of the seven stars listed in H.O. 249 Vol. 1, and decide to work these for our fix.

BODY	GMT	Hs
Canopus	09-41-17	44° 55′.8
Rigel	09-43-01	36° 57′.7
Aldebaran	09-44-40	23° 24′.5
Fomalhaut	09-46-16	50° 19′.2
Hamal	09-47-50	26° 21′.0
Alpheratz	09-49-32	15° 13′.5

What was our December 18 0950 fix?

Problem 5-3

From our December 18, 1993, 0950 fix, at 40° 17′S, 158° 06′E, we continue on 090° at 7.5 knots, I.C. -1′.0, height of eye 8 feet, and plan for A.M. twilight.

	LMT	GMT
Nautical twilight	0311 (12/19)	1634 (12/18)
Civil twilight	0353 (12/19)	1716 (12/18)
Sunrise	0426 (12/19)	1749 (12/18)

The LHA γ at 0353 (12/19) LMT is 146°. Our sighting plan is:

MAG.	BODY	ALT.	AZ.	
1.3	Regulus	38°	007°	
1.2	Spica	34°	075°	
1.1	Acrux	57°	147°	X
0.5	Procyon	36°	320°	
0.5	Betelgeuse	19°	298°	
0.3	Rigel	22°	278°	
-0.9	Canopus	54°	232°	
-1.3	Jupiter	24°	087°	
-1.6	Sirius	45°	288°	

Acrux's azimuth is close to the reciprocal of Procyon's, so we will probably skip it.

Our sights are:

BODY	GMT	Hs
Regulus	17-20-35	37° 34′.3
Spica	17-22-04	34° 25′.8
Procyon	17-23-44	35° 31′.2
Canopus	17-25-16	53° 09′.9
Jupiter	17-26-50	25° 42′.5
Sirius	17-29-01	42° 46′.9

What is our December 18 1729 fix?

Problem 5-4
We plan for a sun-and-moon fix today, noting that the moon has moved closer to the sun since yesterday. Since our December 18, 1993, 1729 fix, at 40°17′S, 159°17′E, we are still on course 090° at 7.5 knots, and our I.C. is still -1.′0 and height of eye 8 feet. We get a few sun shots first, and wait two hours before sighting the moon, so the LOPs will cross at a better angle.

BODY	GMT	Hs
Sun	03-04-10	61° 12.′7
Sun	03-05-50	60° 51.′8
Sun	03-07-41	60° 32.′4
Moon (LL)	05-15-06	51° 55.′6
Moon (LL)	05-16-45	51° 58.′7
Moon (LL)	05-18-29	52° 02.′7

What is our December 19 0308-0518 R. fix?

Problem 5-5
From our December 19, 1993, 0518 fix, at 40°16′S, 161°06′E, we continue on 090°, making 7.5 knots, I.C. -1.′0, height of eye 8 feet, and plan for a round of sights at P.M. twilight. We find the twilight times, as usual, interpolating in the almanac.

	LMT	GMT
Sunset	1929	0842
Civil twilight	2002	0915
Nautical twilight	2044	0957

The LHA γ at 2002 LMT is 29°. Except for the moon, we could use last night's plan, but here is the new one:

MAG.	BODY	ALT.	AZ.	
-	Moon	32°	298°	
-1.6	Sirius	24°	092°	X
-0.9	Canopus	44°	129°	
0.3	Rigel	36°	068°	
1.1	Aldebaran	22°	042°	
1.3	Fomalhaut	53°	272°	
2.1	Peacock	36°	222°	X
2.2	Hamal	27°	003°	
2.2	Alpheratz	17°	336°	

We left Sirius and Peacock in the plan, in case we need them, but will probably not use them, because of azimuth duplications. (We prefer Fomalhaut to Sirius, because we can work the Fomalhaut sight with H.O. 249 Vol. 1.)

At sighting time, there's a haze, and we may not get any stars or planets. We get a good moon sight, then luckily are able to find Sirius and Canopus. No other bodies are visible by the time the horizon has faded.

BODY	GMT	Hs
Moon (LL)	09-10-15	32° 21:0
Sirius	09-20-13	24° 47:4
Canopus	09-21-52	44° 49:6
Moon (LL)	09-24-01	30° 00:8
Sirius	09-25-40	25° 51:5
Canopus	09-27-12	45° 35:3

What is our December 19 0927 fix?

Chapter 6

San Francisco to Kushiro

Problem 6-1
On July 6, 1993, we are near the middle of a long cruise to Japan, with no checkpoints along the way, except for the international date line a couple of weeks from now. We are outside the limits of accurate loran coverage, so we need to rely on celestial navigation. Our 0400 DR position was 39°31′N, 156°53′W. Our course is 270°, speed 6.5, I.C. -1′.5, and height of eye 9 feet. We plan for P.M. twilight:

	LMT	GMT
Sunset	1930 (7/5)	0559 (7/6)
Civil twilight	2002 (7/5)	0631 (7/6)
Nautical twilight	2043 (7/5)	0712 (7/6)

The LHA γ at 2002 (7/5) LMT is 227°. Our plan, from H.O. 249 Vol. 1 and the star-finder, is:

MAG.	BODY	ALT.	AZ.
-1.5	Jupiter	36°	230°
0.9	Altair	21°	096°
1.2	Spica	34°	211°
1.2	Antares	21°	161°
1.3	Deneb	31°	055°
1.7	Mars	25°	262°
2.0	Dubhe	48°	322°
2.1	Polaris	39°	000°
2.1	Rasalhague	48°	120°
2.2	Denebola	40°	254°

Around civil twilight, we get the following sights:

BODY	GMT	Hs
Spica	06-33-17	34°43′.9
Mars	06-34-53	20°11′.8
Deneb	06-35-45	30°25′.8
Dubhe	06-37-25	47°49′.3
Polaris	06-38-50	38°52′.7
Rasalhague	06-40-07	48°02′.6

What is our July 6 0640 fix?

Problem 6-2
From our July 6, 1993, 0640 fix, at 39°31′N, 157°16′W, we continue on course 270°, speed 6.5. Our sextant I.C. is -1.5,

and height of eye 9 feet. We plan for A.M. twilight sights, and interpolate in the almanac for the times:

	LMT	GMT
Nautical twilight	0327	1401
Civil twilight	0408	1442
Sunrise	0439	1513

The LHA γ at 0408 LMT is 346°. Our sighting plan is:

MAG.	BODY	ALT.	AZ	
2.2	Kochab	30°	345°	
2.2	Diphda	28°	153°	
1.3	Fomalhaut	20°	182°	
1.1	Aldebaran	16°	082°	
0.9	Altair	37°	248°	
0.4	Saturn	36°	197°	
0.2	Capella	26°	050°	
0.1	Vega	40°	292°	
-3.9	Venus	24°	087°	X
-	Moon	31°	215°	

If we shoot Aldebaran, we will not need Venus. At sighting time, visibility is excellent, and we get six good shots of stars listed in H.O. 249 Vol. 1.

BODY	GMT	Hs
Kochab	14-35-16	30° 06ʹ.3
Diphda	14-36-51	27° 49ʹ.7
Fomalhaut	14-38-33	20° 52ʹ.0
Aldebaran	14-40-02	16° 10ʹ.9
Capella	14-41-37	26° 01ʹ.0
Vega	14-43-05	38° 47ʹ.7

What is our July 6 1443 fix?

Problem 6-3

We continue on course 270°, 6.5 knots, from our July 6, 1993, 1443 fix, at 39° 35′N, 158° 16′W. Our I.C. is still -1ʹ.5 and height of eye 9 feet. The sky remains clear, the humidity is low, and the boat is moving smoothly with no pounding.

We decide to try for a sun-Venus fix at around 2100. We set up the star-finder, 2102D, to confirm that the Venus and sun LOPs will cross at a suitable angle. We could use the star finder to get the predicted altitude and azimuth, but it's easy (and more accurate) to use H.O. 249 Vol. 2. We look up the GHA of Venus (180° 29′, rounded) in the almanac, and subtract the DR longitude (159° 10′), at 2100, to get the LHA (21° 19′). The declination is N17° 42′, and our DR latitude is 39° 35′N. We enter H.O. 249 Vol. 2 with latitude, declination, and LHA, and take out the predicted altitude, 61° 30′, and azimuth, 226°. We could interpolate for Hp, but it's hardly necessary.

At 2100, we set the sextant to 61°30′, turn to 226° true, and with a little looking spot Venus and get three shots.

BODY	GMT	Hs
Venus	21-03-49	60°47.′5
Venus	21-05-25	60°36.′3
Venus	21-08-01	60°13.′1
Sun	21-09-40	64°01.′1
Sun	21-11-03	64°16.′3
Sun	21-12-42	64°32.′3

What is our July 6 2113 fix?

Problem 6-4
From our July 6, 1993, 2113 fix, at 39°38′N, 159°04′W, we continue on 270°/6.5, with I.C. −1.5 and height of eye 9 feet, and plan for P.M. twilight.

	LMT	GMT
Sunset	1930 (7/6)	0612 (7/7)
Civil twilight	2003 (7/6)	0645 (7/7)
Nautical twilight	2043 (7/6)	0725 (7/7)

The LHA γ at 2003 (7/6) LMT is 227°. Our sighting plan is:

MAG.	BODY	ALT.	AZ.	
−1.5	Jupiter	35°	231°	
0.9	Altair	21°	096°	
1.2	Spica	34°	211°	
1.2	Antares	21°	161°	
1.3	Deneb	31°	055°	X
1.7	Mars	25°	262°	
2.0	Dubhe	48°	322°	
2.1	Polaris	39°	000°	
2.1	Rasalhague	48°	120°	
2.2	Denebola	40°	254°	X

We added Polaris, because we needed a body at its azimuth. We will probably not use Deneb or Denebola, but leave them in the plan.

Our sights are:

BODY	GMT	Hs
Jupiter	06-35-18	36°41.′6
Altair	06-42-13	19°31.′2
Spica	06-43-55	34°21.′8
Antares	06-45-31	21°09.′3
Dubhe	06-47-02	47°48.′2
Polaris	06-48-42	39°02.′2

What is our July 7 0649 fix?

Problem 6-5

After our July 7, 1993, 0649 fix, at 39°42′N, 160°18′W, our course changed to 265° and speed dropped to 6.0. Our I.C. is still -1′.5 and height of eye 9 feet. We plan for A.M. twilight:

	LMT	GMT
Nautical twilight	0327	1413
Civil twilight	0407	1453
Sunrise	0439	1525

The LHA γ at 0407 is 347°. Our sighting plan is:

MAG.	BODY	ALT.	AZ.	
2.2	Kochab	30°	245°	
2.2	Diphda	28°	154°	
1.3	Fomalhaut	20°	183°	
1.1	Aldebaran	17°	082°	
0.9	Altair	36°	249°	
0.2	Capella	26°	051°	
0.1	Vega	39°	292°	
-3.9	Venus	24°	086°	X
-	Moon	39°	205°	

If we get a good shot of Aldebaran, we may skip Venus, because the two are close in azimuth.

During nautical twilight, the visibility worsens, and by the time the horizon becomes usable, only the brightest objects can be seen. But we are able to shoot three bodies:

BODY	GMT	Hs
Vega	15-10-17	35°29′.5
Vega	15-12-05	35°11′.0
Venus	15-15-39	28°56′.4
Venus	15-17-16	29°12′.6
Moon (LL)	15-19-14	36°20′.4
Moon (LL)	15-20-51	36°09′.3

What is our July 7 1521 fix?

Problem 6-6

From our July 7, 1993, 1521 fix, at 39°46′N, 161°20′W, our course and speed are the same: 265°/6.0. Our I.C. is still -1′.5, and height of eye 9 feet. We mark the positions of the sun, moon, and Venus on the star-finder, and note that they are not well positioned for a good three-body fix during the day. A moon-Venus fix might be possible. But during the morning, cirrus clouds move in, precluding a daytime Venus shot. We will try for a running fix on the sun, instead.

Here's what we get:

BODY	GMT	Hs
Sun	21-30-17	65° 04.′5
Sun	21-31-55	65° 19.′7
Sun	23-02-04	72° 23.′8
Sun	23-03-49	72° 19.′4
Sun	00-31-28	63° 00.′2
Sun	00-33-12	62° 44.′9

What is our July 8 0033 fix?

Problem 6-7
From our 0033 July 8, 1993, fix, at 39° 51′N, 162° 34′W, we continue on 265°/6.0, I.C. -1.′5, height of eye 9 feet, and plan for P.M. twilight:

	LMT	GMT
Sunset	1930 (7/7)	0624 (7/8)
Civil twilight	2003 (7/7)	0657 (7/8)
Nautical twilight	2044 (7/7)	0738 (7/8)

The LHA γ at 2003 (7/7) LMT is 227°. Our plan is:

MAG.	BODY	ALT.	AZ.	
-1.5	Jupiter	35°	232°	
0.9	Altair	21°	096°	
1.2	Spica	34°	211°	
1.2	Antares	21°	161°	
1.3	Deneb	31°	055°	X
2.0	Dubhe	48°	322°	
2.1	Polaris	39°	000°	
2.1	Rasalhague	48°	120°	
2.2	Denebola	40°	254°	

We may not use Deneb, whose azimuth is close to the reciprocal of Jupiter's. We included Polaris to give us complete azimuth coverage.

Our shots are:

BODY	GMT	Hs
Jupiter	06-50-30	35° 34.′7
Altair	06-52-08	19° 48.′8
Spica	06-53-49	33° 57.′3
Antares	06-55-31	21° 03.′9
Deneb	06-57-10	31° 29.′7
Polaris	06-58-42	39° 15.′4
Rasalhague	07-00-23	48° 27.′6
Denebola	07-01-59	38° 56.′8

What is our July 8 0702 fix?

Problem 6-8
From 39°56′N, 163°22′W, our July 8, 1993, 0702 fix, we change course to 275°. Our speed is still 6.0, I.C. -1′.5, height of eye 9 feet. We plan for A.M. twilight.

	LMT	GMT
Nautical twilight	0325	1423
Civil twilight	0406	1504
Sunrise	0438	1536

The LHA γ at 0406 LMT is 348°. Our plan is:

MAG.	BODY	ALT.	AZ.	
2.2	Kochab	30°	345°	
2.2	Diphda	28°	155°	
1.3	Fomalhaut	20°	183°	
1.1	Aldebaran	17°	083°	
0.9	Altair	36°	249°	
0.4	Saturn	35°	199°	
0.2	Capella	27°	051°	
0.1	Vega	38°	292°	
-3.9	Venus	24°	086°	X
-	Moon	46°	192°	X

We note that the moon and Saturn are close in azimuth, as are Venus and Aldebaran.

At sighting time, the sky is clear, and we get six good shots:

BODY	GMT	Hs
Diphda	15-11-06	28°55′.9
Fomalhaut	15-12-48	20°13′.6
Aldebaran	15-14-31	19°38′.5
Saturn	15-16-05	34°19′.0
Capella	15-17-41	29°10′.9
Vega	15-19-15	35°31′.7

What is our July 8 1519 fix?

Problem 6-9
After our July 8, 1993, 1519 fix, at 40°01′N, 164°26′W, our course changes to 280° and speed drops to 5.0. Our I.C. is still -1′.5, and height of eye 9 feet. The moon and sun are not well placed for a fix, but if we can spot Venus this morning, we will try for a Venus-sun or Venus-moon fix.

We set up the star-finder, and see that a Venus-moon fix is feasible, so we precompute for Venus's altitude and azimuth at our 1900 DR position: 65°/146°.

We spot Venus a few minutes ahead of time, and get our shots:

BODY	GMT	Hs
Venus	18-55-01	64° 14.'8
Venus	18-57-03	64° 30.'6
Venus	18-59-14	64° 43.'6
Moon (UL)	19-01-45	16° 13.'8
Moon (UL)	19-03-20	15° 59.'1
Moon (UL)	19-04-51	15° 40.'7

What is our July 8 1905 fix?

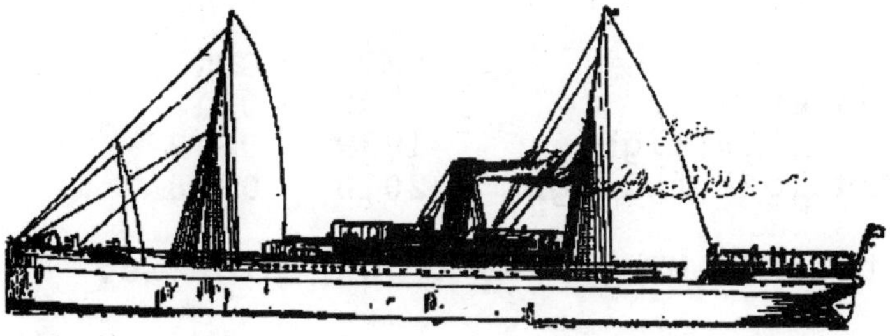

Chapter 7

Petropavlovsk Kamchatskiy to Wake Island

Problem 7-1
We are halfway on our cruise south to Wake Island the evening of May 15, 1993, on a calm sea with very light wind. Our course is 170°, speed 3.5 knots. Our DR position at 0700 May 15 was 40°30′N, 162°30′E. Our I.C. is +1.′0 and height of eye 10 feet. We plan for P.M. twilight:

	LMT	GMT
Sunset	1908	0818
Civil twilight	1939	0849
Nautical twilight	2018	0928

The LHA γ at 1939 LMT is 168°. Our plan is:

MAG.	BODY	ALT.	AZ.
-1.9	Jupiter	46°	155°
0.2	Capella	28°	308°
0.2	Arcturus	46°	104°
0.5	Procyon	31°	249°
1.2	Pollux	46°	271°
1.2	Spica	30°	142°
1.3	Regulus	59°	211°
2.1	Polaris	40°	359°
2.2	Kochab	48°	019°

We skipped Mars, because its azimuth is close to Procyon's, and we added Polaris because we had nothing else near its azimuth.

Our sights are:

BODY	GMT	Hs
Jupiter	08-42-13	45°22.′8
Capella	08-43-49	29°18.′1
Arcturus	08-45-36	44°53.′0
Procyon	08-47-04	31°17.′5
Regulus	08-48-39	58°24.′4
Polaris	08-50-16	39°58.′2

What is our May 15 0850 fix?

Problem 7-2
From our 0850 May 15, 1993, fix, at 40°26′N, 162°34′E, we continue on 170° at 3.5 knots, I.C. still +1.′0. height of eye 10 feet. We plan for A.M. twilight.

	LMT	GMT
Nautical twilight	0338 (5/16)	1647 (5/15)
Civil twilight	0415 (5/16)	1724 (5/15)
Sunrise	0446 (5/16)	1755 (5/15)

The LHA γ at 0415 (5/16) LMT is 297°. Our plan is:

MAG.	BODY	ALT.	AZ.	
2.5	Enif	50°	132°	X
2.2	Alpheratz	37°	080°	
2.2	Kochab	42°	339°	
2.1	Rasalhague	50°	236°	
1.9	Alkaid	29°	312°	
0.9	Altair	59°	179°	
0.6	Saturn	28°	140°	
0.2	Arcturus	17°	281°	
-4.6	Venus	15°	096°	X
-	Moon	28°	117°	

We note the azimuth duplications: Enif-Alkaid, and Venus-Arcturus.

At sighting time, a haze prevents us from seeing any but the brightest bodies. Here is what we get:

BODY	GMT	Hs
Arcturus	17-37-17	14°46:4
Arcturus	17-38-50	14°26:6
Venus	17-41-04	18°07:7
Venus	17-42-39	18°24:1
Moon (LL)	17-46-10	32°04:6
Moon (LL)	17-47-51	32°22:3

Do these sights give us a useful fix?

Problem 7-3

From our May 15, 1993, 1748 fix, at 39°55′N, 162°45′E, we proceed on course 170° at 3.5 knots, with I.C. still +1:0 and height of eye 10 feet. We get some shots for a sun-moon fix:

BODY	GMT	Hs
Moon (LL)	21-03-42	50°31:8
Moon (LL)	21-05-21	50°31:1
Moon (LL)	21-07-01	50°31:5
Sun	21-09-33	35°36:7
Sun	21-11-12	35°51:5
Sun	21-12-44	36°08:6

What is our May 15 2107-2113 R. fix?

Chapter 8

Valencia to Genoa

Problem 8-1
We left Valencia yesterday, and are moving nicely in a moderate sea. There are scattered cumulus, but we hope to get a round of star sights, so we plan for P.M. twilight. It is May 13, 1993, and our latest fix, at 1800, was 39° 35′N, 1° 05′E. Our course is 060°, speed 7.0, I.C. -0.′6, and height of eye 12 feet. We interpolate in the almanac to get the twilight times:

	LMT	GMT
Sunset	1906	1901
Civil twilight	1937	1932
Nautical twilight	2015	2010

The LHA γ at 1937 LMT is 166°. Our plan is:

MAG.	BODY	ALT.	AZ.
-1.9	Jupiter	46°	151°
0.2	Capella	29°	307°
0.2	Arcturus	44°	103°
0.5	Procyon	32°	247°
1.2	Pollux	48°	270°
1.2	Spica	29°	140°
1.3	Regulus	59°	208°
2.1	Polaris	39°	359°
2.2	Kochab	47°	020°

At twilight, the sky is clear, and we get our sights:

BODY	GMT	Hs
Jupiter	19-23-08	45° 00.′8
Capella	19-30-17	29° 41.′4
Arcturus	19-31-58	44° 08.′8
Pollux	19-33-41	32° 31.′5
Regulus	19-35-24	59° 31.′1
Polaris	19-37-05	39° 16.′2

What is our May 13 1937 fix?

Problem 8-2
From our May 13, 1993, 1937 fix, at 39° 41′N, 1° 12′E, we continue toward Genoa on course 060° at 7.0 knots, I.C. still -0.′6, height of eye 12 feet, and plan for A.M. twilight.

	LMT	GMT
Nautical twilight	0337	0328
Civil twilight	0414	0405
Sunrise	0446	0437

The LHA γ at 0414 LMT is 296°. Our plan is:

MAG.	BODY	ALT.	AZ.	
2.5	Enif	50°	130°	X
2.2	Alpheratz	36°	080°	
2.2	Kochab	43°	339°	
2.1	Rasalhague	50°	235°	
1.9	Alkaid	30°	311°	
0.9	Altair	59°	177°	
0.2	Arcturus	18°	280°	
-4.6	Venus	14°	097°	X
-	Moon	33°	137°	X

We plan to skip Enif (a dim star, whose azimuth is close to the reciprocal of Alkaid's), and will not need Venus and the moon, if our other sights are good.

Our results are:

BODY	GMT	Hs
Alpheratz	03-55-17	33° 31′.9
Kochab	03-56-59	43° 37′.6
Rasalhague	03-58-45	51° 52′.9
Alkaid	04-00-25	30° 57′.7
Altair	04-02-03	58° 37′.0
Arcturus	04-03-44	18° 59′.0

We got six good shots, all of which can be worked by H.O. 249 Vol. 1, so we didn't need Venus or the moon, whose sights we would have had to work by H.O. 249 Vol. 2. What is our May 14 0404 fix?

Problem 8-3

From our May 14, 1993, 0404 fix, at 40° 10′N, 2° 09′E, we proceed on 060/7.0, with I.C. still -0′.6 and height of eye 12 feet. We get sights for a morning sun-moon fix:

BODY	GMT	Hs
Sun	09-15-18	50° 59′.8
Sun	09-16-53	51° 18′.4
Sun	09-18-41	51° 34′.8
Moon (UL)	09-21-04	29° 43′.1
Moon (UL)	09-22-39	29° 31′.2
Moon (UL)	09-24-05	29° 17′.0

What is our May 14 0924 fix?

At the time of the last sun sight, a crew member put the shadow pin on the compass and checked the bearing of the sun, for a check on the compass error. The shadow-pin bearing was 115°, and the variation in this locality was 7°W. What was our compass deviation on that heading?

Chapter 9

Hungnam to Akita

Problem 9-1
On the morning of July 6, 1993, we are halfway across the Sea of Japan on a cruise from North Korea to Honshu. We are sailing smoothly at 7.0 knots on course 095°. Our sextant I.C. is -1:0, and height of eye 9 feet. Our DR position at 0100 was 39°50′N, 134°40′E, but we need to confirm this with a fix, because we haven't had clear weather for several days, until this morning.

We get some morning sun sights, then more in the afternoon, for a running fix:

BODY	GMT	Hs
Sun	01-53-32	66°54:5
Sun	01-55-16	67°10:0
Sun	01-56-58	67°21:5
Sun	04-12-03	67°21:1
Sun	04-13-50	67°07:9
Sun	04-15-35	66°50:2

What is our July 6 0157-0416 R. fix?

Problem 9-2
From our July 6, 1993, 0416 fix, at 39°52′N, 135°15′E, we continue on course 095°, at 7.0 knots. Our I.C. is still -1:0, and height of eye 9 feet. We plan for P.M. twilight.

	LMT	GMT
Sunset	1930	1025
Civil twilight	2003	1058
Nautical twilight	2044	1139

The LHA γ at 2003 LMT is 226°. Our plan is:

MAG.	BODY	ALT.	AZ.	
-1.5	Jupiter	36°	230°	X
0.9	Altair	20°	095°	
1.2	Spica	34°	210°	
1.2	Antares	21°	160°	
1.3	Deneb	31°	055°	
2.0	Dubhe	48°	322°	
2.1	Polaris	39°	000°	
2.1	Rasalhague	47°	119°	
2.2	Denebola	41°	254°	

We will skip Jupiter, unless we are unable to shoot Deneb. We omitted Mars, and added Polaris, because of azimuth considerations.

At twilight, the sighting conditions are nearly perfect, and we get the following sights:

BODY	GMT	Hs
Altair	10-57-06	19° 14:7
Spica	10-58-51	34° 19:6
Antares	11-00-38	18° 51:3
Deneb	11-02-14	30° 58:5
Polaris	11-03-50	39° 10:7
Rasalhague	11-05-29	48° 03:8

What is our July 6 1105 fix?

Problem 9-3

From our July 6, 1993, 1105 fix, at 39°50′N, 136°30′E, we continue across the Sea of Japan, now on course 090°, and with our speed increased to 8.5. Our height of eye is still 9 feet.

As we begin our plan for A.M. twilight, the boat rolls, and the sextant bangs to the cabin sole. We examine it, and find a broken horizon mirror. This is not a disaster, since we have a spare mirror, but there isn't time to replace it and adjust the sextant before twilight.

We decide to try something unusual—a sight of the sun, without using a sextant, when the lower limb is exactly on the horizon. (An upper-limb sight would be much harder.)

From the almanac, we find that the time of sunrise is 0438 (7/7) LMT, 1926 (7/6) GMT, and we get ready at about 1920. We hold up the sextant so we can squint through the next-to-strongest sun filter, and when the upper limb breaks the horizon, we wait. The sun looks slightly squashed, but as the lower limb touches the horizon, we record the time.

BODY	GMT	Hs
Sun	19-29-15	00° 00:0

We note the air temperature, 63°F, and the barometer, 30.2 inches Hg (important at low altitudes).

What was the LOP we got from this sight, and what can we do with it?

Problem 9-4

Our sun LOP, at 1929 July 6, 1993, was 1.0 toward 060, with an A.P. of 40°N, 137°53′E. (Our last fix, at 1105 July 6, 1993, was at 39°50′N, 136°30′E.) We continue on 090°, at 8.5 knots, with our height of eye still 9 feet.

We repair and check the sextant, getting a new I.C. of +1:3. We plan for some sun sights later this morning to cross with the 1929 sun LOP for a running fix.

We get:

BODY	GMT	Hs
Sun	02-04-44	70° 16:6
Sun	02-06-26	70° 28:4
Sun	02-08-13	70° 37:9
Sun	02-10-06	70° 45:4

What is our 1929 (7/6)-0210 (7/7) R. fix?

Problem 9-5
Our last fix, at 0210 July 7, 1993, was at 39°51′N, 139°11′E. It depended on the one sun sight we took, without a sextant (an inexact method, because of the uncertainties of refraction at low altitudes). Also, as we approach our landfall, we'd like to check our repaired sextant. We'd like another fix, so we plan for P.M. twilight.

	LMT	GMT
Sunset	1930	1010
Civil twilight	2003	1043
Nautical twilight	2044	1124

The LHA γ at 2003 LMT is 227°. Our plan is:

MAG.	BODY	ALT.	AZ.	
-1.5	Jupiter	35°	231°	
0.9	Altair	21°	096°	
1.2	Spica	34°	211°	
1.2	Antares	21°	161°	
1.3	Deneb	31°	055°	X
2.0	Dubhe	48°	322°	
2.1	Polaris	39°	000°	
2.1	Rasalhague	48°	120°	
2.2	Denebola	40°	254°	

With more bodies available than we need, we will shoot Jupiter, but not bother to work the sight if we get Deneb (whose azimuth is close to Jupiter's reciprocal). We expect to shoot, and work the sights of, Altair, Spica, Antares, Deneb, Dubhe, and Polaris—all listed in H.O. 249 Vol. 1.

At sighting time, there's an unfortunate haze, and we see only the brightest bodies. We shoot Jupiter, Arcturus, and Altair, then repeat the first two. We'd like to get a body to the NE or SW, to give us good azimuth coverage. Just before the horizon fades, we spot Antares, and get one good shot.

BODY	GMT	Hs
Jupiter	10-51-01	34° 03.9'
Arcturus	10-54-09	65° 42.9'
Altair	10-55-48	22° 18.6'
Jupiter	10-57-17	33° 04.0'
Arcturus	10-58-50	65° 12.0'
Antares	11-00-38	21° 59.5'

What is our July 7 1101 fix?

Chapter 10

Barcelona to Tunis

Problem 10-1
We are in the Mediterranean, a few miles NE of Minorca, the afternoon of December 18, 1993, course 120°, speed 6.5. Our 1200 DR position was 40° 30′N, 5° 05′E. Our sextant I.C. is -1′.0, and height of eye 10 feet. We take some sights for a sun-moon fix.

BODY	GMT	Hs
Sun	13-45-23	19° 16′.2
Sun	13-47-02	19° 07′.8
Sun	13-48-43	18° 54′.7
Moon (UL)	13-50-21	34° 58′.5
Moon (UL)	13-51-58	35° 09′.2
Moon (UL)	13-53-37	35° 23′.3

L 40° 23.0′N
λ 5° 17.1′E

What is our December 18 1354 fix?

Problem 10-2
From our December 18, 1993, 1354 fix, at 40°23′N, 5° 17′E, we continue on 120°/6.5, with I.C. still -1′.0, and height of eye 10 feet. We plan for P.M. twilight.

	LMT	GMT
Sunset	1636	1613
Civil twilight	1707	1644
Nautical twilight	1741	1718

The LHA γ at 1707 LMT is 344°. Our sighting plan is:

MAG.	BODY	ALT.	AZ.	
-	Moon	42°	196°	X
0.1	Vega	41°	291°	
0.2	Capella	25°	050°	
0.7	Saturn	34°	198°	X
0.9	Altair	28°	246°	
1.3	Fomalhaut	20°	180°	
2.2	Hamal	47°	098°	
2.2	Diphda	27°	151°	
2.2	Kochab	31°	344°	

This is a big list. We won't need both Saturn and the moon (which are very close in azimuth), and we can skip both if we shoot Fomalhaut.

We get an early moon sight; then, at twilight, we shoot the other bright bodies, and continue down the list. But we work only those included in H.O. 249 Vol. 1.

BODY	GMT	Hs
Vega	16-47-18	40° 42′.2
Capella	16-48-54	25° 25′.3
Altair	16-50-31	37° 19′.6
Fomalhaut	16-52-02	20° 11′.2
Diphda	16-53-40	27° 32′.1
Kochab	16-55-11	30° 24′5

What is our December 18 1655 fix?

Problem 10-3

From our December 18, 1993, 1655 fix, at 40° 14′N, 5° 35′E, we stay on course 120°, but now at 6.0 knots, with I.C. still -1′.0 and height of eye 10 feet. We plan for A.M. twilight.

	LMT	GMT
Nautical twilight	0612	0544
Civil twilight	0646	0618
Sunrise	0716	0648

The LHA γ at 0646 LMT is 189°. Our plan is:

MAG.	BODY	ALT.	AZ.	
2.3	Alphecca	51°	096°	X
2.1	Polaris	39°	000°	
1.3	Regulus	47°	240°	X
1.2	Pollux	30°	283°	
1.2	Spica	38°	165°	
0.2	Arcturus	60°	127°	
0.2	Capella	16°	317°	
0.1	Vega	24°	058°	
-1.3	Jupiter	32°	149°	

H.O. 249 Vol. 1 did not have a perfect round of stars to select from, for 40°N latitude and 189° LHA, so there are some near-duplicates: Vega-Regulus (reciprocals) and Pollux-Alphecca (reciprocals). We add Polaris for better azimuth coverage.

Sighting conditions are good at twilight, and we shoot the following:

BODY	GMT	Hs
Polaris	06-25-13	38° 54′.7
Pollux	06-26-53	28° 17′.1
Spica	06-28-31	38° 41′.2
Arcturus	06-30-11	62° 36′.0
Capella	06-31-50	14° 10′.3
Vega	06-33-07	26° 09′.4

What is our December 19 0633 fix?

Chapter 11

Hilo to Seattle

Problem 11-1
It is late afternoon, May 13, 1993, and we are 600 miles or so off the Northern California coast, on our way from the Hawaiian islands to Seattle. We are heading 040°, making 5.2 knots. Our sextant I.C. is -1'.3, and height of eye 8 feet. Our last DR position was 39°40'N, 134°38'W, at 0400 May 13. It may not be accurate, because we got it from a running fix on the sun, with the last sights shot from a rolling deck. The sea is calm now, and the few cumulus clouds from the afternoon are dissipating—but we need a good fix. We plan for P.M. twilight.

	LMT	GMT
Sunset	1906 (5/12)	0404 (5/13)
Civil twilight	1937 (5/12)	0435 (5/13)
Nautical twilight	2015 (5/12)	0513 (5/13)

The LHA γ at 1937 (5/12) LMT is 165°. Our sighting plan is:

MAG.	BODY	ALT.	AZ.	
-1.9	Jupiter	45°	150°	
0.2	Capella	30°	307°	
0.2	Arcturus	43°	102°	
0.5	Procyon	33°	246°	
1.2	Pollux	48°	269°	
1.2	Spica	29°	139°	
1.3	Regulus	60°	206°	
2.1	Polaris	40°	359°	
2.2	Kochab	46°	020°	X

We omitted Mars (its azimuth is close to Procyon's), and included Polaris. We will probably not sight Kochab (its azimuth is close to Regulus's reciprocal). We would like to get Jupiter, Capella, Arcturus, Procyon, Regulus, and Polaris, for good azimuth distribution.

At twilight, sighting conditions are good, and we get the preferred bodies:

BODY	GMT	Hs
Jupiter	04-27-12	44°55'.7
Capella	04-31-42	30°11'.6
Arcturus	04-33-38	43°32'.1
Procyon	04-35-01	33°10'.2
Regulus	04-36-40	59°56'.6
Polaris	04-38-20	39°12'.4

What is our May 13 0438 fix?

Problem 11-2
From our last fix, at 0438 May 13, 1993, 39°37′N, 134°26′W, we continue on 040°/5.2, with I.C. -1ʹ.3, and height of eye 8 feet. A thin veil of cirrostratus has moved in during the night, but we plan for A.M. twilight, hoping to be able to get some sights.

	LMT	GMT
Nautical twilight	0337	1232
Civil twilight	0414	1309
Sunrise	0445	1340

The LHA γ at 0414 LMT is 294°. We make up our sighting plan.

MAG.	BODY	ALT.	AZ.	
2.5	Enif	49°	128°	X
2.2	Alpheratz	34°	079°	
2.2	Kochab	43°	339°	
2.1	Rasalhague	52°	233°	
1.9	Alkaid	31°	311°	
0.9	Altair	59°	173°	
0.2	Arcturus	20°	279°	
-4.6	Venus	14°	097°	X
-	Moon	33°	145°	

We have omitted Saturn (whose azimuth is close to that of the moon and to the reciprocal of Alkaid's), but included Enif and Venus (azimuths close to reciprocals of Alkaid's and Arcturus's, respectively), although we will probably not use them. With Altair available, we don't need Polaris. We'd like to shoot Alpheratz, Kochab, Rasalhague, Alkaid, Altair, and Arcturus.

As the horizon appears, we can't spot any stars or planets, so we get some moon shots. Then we see Venus, dimly, and quickly shoot it.

BODY	GMT	Hs
Moon (LL)	13-05-12	32°32ʹ.6
Moon (LL)	13-06-53	32°46ʹ.6
Moon (LL)	13-08-35	32°54ʹ.9
Venus	13-15-01	15°01ʹ.0
Venus	13-16-42	15°19ʹ.1
Venus	13-18-20	15°39ʹ.6

These LOPs make a very poor cut. Can we get a usable fix from them?

Problem 11-3
Since our fix, 40° 05′N, 133° 52′W, at 1318 May 13, 1993, we continue on course 040°. The wind has picked up, so we have reduced sail. We are making 5.0 knots. Our I.C. is still -1:3, and height of eye 8 feet. The latest WWVH broadcast reported a storm building to our north, with 35-knot winds and 10-foot seas, moving SW. We want to get out of its way, but before we decide on a new heading, we want a good fix. We got three moon sights an hour ago, and worked them while we waited for the sun to move in azimuth enough to give us a good cut.

Now we get our sun sights.

BODY	GMT	Hs
Moon (UL)	18-45-13	20° 52:4
Moon (UL)	18-46-55	20° 35:0
Moon (UL)	18-48-40	20° 20:7
Sun	19-51-03	64° 32:1
Sun	19-52-50	64° 42:5
Sun	19-54-29	64° 55:1

What is our May 13 1849–1954 R. fix?

Problem 11-4
From our May 13, 1993, 1954 fix, at 40° 25′N, 133° 25′W, we want to change course to the one that will give us the maximum distance from the storm, which is now at 42° 55′N, 133° 25′W, moving SW at 20 knots. We test the boat, and find that we can make 6 knots on any heading between ESE and S.

We figure the best course by the quick method. (See Appendix B.) What is the new course?

After we have turned away from the storm, we work the complete maneuvering-board solution (see Appendix B), which gives us:

The best avoidance course (as above)
The closest point of approach (CPA) of the storm
The time of the CPA

Problem 11-5
From our 1954 May 13, 1993, fix, at 40° 25′N, 133° 25′W, we turned to course 150°, and are making 6.0 knots, as we expected. Our I.C. is still -1:3, and height of eye 8 feet. The boat is taking some pounding, but we plan for P.M. twilight, hoping to get a few sights.

	LMT	GMT
Sunset	1906 (5/13)	0357 (5/14)
Civil twilight	1937 (5/13)	0428 (5/14)
Nautical twilight	2015 (5/13)	0506 (5/14)

The LHA γ at 1937 (5/13) LMT is 166°. Our plan is:

MAG.	BODY	ALT.	AZ.
-1.9	Jupiter	46°	152°
0.2	Capella	29°	307°
0.2	Arcturus	44°	103°
0.5	Procyon	32°	247°
1.2	Pollux	48°	270°
1.2	Spica	29°	140°
1.3	Regulus	59°	208°
2.1	Polaris	39°	359°
2.2	Kochab	47°	020°

At twilight the sea is even rougher, but another crew member helps to steady the navigator, and we get a few fair sights.

BODY	GMT	Hs
Jupiter	04-32-05	46°25!8
Jupiter	04-34-31	46°33!4
Capella	04-37-01	27°49!7
Procyon	04-39-42	30°26!7

What is our May 14 0440 fix?

Problem 11-6
The storm is still a concern, but it's behind us, now, and has not worsened. We decide to change course to ease the boat's motion. From our May 14, 1993, 0440 fix, at 39°40'N, 132°48'W, we head 090°, and make 6.5 knots, even with our reduced sail. Our I.C. is still -1!3, height of eye 8 feet. Through the night there are scattered ragged clouds, but we hope for some sights at A.M. twilight.

	LMT	GMT
Nautical twilight	0340	1225
Civil twilight	0417	1303
Sunrise	0447	1333

The LHA γ at 0417 LMT is 296°. Here is our plan:

MAG.	BODY	ALT.	AZ.	
2.5	Enif	50°	130°	X
2.2	Alpheratz	36°	080°	
2.2	Kochab	43°	339°	
2.1	Rasalhague	50°	235°	
1.9	Alkaid	30°	311°	
0.9	Altair	59°	177°	
0.2	Arcturus	18°	280°	
-4.6	Venus	15°	097°	
-	Moon	32°	132°	

We will probably skip Enif, whose azimuth is close to the moon's and to the reciprocal of Alkaid's. We omitted Saturn and Polaris, also because of azimuth duplications. We would like to get Alpheratz, Kochab, Rasalhague, Alkaid, Altair, and Arcturus—all H.O. 249 Vol. 1 stars, nicely spaced in azimuth.

At sighting time, the sky is surprisingly clear, and we sight the preferred bodies:

BODY	GMT	Hs
Alpheratz	12-58-13	35° 04′.3
Kochab	12-59-52	42° 46′.4
Rasalhague	13-01-30	50° 46′.0
Alkaid	13-03-17	29° 27′.7
Altair	13-04-57	59° 04′.9
Arcturus	13-06-10	17° 21′.3

What is our May 14 1306 fix?

Chapter 12

Tristan da Cunha to Punta Delgada

Problem 12-1
It is September 10, 1993, a mild night near the end of winter, in the middle of a featureless ocean. Our 0600 DR was 39°35′S, 30°40′W, and we are making 6.5 knots on a heading of 270°. Our sextant I.C. is +1ʹ.5, and height of eye 11 feet. We plan for A.M. twilight.

	LMT	GMT
Nautical twilight	0509	0713
Civil twilight	0541	0745
Sunrise	0608	0812

The LHA γ at 0541 LMT is 74°.
Our plan is:

MAG.	BODY	ALT.	AZ.	
2.2	Hamal	15°	320°	
2.2	Diphda	31°	272°	
1.3	Fomalhaut	19°	247°	
1.1	Aldebaran	33°	354°	
1.1	Acrux	26°	152°	
0.6	Achernar	54°	224°	
0.5	Betelgeuse	41°	019°	
-1.6	Sirius	57°	053°	
-	Moon	27°	016°	X

The H.O. 249 Vol. 1 stars leave a big gap in azimuth, which we have filled in with Diphda. We won't bother with the moon if we are able to shoot Betelgeuse, because their azimuths are nearly the same.

The two dimmer stars have faded by the time the horizon is visible, but we get six good sights:

BODY	GMT	Hs
Fomalhaut	07-50-05	17°17ʹ.5
Aldebaran	07-51-41	33°32ʹ.1
Acrux	07-53-19	26°49ʹ.9
Achernar	07-54-57	52°01ʹ.8
Betelgeuse	07-56-41	41°55ʹ.7
Sirius	07-58-32	59°31ʹ.9

What is our September 10 0759 fix?

Problem 12-2
From our September 10, 1993, 0759 fix, at 39°36′S, 30°55′W, we continue on course 270°, speed 6.5, with I.C. still +1.′5, height of eye 11 feet. We check the position of the moon with the star-finder, and decide to get a sun-moon fix this morning.

BODY	GMT	Hs
Moon (UL)	11-15-03	20°25.′6
Moon (UL)	11-17-46	20°09.′9
Moon (UL)	11-19-31	19°56.′2
Sun	11-21-14	31°50.′7
Sun	11-22-50	32°08.′2
Sun	11-24-29	32°23.′1

What is our September 10 1124 fix?

Problem 12-3
From our September 10, 1993, 1124 fix, at 39°38′S, 31°23′W, we stay on course 270°, with speed 6.5, I.C. +1.′5, and height of eye 11 feet. We plan for P.M. twilight.

	LMT	GMT
Sunset	1746	1957
Civil twilight	1813	2024
Nautical twilight	1845	2056

The LHA γ at 1813 LMT is 263°. Our plan is:

MAG.	BODY	ALT.	AZ.	
-1.2	Jupiter	21°	281°	X
0.1	Rigil Kent.	56°	217°	
0.2	Arcturus	15°	312°	
0.4	Saturn	28°	086°	
0.6	Achernar	19°	151°	
0.9	Altair	32°	041°	
1.2	Spica	29°	280°	
1.3	Fomalhaut	25°	109°	
2.1	Rasalhague	37°	001°	

Mars and Spica are lined up on nearly the same azimuth as Jupiter, 3° and 8° higher, respectively. We have omitted Mars from the plan, and will probably not shoot Jupiter.

Sighting conditions are very good at twilight, and we get the following:

BODY	GMT	Hs
Rigil Kent.	20-28-01	55° 39.′2
Arcturus	20-29-43	14° 24.′5
Saturn	20-31-32	29° 57.′3
Achernar	20-33-04	19° 48.′8
Spica	20-34-50	26° 35.′4
Rasalhague	20-36-30	37° 38.′3

What is our September 10 2036 fix?

Problem 12-4

From our September 10, 1993, 2036 fix, at 39° 45′S, 32° 34′W, we continue as before: 270°/6.5, I.C. +1.′5, height of eye 11 feet. We prepare for A.M. twilight.

	LMT	GMT
Nautical twilight	0509	0725
Civil twilight	0541	0757
Sunrise	0608	0824

The LHA γ at 0541 is 75°. Our plan is:

MAG.	BODY	ALT.	AZ.	
2.2	Suhail	45°	116°	
2.2	Diphda	31°	272°	
1.1	Aldebaran	33°	353°	
1.1	Acrux	26°	152°	
0.6	Achernar	53°	225°	X
0.5	Betelgeuse	41°	018°	
-1.6	Sirius	57°	052°	
-	Moon	24°	030°	X

This is a good plan, from which we can pick half a dozen nicely spaced stars, all listed in H.O. 249 Vol. 1. We won't bother with the moon if we shoot Betelgeuse.

Here's what we get:

BODY	GMT	Hs
Suhail	08-05-13	46° 38.′6
Diphda	?	28° 42.′7
Aldebaran	08-08-34	32° 54.′2
Acrux	08-10-15	27° 54.′3
Betelgeuse	08-11-55	41° 54.′4
Sirius	08-13-40	60° 18.′7

When we finish shooting, we are appalled to find that we forgot to record the time for the Diphda sight—and when we go on deck, we can no longer see the star. That's too bad, because we needed a body near that azimuth. Do we have to discard the sight? What is our September 11 0814 fix?

Problem 12-5
From our September 11, 1993, 0814 fix, at 39°56′S, 33°56′W, we continue on 265°/6.0, with I.C. still +1′.5, height of eye 11 feet, and plan for a sun-moon fix this morning.

We get these sights:

BODY	GMT	Hs
Moon (UL)	10-16-25	29°48′.2
Moon (UL)	10-18-01	29°47′.2
Moon (UL)	10-19-38	29°46′.9
Sun	10-22-04	20°44′.5
Sun	10-23-40	21°02′.0
Sun	10-25-10	21°14′.5

What is our September 11 1025 fix?

Problem 12-6
We proceed from our Sept. 11, 1993, 1025 fix, at 39°58′S, 34°09′W, on course 265°, speed 6.0. Our I.C. is still +1′.5 and height of eye 11 feet. We plan for P.M. twilight.

	LMT	GMT
Sunset	1746	2008
Civil twilight	1813	2035
Nautical twilight	1845	2107

The LHA γ at 1813 LMT is 264°. Our sighting plan is:

MAG.	BODY	ALT.	AZ.	
-1.2	Jupiter	21°	280°	X
0.1	Rigil Kent.	56°	217°	
0.2	Arcturus	15°	311°	
0.4	Saturn	29°	085°	
0.6	Achernar	19°	150°	
0.9	Altair	32°	040°	X
1.2	Spica	28°	279°	
1.3	Fomalhaut	26°	108°	
2.1	Rasalhague	37°	000°	

Jupiter and Altair are both redundant in azimuth. The bodies we would prefer are Rigil Kentaurus, Arcturus, Saturn, Achernar, Fomalhaut, and Rasalhague.

At twilight, there is a thin deck of cirrostratus blocking out all but the brightest bodies, but we get a few:

BODY	GMT	Hs
Jupiter	20-41-14	19°27′.8
Rigil Kent.	20-42-50	55°10′.5
Arcturus	20-44-35	13°00′.8
Saturn	20-46-18	31°21′.9
Jupiter	20-48-01	18°09′.4

What is our September 11 2048 fix?

Problem 12-7
From our 2048 fix on September 11, 1993, at 40°13′S, 35°24′W, we continue on course 265°. Our speed, according to the indicator, has increased to 7.8, although some of the crew think we are going no faster than we were yesterday. We prepare for A.M. twilight. Our I.C. is still +1′.5, and height of eye 11 feet.

	LMT	GMT
Nautical twilight	0509	0738
Civil twilight	0541	0810
Sunrise	0608	0837

The LHA γ at 0541 LMT is 76°. Our plan is similar to yesterday's, except for the GMT of civil twilight and the moon prediction.

MAG.	BODY	ALT.	AZ.	
2.2	Suhail	45°	115°	
2.2	Diphda	30°	271°	
1.1	Aldebaran	33°	352°	
1.1	Acrux	27°	152°	
0.6	Achernar	53°	225°	X
0.5	Betelgeuse	41°	017°	
-1.6	Sirius	58°	050°	
-	Moon	20°	044°	X

We expect to skip Achernar and the moon, because of azimuth duplications.

At twilight, viewing conditions are excellent, and we get the following:

BODY	GMT	Hs
Suhail	08-04-17	45°15′.2
Diphda	08-05-59	30°21′.5
Aldebaran	08-07-42	32°51′.1
Acrux	08-09-21	27°26′.7
Betelgeuse	08-11-01	41°13′.1
Sirius	08-12-46	58°55′.4

What is our September 12 0813 fix?

Problem 12-8
Our speed indicator was in error, but it's fixed now, and we are on course 270°, speed 6.0, since our 0813 September 12, 1993, fix, at 40°16′S, 36°51′W. Our I.C. is still +1′.5, and height of eye 11 feet. The moon is closer to the sun than it was yesterday, and simultaneous sun-moon LOPs would not have a very good crossing angle. Therefore we shoot the sun first, and a couple of hours later the moon.

BODY	GMT	Hs
Sun	11-25-06	29°15.′6
Sun	11-26-39	29°28.′5
Sun	11-28-17	29°45.′5
Moon (UL)	13-10-13	26°22.′4
Moon (UL)	13-11-54	26°14.′0
Moon (UL)	13-13-39	26°01.′6

What is our 1128-1314 Sept 12 R. fix?

Problem 12-9
Since our September 12, 1993, 1314 fix, at 40°20′S, 37°26′W, we continue on 270°/6.0, with I.C. +1.′5, and height of eye 11 feet. We plan for P.M. twilight.

	LMT	GMT
Sunset	1746	2020
Civil twilight	1813	2047
Nautical twilight	1845	2119

The γ at 1813 LMT is 265°. Our plan is:

MAG.	BODY	ALT.	AZ.	
-1.2	Jupiter	20°	279°	X
0.1	Rigil Kent.	56°	218°	
0.2	Arcturus	14°	311°	
0.4	Saturn	30°	084°	
0.6	Achernar	20°	150°	
0.9	Altair	33°	039°	X
1.2	Spica	27°	278°	
1.3	Fomalhaut	26°	108°	
1.9	Mars	23°	280°	X
2.1	Rasalhague	37°	358°	

We probably won't sight Mars or Jupiter if we get Spica, or Altair if we get Rigil Kentaurus; also, we may skip Saturn.

Unfortunately, the sky is thinly overcast at twilight, but we get two Jupiter shots, then spot and shoot Rigil Kentaurus and Achernar:

BODY	GMT	Hs
Jupiter	20-55-14	21°35.′6
Jupiter	20-57-35	21°11.′1
Rigil Kent.	21-00-03	54°11.′3
Achernar	21-01-49	21°41.′6

What is our September 12 2102 fix?

Chapter 13

Hamilton to Rotterdam

Problem 13-1
We are well into our cruise from Bermuda to the Netherlands early in the morning of July 6, 1993. The weather is mild, the sky is clear, and the wind is light. We should have good sighting conditions for A.M. twilight. Our 0600 DR position was 39°30′N, 40°00′W. We are drifting along at 3.2 knots on course 070°. Our I.C. is −2′.0, and height of eye 12 feet. We interpolate in the almanac to get the twilight times:

	LMT	GMT
Nautical twilight	0327	0607
Civil twilight	0408	0648
Sunrise	0439	0719

The LHA γ at 0408 LMT is 346°. Our plan is:

MAG.	BODY	ALT.	AZ.	
2.2	Kochab	30°	345°	
2.2	Diphda	28°	153°	
1.3	Fomalhaut	20°	182°	
1.1	Aldebaran	16°	082°	
0.9	Altair	37°	248°	
0.4	Saturn	36°	196°	
0.2	Capella	26°	050°	
0.1	Vega	40°	292°	
−3.9	Venus	24°	087°	X
-	Moon	28°	218°	X

We will probably not shoot Venus (whose azimuth is close to Aldebaran's) or the moon.

Our actual sights are:

BODY	GMT	Hs
Kochab	06-51-13	29°35′.7
Diphda	06-53-59	28°47′.4
Fomalhaut	06-55-45	20°55′.0
Aldebaran	06-57-31	18°11′.0
Capella	06-59-02	?
Vega	07-00-48	36°51′.7

After twilight was over, we saw that we had carelessly forgotten to read the sextant for the Capella sight. What is our July 6 0701 fix from the other sights?

Problem 13-2
From our July 6, 1993, 0701 fix, at 39°29′N, 39°43′W, we continue on 070°/3.2, I.C. -2ʹ.0, and height of eye 12 feet. The moon is out of position for a sun-moon fix, but if we are lucky enough to spot Venus in daylight, we can get a sun-Venus fix. We mark the base of the star-finder with the sun and Venus positions, rotate the transparent disc until we get suitable altitudes for both, and note the LHA γ on the outer scale. In the almanac, we find the GMT for this LHA γ, and at that time go on deck and look for Venus, with the sextant set to the predicted altitude, and facing toward the predicted azimuth. After a little searching, we spot Venus and get our sights.

BODY	GMT	Hs
Venus	13-40-17	55°09ʹ.3
Venus	13-41-59	54°55ʹ.3
Venus	13-43-42	54°38ʹ.5
Sun	13-45-11	69°02ʹ.9
Sun	13-46-50	69°14ʹ.4
Sun	13-48-35	69°29ʹ.3

What is our July 6 1349 fix?

Problem 13-3
From our July 6, 1993, 1349 fix, at 39°37′N, 39°14′W, we continue on course 070°, speed 3.2, with I.C. -2ʹ.0, and height of eye 12 feet. We plan for P.M. twilight.

	LMT	GMT
Sunset	1930	2205
Civil twilight	2003	2238
Nautical twilight	2044	2319

The LHA γ at 2003 LMT is 226°. Our plan includes more bodies than we need:

MAG.	BODY	ALT.	AZ.	
-1.5	Jupiter	36°	231°	X
0.9	Altair	20°	095°	
1.2	Spica	34°	210°	
1.2	Antares	21°	160°	
1.3	Deneb	31°	055°	
1.7	Mars	25°	262°	X
2.0	Dubhe	48°	322°	
2.1	Polaris	39°	000°	
2.1	Rasalhague	47°	119°	
2.2	Denebola	41°	254°	

From this list, we pick Altair, Spica, Antares, Deneb, Dubhe, Polaris, and Rasalhague as the best for our round of sights. But at sighting time, a haze obscures all but the brightest bodies. We get:

BODY	GMT	Hs
Jupiter	22-48-30	33° 52.′1
Jupiter	22-50-02	33° 39.′3
Altair	22-51-54	22° 26.′2
Altair	22-53-39	22° 45.′3
Antares	22-55-17	22° 04.′9
Antares	22-56-51	22° 07.′4

What is our July 6 2257 fix?

Problem 13-4
From our July 7, 1993, 2257 fix, at 39°46′N, 38°34′W, we maintain course 070°, but our speed picks up to 5.0 knots. Our I.C. is still -2.′0, and height of eye 12 feet. The night is cooler, the sky is clear, and many stars are visible as we plan for A.M. twilight.

	LMT	GMT
Nautical twilight	0325	0556
Civil twilight	0406	0637
Sunrise	0438	0709

The LHA γ at 0406 LMT is 347°. We list a lot of bodies in the plan, including Venus, which we will probably not sight, because its azimuth is close to Aldebaran's.

MAG.	BODY	ALT.	AZ.	
2.2	Kochab	30°	345°	
2.2	Diphda	28°	154°	
1.3	Fomalhaut	20°	183°	
1.1	Aldebaran	17°	082°	
0.9	Altair	36°	249°	
0.4	Saturn	36°	197°	
0.2	Capella	26°	051°	
0.1	Vega	39°	292°	
-3.9	Venus	24°	086°	X
-	Moon	36°	208°	

The incredible sighting conditions hold up, the boat is a steady platform, and we sight everything on the list:

BODY	GMT	Hs
Kochab	06-25-09	30° 48.′2
Diphda	06-26-43	27° 03.′5
Fomalhaut	06-28-21	20° 30.′0
Aldebaran	06-30-01	15° 18.′9
Altair	06-31-39	37° 23.′1
Saturn	06-33-18	35° 43.′3
Capella	06-34-55	26° 04.′0
Vega	06-36-30	39° 14.′6
Venus	06-38-08	24° 32.′7
Moon (LL)	06-40-02	35° 34.′4

What is our July 7, 0640 fix?

Problem 13-5

Since our July 7, 1993, 0640 fix, at 40°01′N, 37°37′W, we continue on 070°/5.0, with I.C. -2′.0 and height of eye 12 feet. There are scattered cumulus, and we know we won't find Venus. We will be lucky to get a running fix on the sun.

We get three good sights in the late morning, and wait a little over three hours for the sun to get to a good azimuth to cross with our morning LOPs. Clouds obscure the lower part of the sun for the last two shots, and we use the upper limb.

BODY	GMT	Hs
Sun (LL)	13-31-07	67°57′.2
Sun (LL)	13-34-16	68°21′.2
Sun (LL)	13-37-05	68°40′.6
Sun (LL)	15-51-02	65°18′.3
Sun (UL)	15-56-27	65°01′.3
Sun (UL)	15-58-11	64°43′.6

What is our 1337-1558 R. fix?

Problem 13-6

We continue on 070°/5.0 from our July 7, 1993, 1558 fix, at 40°19′N, 36°30′W. Our I.C. is still -2′.0, and height of eye 12 feet. We plan for P.M. twilight.

	LMT	GMT
Sunset	1933	2156
Civil twilight	2006	2229
Nautical twilight	2048	2311

The LHA γ at 2006 LMT is 227°. Our sighting plan is:

MAG.	BODY	ALT.	AZ.	
-1.5	Jupiter	34°	232°	X
0.9	Altair	21°	096°	
1.2	Spica	34°	211°	
1.2	Antares	21°	161°	
1.3	Deneb	31°	055°	
1.7	Mars	24°	262°	X
2.0	Dubhe	48°	322°	
2.1	Polaris	40°	000°	
2.1	Rasalhague	48°	120°	
2.2	Denebola	40°	254°	

We will probably skip Jupiter and Mars, because of azimuth duplications, unless we need them to complete our round of sights.

The sky is clear at twilight, and we get:

BODY	GMT	Hs
Altair	22-20-35	19°22.′8
Spica	22-22-12	33°41.′2
Antares	22-23-47	20°23.′8
Deneb	22-25-08	31°24.′5
Dubhe	22-26-41	48°21.′8
Polaris	22-28-20	39°49.′3
Rasalhague	22-30-05	48°03.′2

What is our July 7 2230 fix?

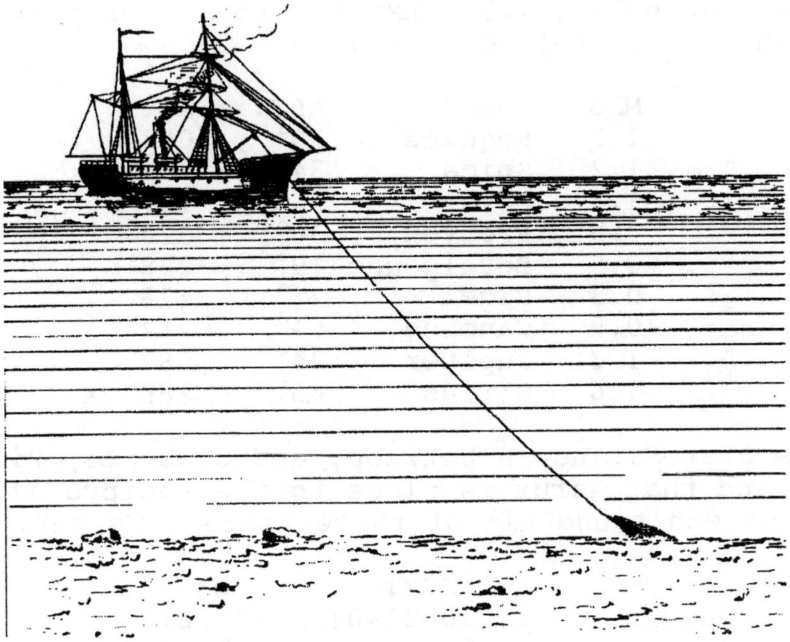

Chapter 14

Crozet Islands to Melbourne

Problem 14-1
We are approaching Australia's southern coast in a catamaran early in the morning of December 19 (GMT), 1993 (which is December 20, LMT). Our course is 090°, speed 14.0 knots, sextant I.C. -3.′0, and height of eye 10 feet. Our DR position at 2000 December 19 was 40°30′S, 110°00′E. We plan for A.M. twilight.

	LMT	GMT
Nautical twilight	0309 (12/20)	1948 (12/19)
Civil twilight	0352 (12/20)	2031 (12/19)
Sunrise	0425 (12/20)	2104 (12/19)

The LHA γ at 0352 (12/20) LMT is 146°. Our plan, from H.O. 249 Vol. 1 and the star-finder, is:

MAG.	BODY	ALT.	AZ.	
1.3	Regulus	38°	007°	
1.2	Spica	34°	075°	
1.1	Acrux	57°	147°	
0.5	Procyon	36°	320°	
0.5	Betelgeuse	19°	298°	
0.3	Rigel	22°	278°	
-0.9	Canopus	54°	232°	
-1.3	Jupiter	24°	087°	
-1.6	Sirius	45°	288°	X

We note that Sirius is between, and close to, Rigel and Betelgeuse, and that Acrux is close to the reciprocal of Procyon—so we won't use all of these stars. We get:

BODY	GMT	Hs
Regulus	20-27-01	37°16.′5
Spica	20-28-42	33°28.′7
Acrux	20-30-19	57°27.′8
Betelgeuse	20-32-52	18°35.′9
Rigel	20-33-35	21°45.′5
Canopus	20-35-08	53°31.′5

What is our December 19 2035 fix?

Problem 14-2
From our December 19, 1993, 2035 fix, at 40°30′S, 110°17′E, we continue moving nicely at 14.0 knots on course 090°. Our I.C. is

still -3:0, and height of eye 10 feet. We'd like a running fix on the sun, or a sun-moon fix.

Because the moon is a little too far from the sun for an optimum cut, we decide to shoot the sun first, then wait a couple of hours before getting our moon sights. The results are:

BODY	GMT	Hs
Sun	06-02-10	63° 08:5
Sun	06-03-48	62° 54:1
Sun	06-05-32	62° 34:7
Moon (UL)	08-12-13	42° 20:0
Moon (UL)	08-14-00	42° 30:5
Moon (UL)	08-15-39	42° 42:5

What is our December 20 0606-0816 R. fix?

Problem 14-3

From our last fix, at 0816 December 20, at 40° 20′S, 114° 06′E, we maintain our 090° course, at 14.0 knots, with I.C. still -3:0 and height of eye 10 feet. We plan for P.M. twilight.

	LMT	GMT
Sunset	1929	1148
Civil twilight	2002	1221
Nautical twilight	2044	1303

The LHA γ at 2002 LMT is 30°. Our plan is:

MAG.	BODY	ALT.	AZ.	
-	Moon	35°	312°	X
-1.6	Sirius	25°	091°	
-0.9	Canopus	44°	129°	
0.3	Rigel	36°	067°	
1.1	Aldebaran	23°	041°	
1.3	Fomalhaut	52°	271°	X
2.1	Peacock	36°	222°	X
2.2	Hamal	27°	002°	

There are a number of azimuth duplications in this list: the moon and Canopus; Saturn (which we didn't bother to list), Fomalhaut, and Sirius; Peacock and Aldebaran.

We get the following sights:

BODY	GMT	Hs
Sirius	12-15-42	23° 51:6
Canopus	12-17-21	44° 04:1
Rigel	12-18-59	35° 55:1
Aldebaran	12-20-39	22° 41:5
Hamal	12-22-03	26° 25:3

What is our December 20 1222 fix?

Chapter 15

New York to Lisbon

Problem 15-1

The passenger airship has been revived, as a leisurely but reasonably fast means of ocean travel, and we are navigating to Portugal in one. It is early in the morning of May 15, 1993, and our last DR position was 40°15′N, 59°10′W, at 0700 today. Our course is 095°, speed 75 knots. The airship has an observation platform set into a small depression on top, and we hope to get a round of A.M. twilight sights from there, using an ordinary marine sextant (I.C. -2.′0). From higher altitudes, the horizon is often obscured by haze, so we will descend to 1000 feet for the sights. The *Nautical Almanac* "Explanations" section gives the formula for height-of-eye corrections as 0.97 $\sqrt{\text{H.E.}}$ (in feet). We find the twilight times as usual:

	LMT	GMT
Nautical twilight	0337	0727
Civil twilight	0415	0805
Sunrise	0446	0836

The LHA γ at 0415 LMT is 297°. Our sighting plan is:

MAG.	BODY	ALT.	AZ.	
2.5	Enif	50°	132°	X
2.2	Alpheratz	37°	080°	
2.2	Kochab	42°	339°	
2.1	Rasalhague	50°	236°	
1.9	Alkaid	29°	312°	
0.9	Altair	59°	179°	
0.6	Saturn	28°	140°	
0.2	Arcturus	17°	281°	
-4.6	Venus	15°	096°	X
-	Moon	30°	122°	

We note that Enif has nearly the same azimuth as the reciprocal of Alkaid, and ditto for Venus and Arcturus. We'd like to get Alpheratz, Kochab, Rasalhague, Alkaid, Altair, and Arcturus.

But at twilight we find that observation conditions are far from optimum. We do see the brightest bodies, however, and shoot the following:

BODY	GMT	Hs
Altair	08-15-32	59°05.9
Saturn	08-17-19	29°45.3
Venus	08-19-02	18°07.4
Moon (LL)	08-21-05	32°45.7

What is our May 15 0821 fix?

Problem 15-2

From our May 15, 1993, 0821 fix, at 40°16′N, 57°08′W, we continue on course 095° at 75 knots. Our I.C. is still -2.0, and for the next group of sights we will drop to 800 feet. We make plans for a sun-moon fix.

When the bodies have moved into position, we get the following sights:

BODY	GMT	Hs
Moon (UL)	11-17-31	49°06.9
Moon (UL)	11-19-08	49°02.5
Moon (UL)	11-20-45	49°00.3
Sun	11-24-03	34°45.5
Sun	11-25-38	35°04.9
Sun	11-27-22	35°28.3

What is our May 15 1127 fix?

Problem 15-3

We expect to overfly the Azores during the night. From our May 15, 1993, 1127 fix, at 40°09′N, 52°13′W, we change course to 092°, maintaining 75 knots, and for our P.M. twilight sights we will fly at 1500 feet. Our I.C. is still -2.0.

	LMT	GMT
Sunset	1906	2126
Civil twilight	1937	2157
Nautical twilight	2015	2235

The LHA γ at 1937 LMT is 168°. Our plan is:

MAG.	BODY	ALT.	AZ.
-1.9	Jupiter	47°	155°
0.2	Capella	28°	308°
0.2	Arcturus	46°	104°
0.5	Procyon	31°	249°
1.2	Pollux	46°	271°
1.2	Spica	30°	142°
1.3	Regulus	59°	211°
2.1	Polaris	40°	359°
2.2	Kochab	48°	019°

The weather is fair at sighting time, and we get six clear sights:

BODY	GMT	Hs
Capella	21-56-40	28° 26!4
Arcturus	21-58-16	47° 15!8
Procyon	21-59-57	30° 23!5
Spica	22-01-32	31° 52!5
Regulus	22-03-12	58° 03!0
Polaris	22-05-01	40° 18!2

What is our May 15 2205 fix?

Chapter 16

Oporto to Dover

Problem 16-1
We are sailing from Portugal to Delaware, early in the morning of July 8, 1993. After a week of overcast weather, a cold front has passed, shaking up the boat a little, but leaving us with clear skies. We are moving nicely at 7.0 knots on course 270°. Our DR is doubtful, somewhere near 40°N, 40°W, at 0700. Our sextant I.C. is -1:0, and height of eye 9 feet. We prepare for A.M. twilight.

	LMT	GMT
Nautical twilight	0325	0605
Civil twilight	0406	0646
Sunrise	0438	0718

The LHA γ at 0406 LMT is 348°. Our plan is:

MAG.	BODY	ALT.	AZ.	
2.2	Kochab	30°	345°	
2.2	Diphda	28°	155°	
1.3	Fomalhaut	20°	183°	
1.1	Aldebaran	17°	083°	
0.9	Altair	36°	249°	
0.2	Capella	27°	051°	
0.1	Vega	38°	292°	
-3.9	Venus	24°	086°	X
-	Moon	44°	197°	

We note that Saturn will be visible just below the moon. If we shoot enough of the other bodies, we won't bother with Venus, whose azimuth is close to Aldebaran's.

The air is clear and cool as twilight approaches, and we get some good sights:

BODY	GMT	Hs
Kochab	06-40-13	30°47:1
Fomalhaut	06-41-52	20°02:5
Aldebaran	06-43-39	16°29:4
Altair	06-45-15	36°10:4
Capella	06-47-00	26°58:8
Vega	06-48-50	38°29:6

What is our July 8 0649 fix?

Problem 16-2

From our July 8, 1993, 0649 fix, at 40°25′N, 40°31′W, we continue on course 270° at 7.0 knots. Our I.C. is still -1′.0, and height of eye 9 feet. We mark the star-finder base with the current positions of the sun, moon, and Venus, and check to see whether we can get a daylight shot of Venus (if the sky stays clear) and at least one other body.

For a better cut of LOPs, we decide to get our moon sights at about 1000, then wait an hour and a quarter for the sun and Venus to move to better azimuths. The sky is clear, and we are lucky to spot Venus.

BODY	GMT	Hs
Moon (UL)	10-05-45	19°23′.0
Moon (UL)	10-07-30	19°06′.8
Venus	11-14-38	66°56′.4
Venus	11-17-03	67°02′.9
Sun	11-20-05	42°27′.5
Sun	11-21-48	42°50′.0

What is our 1008-1122 R. fix?

Problem 16-3

From our July 8, 1993, 1122 fix, at 40°23′N, 41°06′W, we continue on course 270° at 7.0 knots. Our height of eye is still 9 feet. We have not checked the sextant I.C. recently, since we found it to be -1′.0. We plan for P.M. twilight.

	LMT	GMT
Sunset	1932	2223
Civil twilight	2006	2257
Nautical twilight	2047	2338

The LHA γ at 2006 LMT is 228°. Our plan is:

MAG.	BODY	ALT.	AZ.	
-1.5	Jupiter	34°	232°	X
0.9	Altair	21°	096°	
1.2	Spica	33°	212°	
1.2	Antares	21°	162°	
1.3	Deneb	32°	055°	
1.7	Mars	23°	262°	X
2.0	Dubhe	47°	322°	
2.1	Polaris	40°	000°	
2.1	Rasalhague	48°	121°	
2.2	Denebola	39°	255°	

Azimuth duplicates are Jupiter-Deneb (reciprocal) and Mars-Denebola.

The sky is clear at twilight, and we get the following sights:

BODY	GMT	Hs
Altair	23-01-04	22° 31.'5
Spica	23-02-41	32° 10.'9
Antares	23-04-19	21° 35.'5
Deneb	23-05-55	33° 55.'5
Dubhe	23-07-45	46° 06.'5
Polaris	23-09-20	39° 36.'7

What is our 2309 fix?

Chapter 17

The Great Australian Bight

Problem 17-1

We've just been through a terrible storm. Our boat sank, but all six of our crew made it into the life raft, with fresh water, one bottle of champagne, and a carton of sardines. The EPIRB emergency transmitter was lost, but the good sextant was saved, with two of the navigational watches, a copy of H.O. 211, the *Nautical Almanac*, one chart, and pencils and paper. There is no privacy in our eight-person raft, and we are crammed in with the equipment we salvaged. We have seen jet trails, but no search aircraft. After three days, we have five gallons of water left, 30 small tins of sardines, and the champagne.

Today, the seas are down to five or six feet, and no one is seasick. It is 0700 December 18, 1993. We think we are close to 40°S, 138°E, and we hope to get a sun-moon fix this afternoon. Our height of eye will be 3 feet, and our sextant I.C. is -3.'5. The wind is moderate, and we have rigged a sail. With the eastward-setting current, we estimate our course and speed over the bottom at 095°/3.0.

We take some moon shots, then wait an hour, for a better cut, before shooting the sun.

BODY	GMT	Hs
Moon (UL)	07-03-15	57° 39.'9
Moon (UL)	07-05-01	57° 39.'2
Moon (UL)	07-06-42	57° 38.'2
Sun	08-15-03	19° 52.'3
Sun	08-16-40	19° 30.'2
Sun	08-18-05	19° 15.'4

What is our December 18 0707-0818 R. fix?

Problem 17-2

After our December 18, 1993, 0818 fix, at 39° 38'S, 138° 13'E, the wind shifts to NE, and picks up to 20 knots, so we hastily drop the sail and rig a sea anchor. We estimate the current here at 105/1.0. The raft is reasonably steady now, and we make plans for P.M. twilight sights after our evening rations of one sardine and a sip of water each.

	LMT	GMT
Sunset	1927	1014
Civil twilight	2000	1047
Nautical twilight	2040	1127

The LHA γ at 2000 LMT is 27°. Our plan is:

MAG.	BODY	ALT.	AZ.	
-	Moon	29°	286°	
-1.6	Sirius	22°	094°	X
-0.9	Canopus	43°	130°	
0.3	Rigel	34°	070°	
1.1	Aldebaran	21°	043°	
1.3	Fomalhaut	54°	273°	
2.1	Peacock	37°	223°	X
2.2	Alpheratz	17°	337°	
2.2	Hamal	26°	005°	

We note the azimuth duplications: Saturn (which we didn't list)-Sirius-Fomalhaut, and Aldebaran-Peacock.

We get the following:

BODY	GMT	Hs
Canopus	10-53-16	43° 30:7
Rigel	10-54-51	35° 57:2
Aldebaran	10-56-40	23° 03:1
Fomalhaut	10-58-31	51° 40:7
Alpheratz	11-00-12	16° 27:3
Hamal	11-01-47	27° 02:6

What is our 1102 fix?

Problem 17-3

From our December 18, 1993, 1102 fix, at 39°40'S, 138°16'E, we estimate our course and speed over the bottom to be 110/1.4. Height of eye is still 3 feet, I.C. -3.5. The waves are higher, and our motion is worse, but we will plan for a round of A.M. twilight sights.

	LMT	GMT
Nautical twilight	0313 (12/19)	1759 (12/18)
Civil twilight	0355 (12/19)	1841 (12/18)
Sunrise	0427 (12/19)	1913 (12/18)

The LHA γ at 0355 (12/19) LMT is 146°. Our plan is:

MAG.	BODY	ALT.	AZ.
1.3	Regulus	38°	007°
1.2	Spica	34°	075°
1.1	Acrux	57°	147°
0.5	Procyon	36°	320°
0.5	Betelgeuse	19°	298°
0.3	Rigel	22°	278°
-0.9	Canopus	54°	232°
-1.3	Jupiter	24°	087°

At twilight, the raft is no steadier, and we can't spot any of the first four stars on the list. But we manage to shoot Betelgeuse, then the other bright bodies.

BODY	GMT	Hs
Betelgeuse	18-50-31	17° 36′.4
Rigel	18-52-00	20° 23.0
Canopus	18-54-12	51° 50′.9
Jupiter	18-56-35	27° 04′.1
Jupiter	18-58-15	27° 23′.8

What is our December 18 1858 fix?

We note that we have been running along the coast, about 120 miles off, and are close to Mount Gambier. If the wind shifts back to W, or to S, we will hoist the sail and try to get closer to the mainland. If we get into the Bass Strait, we will have our chose of trying for the mainland or for Tasmania.

As we are planning our morning sights, a low-flying airplane is heard, and we try to flash it with our signalling mirrors. They see us, and make a low pass, waggling their wings. They come back and drop a package, and we vote to drink the champagne and have rations from the drop (no sardines) for lunch.

Chapter 18

Tortoli to Naples

Problem 18-1
We are sailing from Sardinia to the Italian mainland early in the morning of September 11, 1993, on a course of 070° at 4.5 knots. Our DR position was 40°05′N, 10°20′E, at 0400. Our sextant I.C. is +1′.0, and height of eye 10 feet. We plan for A.M. twilight.

	LMT	GMT
Nautical twilight	0437	0355
Civil twilight	0510	0428
Sunrise	0537	0455

The LHA γ at 0510 LMT is 68°. Our plan is:

MAG.	BODY	ALT.	AZ.	
2.5	Schedar	49°	314°	
2.2	Hamal	55°	253°	
2.1	Polaris	41°	000°	
2.0	Dubhe	31°	033°	
1.2	Pollux	49°	092°	
0.5	Procyon	36°	117°	
0.3	Rigel	41°	166°	
-1.6	Sirius	25°	145°	
-3.4	Venus	24°	088°	X
-	Moon	54°	114°	X

Our sights are:

BODY	GMT	Hs
Polaris	04-35-18	40°47′.9
Dubhe	04-37-05	32°06′.4
Pollux	04-38-46	50°35′.9
Procyon	04-40-12	37°21′.0
Rigel	04-41-55	41°16′.6
Sirius	04-43-47	26°53′.8

What is our September 11 0444 fix?

Problem 18-2
From our September 11, 1993, 0444 fix, at 40°06′N, 10°24′E, we continue on 070°/4.5, with I.C. +1′.0, and height of eye 10 feet. Conditions are a little rough, but we decide to try a sun-moon fix this morning. The results are:

BODY	GMT	Hs
Moon (UL)	08-55-13	56° 51.'8
Moon (UL)	08-57-04	56° 29.'6
Moon (UL)	08-58-48	56° 14.'5
Sun	09-01-05	43° 21.'1
Sun	09-02-47	43° 34.'2
Sun	09-04-22	43° 43.'2

What is our September 11 0904 fix?

Problem 18-3
Since our September 11, 1993, 0904 fix, at 40° 11′N, 10° 48′E, we continue on 070°/4.5, with I.C. +1.'0, and height of eye 10 feet. We prepare for P.M. twilight, hoping the sea will be calmer by then.

	LMT	GMT
Sunset	1816	1730
Civil twilight	1843	1757
Nautical twilight	1915	1829

The LHA γ at 1843 LMT is 272°. Our plan is:

MAG.	BODY	ALT.	AZ.	
0.2	Arcturus	36°	265°	
0.4	Saturn	14°	124°	X
0.9	Altair	51°	137°	
1.2	Antares	20°	204°	
1.9	Alkaid	44°	305°	
2.1	Polaris	40°	000°	
2.1	Nunki	23°	169°	
2.2	Alpheratz	18°	067°	
2.2	Kochab	49°	342°	

We may skip Saturn, because its azimuth is close to the reciprocal of Aklaid's.

But at sighting time, the sky and horizon are so clear that we decide to shoot everything we can, for the practice. In fact, we get every body in the plan:

BODY	GMT	Hs
Arcturus	17-52-03	37° 42.'6
Saturn	17-53-42	13° 41.'3
Altair	17-55-13	50° 42.'8
Antares	17-56-57	19° 36.'8
Alkaid	17-58-39	44° 28.'2
Polaris	18-00-19	39° 57.'5
Nunki	18-02-02	22° 39.'7
Alpheratz	18-03-48	19° 16.'1
Kochab	18-05-29	48° 56.'7

What is our September 11 1805 fix?

Chapter 19

Capetown to Gough Island

Problem 19-1

We are sailing from South Africa to the tiny island of Gough (Diego Alvarez) in the Tristan da Cunha Group, late in the afternoon of December 18, 1993. Our course is 250°, speed 7.2 knots, sextant I.C. -2.′1, and height of eye 8 feet. Our 1900 DR position is 39°35′S, 7°39′W. A good fix is important as we approach the island. Its height is shown as 2986 feet, so it should be visible from 65 miles away, but we want to be prepared in case the visibility is poor. We prepare for P.M. twilight.

	LMT	GMT
Sunset	1927	1959
Civil twilight	2000	2032
Nautical twilight	2040	2112

The LHA γ at 2000 LMT is 27°. Our plan is:

MAG.	BODY	ALT.	AZ.	
-	Moon	30°	291°	
-1.6	Sirius	23°	093°	X
-0.9	Canopus	43°	130°	
0.3	Rigel	34°	070°	
1.1	Aldebaran	21°	043°	
1.3	Fomalhaut	54°	273°	
2.1	Peacock	37°	223°	X
2.2	Hamal	26°	005°	
2.2	Alpheratz	17°	337°	

This gives us more bodies than we need, and Sirius-Fomalhaut and Aldebaran-Peacock are azimuth duplicates (their reciprocals). We'd like to get Canopus, Rigel, Aldebaran, Fomalhaut, Hamal, and Alpheratz.

But the sky is hazy at twilight time, and we shoot only the three brightest bodies:

BODY	GMT	Hs
Moon (LL)	20-27-03	30°42.′5
Moon (LL)	20-28-41	30°25.′6
Sirius	20-30-22	22°50.′9
Sirius	20-32-04	23°11.′4
Canopus	20-33-48	43°06.′7
Canopus	20-35-33	43°25.′1

What is our December 18 2036 fix?

Problem 19-2

From our December 18, 1993, 2036 fix, at 39°34'S, 7°51'W, we continue on 250°/7.2, with I.C. still -2!1, and height of eye 8 feet. We'd like to get a sun-moon fix after the moon rises, but we're concerned that the visibility will be worse by then. We decide to get some sun sights now—possibly to use for a running fix.

BODY	GMT	Hs
Sun	06-58-12	19°00!3
Sun	06-59-48	19°20!4
Sun	07-02-01	19°45!7
Sun	07-03-44	19°57!7
Sun	07-05-20	20!19!9

These sights were not great, but we did the best we could, with a sky already hazy and an indistinct horizon.

We plot the sights, and are pleased to find that they are reasonably close to our DR position (we are heading for a point several miles north of Gough)—but we want to give ourselves a better chance. We draw a line on the chart from Gough (at 40°19'S, 9°55'W) northward, parallel to the 07-05-20 sun LOP.

At 0745, after we finish working and plotting the sights, we turn to 284° (to get more distance to the north, in case we are farther south than our DR shows us to be), and make 7.0 knots on this heading. When we intersect the line we drew from Gough parallel to the sun LOP, we turn onto that line—that is, to 194°—and expect to make 6.0 knots on the new heading.

When do we turn to 194°, and what time do we expect to arrive at Gough?

Appendix A

Answers to Problems

Appendix A

Answers to Problems

1-1 Our 0744 fix is 40° 14′N, 49° 58′W, with good reliability.

1-2 Our 1229 fix is 40° 13′N, 49° 19′W. The LOPs have more spread than we would like, but we rate the reliability as good.

1-3 Our 1222-1528 R. fix is 40° 10′N, 48° 58′W. This is a poor fix, because of the excessive spread in the LOPs. In fact, there was little point in getting it, in such bad conditions, because we had a good sun-moon fix 3 hours ago.

1-4 Our 2252 fix is 40° 09′N, 48° 02′W. Three of the LOPs form a good (if not perfect) triangle, and the other is borderline. But we will use the fix, since it's all we will have until morning twilight—or later, if the visibility doesn't improve.

1-5 Our 0727 fix is 40° 09′N, 46° 46′W. This is a good fix. We will use it to update our DR.

1-6 Our 1225 fix is 40° 09′N, 46° 04′W. This is a good fix. Only the first moon sight was far from the central point.

1-7 Our noonsight-determined position is 40° 08′N, 45° 42′W. The latitude is very reliable, but the longitude is not. The top of our curve was easily determined, but we could have drawn it slightly to the left or right, and it would have seemed to match the points just as well. Crewmember C was right—it would have been better to get a running fix on the sun, with sights before and after local apparent noon.

1-8 No. 1's 2243 fix is 40° 07′N, 44° 37′W. No. 2's 2240 fix is 40° 10′N, 44° 38′W. No. 2's fix was good; No. 1's was excellent, because of its tight grouping of LOPs.

1-9 Our 0728 fix is 40° 08′N, 43° 24′W. This is a poor fix, because all of the LOPs are within 43° of azimuth. But it's close to our DR, and we decide to use it.

1-10 Our 1238 fix is 40° 07′N, 42° 41′W. This is an excellent fix—the best we've had on this cruise.

1-11 Our latitude at LAN is 40° 07′N. We consider it very reliable. It was true that we worked the noonsight without knowing the exact time (using the formula $L = 90° - Ho + d$). We needed only the approximate time to look up the sun's declination. But it would have been better to record the times, so we could plot Hs against GMT, draw a curve, and pick the highest point. (Also, the time is needed to plot the resulting LOP.) Still better would have been to get a running fix, with sun sights before and after LAN—or in the actual situation to get a sun-moon fix while the sun was still visible.

1-12 Our 2238 fix is 40°14′N, 41°47′W. We consider it very reliable. The Regulus LOP is a little off, but the others form a tight pattern, and are well spaced in azimuth.

2-1 Our 0541 fix is 40°35′S, 178°18′E. It appears reliable, but it would have been better to get the sun-moon fix a few hours earlier, when the crossing angle would have been closer to 90°.

2-2 Our 0810 fix is 40°30′S, 178°41′E. The pattern of LOPs is reasonably tight, and the fix appears reliable.

2-3 Our 1552 fix is 40°10′S, 179°53′E. We consider it very reliable. It is close to our DR position, and the LOPs make a tight pattern.

2-4 Our 2208-0333 running fix is 39°48′S, 178°49′W. It looks like an excellent fix.

2-5 Our 0222 fix is 39°44′S, 178°31′W. This is a good fix, showing moderate, apparently random, sighting errors.

2-6 Our 0752 fix is 39°46′S, 177°45′W. It's a good fix, considering the poor sighting conditions.

2-7 Our 1549 fix is 39°38′S, 176°27′W. This is a very good result, with all of the LOPs falling within a couple of miles of the fix point.

2-8 Our 0122 fix is 39°45′S, 175°02′W, and looks very reliable.

2-9 Something is obviously wrong with the Rigel sight. We found the star at about the predicted azimuth, but at a much lower altitude. When we check the star finder, we can see that we have shot Betelgeuse—not Rigel. We make this correction, and get a good plot. Our 0749 fix is 39°53′S, 174°28′W.

2-10 Our 1542 fix is 40°02′S, 173°41′W. This fix is only fair. The spread could be better, and we'd like at least one body near 000° or 180° azimuth. But it's all we have—and not too far from our DR position—so we decide to use it.

3-1 The unknown body was not Pollux, but Procyon. Our 0602 fix is 40°31′N, 153°08′W. We consider it good, because of the closely grouped LOPs and compatibility with the DR position.

3-2 Our 1440 fix is 40°01′N, 153°09′W. It appears reliable.

3-3 Our 2010 fix is 39°42′N, 153°11′W. This looks like a good fix.

4-1 Our 0800 fix is 40°11′N, 50°31′W. The LOP intercepts were long, because we were so far from our DR position—at least we hope that's the reason. This fix is only fair, as we had suspected, considering the poor sighting conditions. We will get another fix as soon as possible.

4-2 Our 1407 sun-Venus fix is 40°10′N, 49°27′W—an excellent fix, confirming this morning's twilight fix.

4-3 Our 2319 fix is 40°08′N, 47°50′W. At first glance, this looks almost useless, because of the poor cut of the LOPs, but we see that the first three sights form a very small triangle, and the Arcturus and Vega lines cross at 45°. The second Jupiter sight helps to confirm the others. This is a usable fix.

4-4 Our 0722 fix is 40°04′N, 46°36′W. The grouping is good, except for the Vega LOP, and we consider the fix reliable.

4-5 Our 0952 fix is 40°03′N, 46°12′W. The fix looks reliable. Although we could have used a simple two-body fix, the sun LOPs give us some additional assurance.

4-6 Our 2304 fix is 39°59′N, 44°12′W. We were lucky to have shot so many sights. Eight of them look good, but the Spica and Deneb LOPs are too far from the center, so we delete these and rely on the others.

4-7 Our 0719 fix is 40°00′N, 42°58′W. The azimuth distribution is poor, and the Saturn and moon azimuths are the same, but the grouping of the LOPs is good, and the fix is usable.

4-8 Our 1413–1705 R. fix is 39°59′N, 41°27′W. It appears reliable.

4-9 Our 2249 fix is 39°54′N, 40°36′W. It appears reliable.

5-1 Our 0524 fix, at 40°18′S, 157°26′E, looks like a very good one.

5-2 Our 0950 fix is 40°17′S, 158°06′E. It appears reliable.

5-3 Our 1729 fix is 40°17′S, 159°17′E. Except for Spica, the LOPs make a tight pattern. But it might be the Canopus sight that's off, so we don't exclude the Spica LOP.

5-4 Our 0308–0518 R. fix is 40°16′S, 161°06′E. Except for the first sun sight, the LOPs are reasonably close together. This looks like a good fix.

5-5 Our 0927 fix is 40°17′S, 161°43′E. The azimuth coverage is not good, but, except for Canopus, each pair of LOPs is very close. The fix is mediocre, but usable.

6-1 Four of the LOPs meet in a point, and one is close—but there is something wrong with Mars: its altitude is about 5° off our predicted altitude. We recheck the computations and find nothing wrong. Then we notice, in the plan, that Regulus is close to Mars in altitude and azimuth. We rework the sight as Regulus, and it comes out right. Our 0640 fix is 39°31′N, 157°16′W. It looks reliable.

6-2 Our 1443 fix is 39°35′N, 158°16′W. The LOPs are grouped well, not too far from the fix point.

6-3 Our 2113 fix is 39°38′N, 159°04′W—apparently a good fix.

6-4 Our 0649 fix is 39°42′N, 160°18′W. The LOPs all look good, except for Spica's, which is slightly off.

6-5 Our 1521 fix is 39°46′N, 161°20′W. It appears reliable, in spite of the poor sighting conditions.

6-6 Our 0033 fix is 39°51′N, 162°34′W. It looks like a very good fix.

6-7 Our 0702 fix, 39°56′N, 163°22′W, looks very good. All eight LOPs are within 2 miles of the fix point.

6-8 Our 1519 fix is 40°01′N, 164°26′W. The Diphda LOP is slightly off, but the others look excellent.

6-9 Our 1905 fix is 40°02′N, 164°54′W. It appears reliable.

7-1 Our 0850 fix is 40°26′N, 162°34′E. The Capella LOP is a little off, but the others are nicely grouped. It looks like a good fix.

7-2 Our sights were good, in spite of the poor visibility, but the LOPs cross at such a small angle that the fix is of little value. The 1748 fix (for what it's worth) is 39°55′N, 162°45′E. We need to replace this with a more reliable fix as soon as possible.

7-3 Our 2113 fix is 39°46′N, 162°51′E. The first sun sight is slightly off, but the fix looks good.

8-1 Our 1937 fix is 39°41′N, 1°12′E. This looks like an excellent fix.

8-2 Our 0404 fix is 40°10′N, 2°09′E. The Kochab LOP is slightly off, but it looks like a very good fix.

8-3 Our 0924 fix is 40°27N, 2°43′E. It appears to be very reliable. The compass deviation on this heading is 6°E.

9-1 Our 0416 fix is 39°52′N, 135°15′E. It looks reliable.

9-2 Something is wrong with the Antares sight, even after double-checking. It seemed to be at the right altitude and azimuth when we shot it; the only clue to something odd is that it looked too dim for 1.2 magnitude. We suspect a mis-identification—and when we check the list of stars for 1993, in the back of the almanac, we find a 2.9-magnitude star, Tau Scorpii, close to Antares, with SHA 111°07′.2 and declination S28°12′.2. We work the sight with these figures, and get a reasonable LOP. The 1105 fix, at 39°50′N, 136°30′E, now looks excellent.

9-3 From an A.P. of 40°N, 137°53′E, we get an intercept of 1.0 toward 060°. We need to be careful about algebraic signs here; note that if the Hc is -21′ and the Ho is -20′, the Ho is *greater*, and the intercept is *toward*. Also, note that this sight may not be very reliable, because of uncertain refraction at low altitudes.

9-4 Combining our morning sun sights with the sunrise sight, we get a 1929 (7/6)-0210 (7/7) running fix at 39°51′N, 139°11′E. This fix is no more reliable than our single sunrise LOP.

9-5 Our 1101 fix is 39°49′N, 139°54′E. It looks good, in spite of the poor sighting conditions.

10-1 Our 1354 fix is 40°23′N, 5°17′E. It looks good.

10-2 Our 1655 fix, at 40°14′N, 5°35′E, looks very good.

10-3 Our 0633 fix, at 39°31′N, 6°55′E, looks very good.

11-1 Our 0438 fix is 39°37′N, 134°26′W. It looks like an excellent fix.

11-2 Our 1318 fix is 40°05′N, 133°52′W. Except for the second moon shot, the LOPs are tightly grouped. The angle of cut, 46°, is at about the limit for a reliable fix, but we consider this one usable.

11-3 Our 1954 fix is 40°25′N, 133°25′W, and it looks like a very good one.

11-4 The quick solution to our new course is 72 ° from the storm's course (on the opposite side), or 152 °. The complete maneuvering-board solution confirms this, and also gives the CPA as 133 miles, at 2332 May 13. (See Appendix B.)

11-5 Our 0440 fix is 39°40′N, 132°48′W. At least one of the Jupiter LOPs is off (and possibly the star sights), but the fix looks fairly good, considering the difficult sighting conditions.

11-6 Our 1306 fix is 39°49′N, 131°47′W. It looks like an excellent fix.

12-1 Our 0759 fix is 39°36′S, 30°55′W. It looks very good.

12-2 Our 1124 fix is 39°38′S, 31°23′W. The last sun sight is a little off; otherwise, it looks like a very good fix.

12-3 Something is wrong with the Spica LOP. It's possible that we misread the sextant scale as 35′.4, instead of 25′.4. We delete this LOP and plot the others to get what looks like a good 2036 fix at 39°45′S, 32°34′W.

12-4 We note the times for the Suhail and Aldebaran sights, and take the time halfway between (08-06-54) for the Diphda sight. We could hardly be more than a few seconds off, because we were shooting as fast as we could, and the time between Suhail and Aldebaran was only 3 minutes 21 seconds. Our 0814 fix is 39°56′S, 33°56′W, and it looks like a good one.

12-5 Our 1025 fix, 39°58′S, 34°09′W, looks very good.

12-6 Our 2048 fix is 40°13′S, 35°24′W. It appears reliable.

12-7 Our 0813 fix is 40°16′S, 36°51′W. It shows that our speed indicator registered too high, as we had suspected.

12-8 Our 1314 fix is 40°20′S, 37°26′W. It seems to be a very good fix.

12-9 Something is wrong with our Jupiter sights. We note that Jupiter, Mars, and Spica are all near 279° azimuth, and very close in altitude. We rework the "Jupiter" sights as Mars sights, and get a good 2102 fix, 40°23′S, 38°22′W.

13-1 Our 0701 fix, 39°29′N, 39°43′W, looks good, in spite of the loss of the Capella sight.

13-2 Our 1349 fix is 39°37′N, 39°14′W. It appears to be reliable.

13-3 Our 2257 fix, at 39°46′N, 38°34′W, looks very good.

13-4 Our 0640 fix is 40°01′N, 37°37′W. The Altair LOP is a little off, but the fix looks very good.

13-5 Our 1558 fix looks excellent. It is 40°19′N, 36°30′W.

13-6 Our 2230 fix is 40°30′N, 35°37′W. All the sights except Altair and Polaris meet almost in a point.

14-1 Something is wrong with the Betelgeuse LOP, but when we recheck the computations, we get the same answer. Then we note that the time from the Acrux sight to the Betelgeuse one was 2m 33s, while the time from Betelgeuse to Rigel was only 0m 43s. It isn't possible to read the sextant, record the time and altitude, reset the sextant, and shoot another star in such a short time—so we conclude that we wrote down the GMT for Betelgeuse a minute later than the correct time. We rework the sight using 20-31-52, and get a reasonable LOP. Our 2035 fix is 40°30'S, 110°17'E.

14-2 Our 0816 fix is 40°20'S, 114°06'E. It looks reliable.

14-3 Our 1222 fix is 40°16'S, 115°28'E. This looks like an excellent fix.

15-1 In spite of the poor sighting conditions, our 0821 fix, 40°16'N, 57°08'W, looks good.

15-2 Our 1127 fix is 40°09'N, 52°13'W. It looks very good.

15-3 The intercepts are long, but our 2205 fix, 40°10'N, 33°44'W, looks like a good one.

16-1 Our 0649 fix, at 40°25'N, 40°31'W, looks only fair (Aldebaran is off).

16-2 Our 1122 fix, at 40°23'N, 41°06'W, looks fair.

16-3 Something is wrong with these sights. The LOPs make a circle around an imaginary point, about 10' away. We conclude that our I.C. (which we forgot to check before or after the twilight sights) must be off. We rework the sights, using an I.C. of +10'.0, and find that we get a nice 2309 fix, at 40°17'N, 42°35'W. We also need to check the sextant to see if anything is loose or broken.

17-1 Our 0818 fix, at 39°38'S, 138°13'E, looks good.

17-2 Our 1102 fix, at 39°40'S, 138°16'E, is fair (Alpheratz is off).

17-3 Our 1858 fix is 39°46'S, 138°32'E. It's a poor fix (not surprising, considering the rough sea), but usable.

18-1 Our 0444 fix, at 40°06'N, 10°24'E, appears reliable.

18-2 Our 0904 fix is 40°11'N, 10°48'E. The LOPs are not closely grouped, but we consider the fix fairly good, considering the poor conditions we had for sighting.

18-3 At last we have a round of sights for which every LOP passes through the same point. The 1805 fix is 40°20'N, 11°32°E.

19-1 These LOPs make a poor cut, but our 2036 fix, such as it is, is at 39°34'S, 7°51'W. It looks plausible, compared to our DR position, so we decide to use it.

19-2 We turn to 194° at 0955 (at DR position 39°58'S, 9°48'W). Our ETA Gough is 1334.

Appendix B

Procedures

Appendix B

Procedures

A.M. Twilight

A navigator should not just go on deck at twilight and start shooting. Proper planning makes star identification easy, and is important for spotting planets. It also speeds up the work at sighting time, so more bodies can be observed.

1. On the right-hand daily page of the almanac, enter the upper half of the table (A.M.) with the expected DR latitude at the time of twilight. (Interpolate for latitude, either by eye or mathematically.) Take out the times of nautical twilight (time to get ready to sight), civil twilight (approximate time of sights), and sunrise (end of sighting period). These times are LMT. Convert each LMT to GMT by adding (if in W longitude) or subtracting (if in E longitude) the time equivalent of the expected DR longitude—found on the first yellow page in the back of the almanac.
2. Find the LHA γ at the LMT of civil twilight. The easiest way to do this is to enter the first column [labelled "UT/(GMT)"] of the left-hand almanac daily page with the *LMT* of civil twilight, and from the next column (labelled "Aries/G.H.A."), take out the *LHA* γ. Add the correction for minutes of LMT, from the yellow pages.
3. Enter H.O. 249 Vol. 1 with the nearest whole degree of latitude and LHA γ and take out the name, altitude (rounded to the nearest whole degree), and azimuth of each of the seven stars. List them in order of magnitude, from dimmest to brightest—the order in which they will be sighted.
4. Set up the star finder, 2102-D, by marking the positions of the moon and planets on the base, for the day in question. It helps, also, to mark the sun's position, for reference. (The RA, in degrees, equals the GHA γ minus the GHA of the body. Use the RA and the declination, with the red template, to mark the base.) Then use the LHA γ found in 2, above, to set the blue template, and read the altitude and azimuth of each available planet, and the moon, and add them to the list made in 3, above. Include Polaris if you need a body at that azimuth. (Approximate altitude equals latitude; azimuth is about 000°.)
5. Now, we have more bodies than we need for a good round of sights. Circle six or seven of the best ones, and put an "X" next to each one whose azimuth is the same as another body's (or its reciprocal). Considerations are: H.O. 249 Vol. 1 stars are preferred; bright bodies are preferred;

azimuth duplicates should be avoided, as should those objects below about 15° and above about 70°.

P.M. Twilight
Preparations are similar to those for A.M. twilight, except that now we list the LMT and GMT of sunset, civil twilight, and nautical twilight (from the bottom of the right-hand daily page), and we make up our sighting plan listing bodies from brightest to dimmest—the way we will sight them.

Sun-Moon Fix
When the moon is close to first quarter (age 7 days) or last quarter (age 21 days), the navigator should plan for a sun-moon fix in the morning (near last quarter) or afternoon (near first quarter). Mark the positions of the sun and moon on the star-finder base, as was done with the moon and planets for twilight preparations, and rotate the blue template until the bodies are suitably placed in altitude. Read the LHA γ on the outer scale, and find its LMT from the left-hand daily page of the almanac; add the minutes correction from the yellow pages. Convert this to GMT: LMT + W longitude (or - east longitude) = GMT.

Venus in Daylight
A rare navigation feat is to spot and shoot Venus in daylight. It can be done only when Venus is about 30° or more (2 hours) from the sun in GHA. (This is shown graphically on p. 9 of the *Nautical Almanac*.) Also, the sky must be clear and the boat steady, and the navigator must know where to look. The star-finder, or H.O. 249 Vol. 1 or 2 (or another method), can be used to find the expected altitude and azimuth for the current GMT, latitude, and longitude. Then the sextant is preset to the computed altitude, and the navigator turns to the computed azimuth and looks. It helps to know Venus is there, and to relax one's eyes so they can focus for distance.

The Noonsight
The noonsight is an anachronism—it is not recommended as useful for practical navigation. It's better to get a running fix of the sun, with sights before and after local apparent noon. If you want to do it anyway, as a connection with your pre-chronometer ancestors, or just for the fun of it, here's how:

1. Find the predicted GMT of LAN: The local apparent time (LAT) of local apparent noon (LAN) is 1200 (by definition). Convert the DR longitude to time (first yellow page in the almanac), and add (for W longitude) or subtract (for E longitude). Then add or subtract the equation of time (bottom right of the right-hand daily page): *Add* it if the sun's meridian passage is *after* 1200; if before, subtract. The result is the predicted GMT of LAN.
2. Find the predicted altitude (needed in cloudy weather): Solve the proper formula (No. 4, below) for Hp (instead of Ho), using the DR latitude and the sun's declination.

3. Start sighting well before the time of LAN, and continue for eight or ten sights after the sun's altitude has increased to its peak and started to drop. Plot the points on graph paper (GMT on the x axis, Hs on the y axis), draw a smooth curve through them, and select the peak Hs and its GMT (not necessarily one of the actual sights).
4. Find the latitude with the appropriate formula:
   d contrary to L: $L = 90° - Ho - d$
   d same as L, and <L: $L = 90° - Ho + d$
   d same as L, and >L: $L = Ho + d - 90°$
5. Some recklessly inclined navigators go on from there to deducing the longitude (a highly inaccurate procedure). If everything is perfect, the longitude, if W, equals the GHA of the sun at the time of LAN (when its altitude was highest). E longitude equals 360° minus the GHA of the sun at LAN. But everything won't be perfect. Expect a big longitude error if you try this.

The Backsight

The backsight is rarely needed (and can't always be used), but can be handy if a body is visible, but the horizon under it is obscured. It's described in *Bowditch* and other texts. The way you do it is to turn your back to the sun (or moon, planet, or star) and measure the angle from the opposite horizon up and over to the body.

There are two things to note. First, the result of 180° minus the actual altitude must not exceed the range of your sextant (probably something like 120°). Second, if the sight looks like an upper-limb one, it's lower limb, and vice-versa. A rough diagram should make both of these things clear.

Working the sight is not complicated, but it has some traps. There are two ways to find Ho. The one the author likes better is:

(a) Apply the normal I.C. and dip corrections to Hs.
(b) Subtract the result from 180°.
(c) If it looked like a normal sun or moon LL sight, apply the UL corrections, and vice versa. For a star or planet, apply the normal refraction correction.

Estimating Speed

Speed can be estimated fairly well by throwing a bio-degradable floating object (like a crumpled paper towel) overboard at the bow, and timing its passage to the stern; using the boat's length for the distance traveled; then converting feet per second to miles per hour. The formula is:

Knots = .592 x feet traveled ÷ seconds

Storm Avoidance

WWV, in Fort Collins, Colorado, broadcasts brief storm warnings for the North Atlantic and the Caribbean, with its time signals, on 2500, 5000, 10000, 15000, and 20000 KHz, at 8, 9, and 10 minutes after each hour, using information obtained at 0500, 1100, 1700, and 2300 GMT. WWVH, in Hawaii, does the same for the

North Pacific, at 48, 49, and 50 minutes, using data obtained at 0000, 0600, 1200, and 1800 GMT. Other stations broadcast weather forecasts and storm warnings on various frequencies, in Morse code or plain language.

A vessel that finds itself close to a storm's path has two problems: finding the best avoidance course in a hurry, then calculating the storm's closest approach and the time it will be there.

The first step is simple, if you have trig tables or a scientific calculator, and the second step can be done with a maneuvering board or a diagram. Let's use Problem 11-4 as an example. Given: A developing storm is 150 miles, 000° from the boat, moving SW (225°) at 20 knots. The boat can make 6 knots on a heading between ESE and S. Find (a) the boat's best course to avoid the storm; (b) the storm's closest point of approach (CPA), in miles, and the time it will be there.

(a) The difference between the storm's course and the boat's best course is:

$$C = \arccos (6 \div 20)$$
$$C = 72.5$$

The best course is 225 - 72.5 = 152.5

(b) The maneuvering board solution of the complete problem is:

1. From the boat's position, R, at the center, draw the storm vector, RA, 225°/20.
2. From A, draw a line tangent to the 6-knot circle, on the side opposite the storm—in this case, on the south side. This line is AB extended.
3. From R, draw a line perpendicular to AB extended, locating point B. RB is the boat's vector, measured as 152°/6. (This confirms the result in a, above.)
4. BA is the relative-movement vector, measured as 242°/19.
5. On another plot, locate C, 000°, 150 miles from R.
6. Draw a line perpendicular to CD extended, locating point D. Measure CD: 69 miles, the miles of relative movement.
7. Measure RD, 133 miles, the closest point of approach.
8. Find the time of the CPA: 69 ÷ 19 = 3.63 hours, or 3 hours 38 minutes—that is, at 2332.

Appendix C

Nautical Almanac Excerpts

A2 ALTITUDE CORRECTION TABLES 10°-90°—SUN, STARS, PLANETS

SUN

OCT.—MAR. App. Alt.	Lower Limb	Upper Limb	APR.—SEPT. App. Alt.	Lower Limb	Upper Limb
° ′	′	′	° ′	′	′
9 34	+10·8	−21·5	9 39	+10·6	−21·2
9 45	+10·9	−21·4	9 51	+10·7	−21·1
9 56	+11·0	−21·3	10 03	+10·8	−21·0
10 08	+11·1	−21·2	10 15	+10·9	−20·9
10 21	+11·2	−21·1	10 27	+11·0	−20·8
10 34	+11·3	−21·0	10 40	+11·1	−20·7
10 47	+11·4	−20·9	10 54	+11·2	−20·6
11 01	+11·5	−20·8	11 08	+11·3	−20·5
11 15	+11·6	−20·7	11 23	+11·4	−20·4
11 30	+11·7	−20·6	11 38	+11·5	−20·3
11 46	+11·8	−20·5	11 54	+11·6	−20·2
12 02	+11·9	−20·4	12 10	+11·7	−20·1
12 19	+12·0	−20·3	12 28	+11·8	−20·0
12 37	+12·1	−20·2	12 46	+11·9	−19·9
12 55	+12·2	−20·1	13 05	+12·0	−19·8
13 14	+12·3	−20·0	13 24	+12·1	−19·7
13 35	+12·4	−19·9	13 45	+12·2	−19·6
13 56	+12·5	−19·8	14 07	+12·3	−19·5
14 18	+12·6	−19·7	14 30	+12·4	−19·4
14 42	+12·7	−19·6	14 54	+12·5	−19·3
15 06	+12·8	−19·5	15 19	+12·6	−19·2
15 32	+12·9	−19·4	15 46	+12·7	−19·1
15 59	+13·0	−19·3	16 14	+12·8	−19·0
16 28	+13·1	−19·2	16 44	+12·9	−18·9
16 59	+13·2	−19·1	17 15	+13·0	−18·8
17 32	+13·3	−19·0	17 48	+13·1	−18·7
18 06	+13·4	−18·9	18 24	+13·2	−18·6
18 42	+13·5	−18·8	19 01	+13·3	−18·5
19 21	+13·6	−18·7	19 42	+13·4	−18·4
20 03	+13·7	−18·6	20 25	+13·5	−18·3
20 48	+13·8	−18·5	21 11	+13·6	−18·2
21 35	+13·9	−18·4	22 00	+13·7	−18·1
22 26	+14·0	−18·3	22 54	+13·8	−18·0
23 22	+14·1	−18·2	23 51	+13·9	−17·9
24 21	+14·2	−18·1	24 53	+14·0	−17·8
25 26	+14·3	−18·0	26 00	+14·1	−17·7
26 36	+14·4	−17·9	27 13	+14·2	−17·6
27 52	+14·5	−17·8	28 33	+14·3	−17·5
29 15	+14·6	−17·7	30 00	+14·4	−17·4
30 46	+14·7	−17·6	31 35	+14·5	−17·3
32 26	+14·8	−17·5	33 20	+14·6	−17·2
34 17	+14·9	−17·4	35 17	+14·7	−17·1
36 20	+15·0	−17·3	37 26	+14·8	−17·0
38 36	+15·1	−17·2	39 50	+14·9	−16·9
41 08	+15·2	−17·1	42 31	+15·0	−16·8
43 59	+15·3	−17·0	45 31	+15·1	−16·7
47 10	+15·4	−16·9	48 55	+15·2	−16·6
50 46	+15·5	−16·8	52 44	+15·3	−16·5
54 49	+15·6	−16·7	57 02	+15·4	−16·4
59 23	+15·7	−16·6	61 51	+15·5	−16·3
64 30	+15·8	−16·5	67 17	+15·6	−16·2
70 12	+15·9	−16·4	73 16	+15·7	−16·1
76 26	+16·0	−16·3	79 43	+15·8	−16·0
83 05	+16·1	−16·2	86 32	+15·9	−15·9
90 00			90 00		

STARS AND PLANETS

App. Alt.	Corrⁿ
° ′	′
9 56	−5·3
10 08	−5·2
10 20	−5·1
10 33	−5·0
10 46	−4·9
11 00	−4·8
11 14	−4·7
11 29	−4·6
11 45	−4·5
12 01	−4·4
12 18	−4·3
12 35	−4·2
12 54	−4·1
13 13	−4·0
13 33	−3·9
13 54	−3·8
14 16	−3·7
14 40	−3·6
15 04	−3·5
15 30	−3·4
15 57	−3·3
16 26	−3·2
16 56	−3·1
17 28	−3·0
18 02	−2·9
18 38	−2·8
19 17	−2·7
19 58	−2·6
20 42	−2·5
21 28	−2·4
22 19	−2·3
23 13	−2·2
24 11	−2·1
25 14	−2·0
26 22	−1·9
27 36	−1·8
28 56	−1·7
30 24	−1·6
32 00	−1·5
33 45	−1·4
35 40	−1·3
37 48	−1·2
40 08	−1·1
42 44	−1·0
45 36	−0·9
48 47	−0·8
52 18	−0·7
56 11	−0·6
60 28	−0·5
65 08	−0·4
70 11	−0·3
75 34	−0·2
81 13	−0·1
87 03	0·0
90 00	

App. Alt. / Additional Corrⁿ

1993

VENUS

Jan. 1–Feb. 2 May 28–July 15	
°	′
0	+0·2
41	+0·1
76	

Feb. 3–Feb. 26 May 5–May 27	
°	′
0	+0·3
34	+0·2
60	+0·1
80	

Feb. 27–Mar. 14 Apr. 19–May 4	
°	′
0	+0·4
29	+0·3
51	+0·2
68	+0·1
83	

Mar. 15–Apr. 18	
°	′
0	+0·5
26	+0·4
46	+0·3
60	+0·2
73	+0·1
84	

July 16–Dec. 31	
°	′
0	+0·1
60	

MARS

Jan. 1–Mar. 7	
°	′
0	+0·2
41	+0·1
76	

Mar. 8–Dec. 31	
°	′
0	+0·1
60	

DIP

Ht. of Eye (m)	Corrⁿ	Ht. of Eye (ft.)
2·4	−2·8	8·0
2·6	−2·9	8·6
2·8	−3·0	9·2
3·0	−3·1	9·8
3·2	−3·2	10·5
3·4	−3·3	11·2
3·6	−3·4	11·9
3·8	−3·5	12·6
4·0	−3·6	13·3
4·3	−3·7	14·1
4·5	−3·8	14·9
4·7	−3·9	15·7
5·0	−4·0	16·5
5·2	−4·1	17·4
5·5	−4·2	18·3
5·8	−4·3	19·1
6·1	−4·4	20·1
6·3	−4·5	21·0
6·6	−4·6	22·0
6·9	−4·7	22·9
7·2	−4·8	23·9
7·5	−4·9	24·9
7·9	−5·0	26·0
8·2	−5·1	27·1
8·5	−5·2	28·1
8·8	−5·3	29·2
9·2	−5·4	30·4
9·5	−5·5	31·5
9·9	−5·6	32·7
10·3	−5·7	33·9
10·6	−5·8	35·1
11·0	−5·9	36·3
11·4	−6·0	37·6
11·8	−6·1	38·9
12·2	−6·2	40·1
12·6	−6·3	41·5
13·0	−6·4	42·8
13·4	−6·5	44·2
13·8	−6·6	45·5
14·2	−6·7	46·9
14·7	−6·8	48·4
15·1	−6·9	49·8
15·5	−7·0	51·3
16·0	−7·1	52·8
16·5	−7·2	54·3
16·9	−7·3	55·8
17·4	−7·4	57·4
17·9	−7·5	58·9
18·4	−7·6	60·5
18·8	−7·7	62·1
19·3	−7·8	63·8
19·8	−7·9	65·4
20·4	−8·0	67·1
20·9	−8·1	68·8
21·4		70·5

Ht. of Eye	Corrⁿ
m	′
1·0	−1·8
1·5	−2·2
2·0	−2·5
2·5	−2·8
3·0	−3·0
See table ←	
m	′
20	−7·9
22	−8·3
24	−8·6
26	−9·0
28	−9·3
30	−9·6
32	−10·0
34	−10·3
36	−10·6
38	−10·8
40	−11·1
42	−11·4
44	−11·7
46	−11·9
48	−12·2
ft.	′
2	−1·4
4	−1·9
6	−2·4
8	−2·7
10	−3·1
See table ←	
ft.	′
70	−8·1
75	−8·4
80	−8·7
85	−8·9
90	−9·2
95	−9·5
100	−9·7
105	−9·9
110	−10·2
115	−10·4
120	−10·6
125	−10·8
130	−11·1
135	−11·3
140	−11·5
145	−11·7
150	−11·9
155	−12·1

App. Alt. = Apparent altitude = Sextant altitude corrected for index error and dip.

App. Alt.	SUN OCT.–MAR. Lower Limb	SUN OCT.–MAR. Upper Limb	SUN APR.–SEPT. Lower Limb	SUN APR.–SEPT. Upper Limb	STARS PLANETS
° ′	′	′	′	′	′
0 00	−18·2	−50·5	−18·4	−50·2	−34·5
03	17·5	49·8	17·8	49·6	33·8
06	16·9	49·2	17·1	48·9	33·2
09	16·3	48·6	16·5	48·3	32·6
12	15·7	48·0	15·9	47·7	32·0
15	15·1	47·4	15·3	47·1	31·4
0 18	−14·5	−46·8	−14·8	−46·6	−30·8
21	14·0	46·3	14·2	46·0	30·3
24	13·5	45·8	13·7	45·5	29·8
27	12·9	45·2	13·2	45·0	29·2
30	12·4	44·7	12·7	44·5	28·7
33	11·9	44·2	12·2	44·0	28·2
0 36	−11·5	−43·8	−11·7	−43·5	−27·8
39	11·0	43·3	11·2	43·0	27·3
42	10·5	42·8	10·8	42·6	26·8
45	10·1	42·4	10·3	42·1	26·4
48	9·6	41·9	9·9	41·7	25·9
51	9·2	41·5	9·5	41·3	25·5
0 54	− 8·8	−41·1	− 9·1	−40·9	−25·1
0 57	8·4	40·7	8·7	40·5	24·7
1 00	8·0	40·3	8·3	40·1	24·3
03	7·7	40·0	7·9	39·7	24·0
06	7·3	39·6	7·5	39·3	23·6
09	6·9	39·2	7·2	39·0	23·2
1 12	− 6·6	−38·9	− 6·8	−38·6	−22·9
15	6·2	38·5	6·5	38·3	22·5
18	5·9	38·2	6·2	38·0	22·2
21	5·6	37·9	5·8	37·6	21·9
24	5·3	37·6	5·5	37·3	21·6
27	4·9	37·2	5·2	37·0	21·2
1 30	− 4·6	−36·9	− 4·9	−36·7	−20·9
35	4·2	36·5	4·4	36·2	20·5
40	3·7	36·0	4·0	35·8	20·0
45	3·2	35·5	3·5	35·3	19·5
50	2·8	35·1	3·1	34·9	19·1
1 55	2·4	34·7	2·6	34·4	18·7
2 00	− 2·0	−34·3	− 2·2	−34·0	−18·3
05	1·6	33·9	1·8	33·6	17·9
10	1·2	33·5	1·5	33·3	17·5
15	0·9	33·2	1·1	32·9	17·2
20	0·5	32·8	0·8	32·6	16·8
25	− 0·2	32·5	0·4	32·2	16·5
2 30	+ 0·2	−32·1	− 0·1	−31·9	−16·1
35	0·5	31·8	+ 0·2	31·6	15·8
40	0·8	31·5	0·5	31·3	15·5
45	1·1	31·2	0·8	31·0	15·2
50	1·4	30·9	1·1	30·7	14·9
2 55	1·6	30·7	1·4	30·4	14·7
3 00	+ 1·9	−30·4	+ 1·7	−30·1	−14·4
05	2·2	30·1	1·9	29·9	14·1
10	2·4	29·9	2·1	29·7	13·9
15	2·6	29·7	2·4	29·4	13·7
20	2·9	29·4	2·6	29·2	13·4
25	3·1	29·2	2·9	28·9	13·2
3 30	+ 3·3	−29·0	+ 3·1	−28·7	−13·0

App. Alt.	SUN OCT.–MAR. Lower Limb	SUN OCT.–MAR. Upper Limb	SUN APR.–SEPT. Lower Limb	SUN APR.–SEPT. Upper Limb	STARS PLANETS
° ′	′	′	′	′	′
3 30	+ 3·3	−29·0	+ 3·1	−28·7	−13·0
35	3·6	28·7	3·3	28·5	12·7
40	3·8	28·5	3·5	28·3	12·5
45	4·0	28·3	3·7	28·1	12·3
50	4·2	28·1	3·9	27·9	12·1
3 55	4·4	27·9	4·1	27·7	11·9
4 00	+ 4·5	−27·8	+ 4·3	−27·5	−11·8
05	4·7	27·6	4·5	27·3	11·6
10	4·9	27·4	4·6	27·2	11·4
15	5·1	27·2	4·8	27·0	11·2
20	5·2	27·1	5·0	26·8	11·1
25	5·4	26·9	5·1	26·7	10·9
4 30	+ 5·6	−26·7	+ 5·3	−26·5	−10·7
35	5·7	26·6	5·5	26·3	10·6
40	5·9	26·4	5·6	26·2	10·4
45	6·0	26·3	5·8	26·0	10·3
50	6·2	26·1	5·9	25·9	10·1
4 55	6·3	26·0	6·0	25·8	10·0
5 00	+ 6·4	−25·9	+ 6·2	−25·6	− 9·9
05	6·6	25·7	6·3	25·5	9·7
10	6·7	25·6	6·4	25·4	9·6
15	6·8	25·5	6·6	25·2	9·5
20	6·9	25·4	6·7	25·1	9·4
25	7·1	25·2	6·8	25·0	9·2
5 30	+ 7·2	−25·1	+ 6·9	−24·9	− 9·1
35	7·3	25·0	7·0	24·8	9·0
40	7·4	24·9	7·2	24·6	8·9
45	7·5	24·8	7·3	24·5	8·8
50	7·6	24·7	7·4	24·4	8·7
5 55	7·7	24·6	7·5	24·3	8·6
6 00	+ 7·8	−24·5	+ 7·6	−24·2	− 8·5
10	8·0	24·3	7·8	24·0	8·3
20	8·2	24·1	8·0	23·8	8·1
30	8·4	23·9	8·1	23·7	7·9
40	8·6	23·7	8·3	23·5	7·7
6 50	8·7	23·6	8·5	23·3	7·6
7 00	+ 8·9	−23·4	+ 8·6	−23·2	− 7·4
10	9·1	23·2	8·8	23·0	7·2
20	9·2	23·1	9·0	22·8	7·1
30	9·3	23·0	9·1	22·7	7·0
40	9·5	22·8	9·2	22·6	6·8
7 50	9·6	22·7	9·4	22·4	6·7
8 00	+ 9·7	−22·6	+ 9·5	−22·3	− 6·6
10	9·9	22·4	9·6	22·2	6·4
20	10·0	22·3	9·7	22·1	6·3
30	10·1	22·2	9·8	22·0	6·2
40	10·2	22·1	10·0	21·8	6·1
8 50	10·3	22·0	10·1	21·7	6·0
9 00	+10·4	−21·9	+10·2	−21·6	− 5·9
10	10·5	21·8	10·3	21·5	5·8
20	10·6	21·7	10·4	21·4	5·7
30	10·7	21·6	10·5	21·3	5·6
40	10·8	21·5	10·6	21·2	5·5
9 50	10·9	21·4	10·6	21·2	5·4
10 00	+11·0	−21·3	+10·7	−21·1	− 5·3

Additional corrections for temperature and pressure are given on the following page.

For bubble sextant observations ignore dip and use the star corrections for Sun, planets, and stars.

ADDITIONAL REFRACTION CORRECTIONS FOR NON-STANDARD CONDITIONS

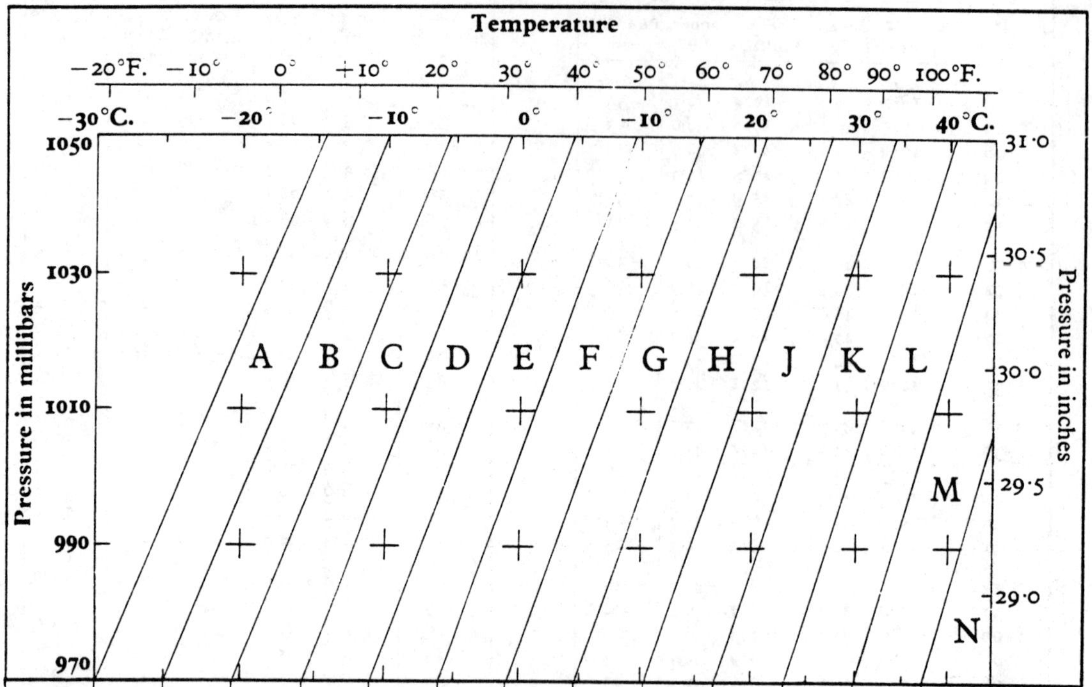

App. Alt.	A	B	C	D	E	F	G	H	J	K	L	M	N	App. Alt.
° ′	′	′	′	′	′	′	′	′	′	′	′	′	′	° ′
0 00	−6·9	−5·7	−4·6	−3·4	−2·3	−1·1	0·0	+1·1	+2·3	+3·4	+4·6	+5·7	+6·9	0 00
0 30	5·2	4·4	3·5	2·6	1·7	0·9	0·0	0·9	1·7	2·6	3·5	4·4	5·2	0 30
1 00	4·3	3·5	2·8	2·1	1·4	0·7	0·0	0·7	1·4	2·1	2·8	3·5	4·3	1 00
1 30	3·5	2·9	2·4	1·8	1·2	0·6	0·0	0·6	1·2	1·8	2·4	2·9	3·5	1 30
2 00	3·0	2·5	2·0	1·5	1·0	0·5	0·0	0·5	1·0	1·5	2·0	2·5	3·0	2 00
2 30	−2·5	−2·1	−1·6	−1·2	−0·8	−0·4	0·0	+0·4	+0·8	+1·2	+1·6	+2·1	+2·5	2 30
3 00	2·2	1·8	1·5	1·1	0·7	0·4	0·0	0·4	0·7	1·1	1·5	1·8	2·2	3 00
3 30	2·0	1·6	1·3	1·0	0·7	0·3	0·0	0·3	0·7	1·0	1·3	1·6	2·0	3 30
4 00	1·8	1·5	1·2	0·9	0·6	0·3	0·0	0·3	0·6	0·9	1·2	1·5	1·8	4 00
4 30	1·6	1·4	1·1	0·8	0·5	0·3	0·0	0·3	0·5	0·8	1·1	1·4	1·6	4 30
5 00	−1·5	−1·3	−1·0	−0·8	−0·5	−0·2	0·0	+0·2	+0·5	+0·8	+1·0	+1·3	+1·5	5 00
6	1·3	1·1	0·9	0·6	0·4	0·2	0·0	0·2	0·4	0·6	0·9	1·1	1·3	6
7	1·1	0·9	0·7	0·6	0·4	0·2	0·0	0·2	0·4	0·6	0·7	0·9	1·1	7
8	1·0	0·8	0·7	0·5	0·3	0·2	0·0	0·2	0·3	0·5	0·7	0·8	1·0	8
9	0·9	0·7	0·6	0·4	0·3	0·1	0·0	0·1	0·3	0·4	0·6	0·7	0·9	9
10 00	−0·8	−0·7	−0·5	−0·4	−0·3	−0·1	0·0	+0·1	+0·3	+0·4	+0·5	+0·7	+0·8	10 00
12	0·7	0·6	0·5	0·3	0·2	0·1	0·0	0·1	0·2	0·3	0·5	0·6	0·7	12
14	0·6	0·5	0·4	0·3	0·2	0·1	0·0	0·1	0·2	0·3	0·4	0·5	0·6	14
16	0·5	0·4	0·3	0·3	0·2	0·1	0·0	0·1	0·2	0·3	0·3	0·4	0·5	16
18	0·4	0·4	0·3	0·2	0·2	0·1	0·0	0·1	0·2	0·2	0·3	0·4	0·4	18
20 00	−0·4	−0·3	−0·3	−0·2	−0·1	−0·1	0·0	+0·1	+0·1	+0·2	+0·3	+0·3	+0·4	20 00
25	0·3	0·3	0·2	0·2	0·1	−0·1	0·0	+0·1	0·1	0·2	0·2	0·3	0·3	25
30	0·3	0·2	0·2	0·1	0·1	0·0	0·0	0·0	0·1	0·1	0·2	0·2	0·3	30
35	0·2	0·2	0·1	0·1	0·1	0·0	0·0	0·0	0·1	0·1	0·1	0·2	0·2	35
40	0·2	0·1	0·1	0·1	−0·1	0·0	0·0	0·0	+0·1	0·1	0·1	0·1	0·2	40
50 00	−0·1	−0·1	−0·1	−0·1	0·0	0·0	0·0	0·0	0·0	+0·1	+0·1	+0·1	+0·1	50 00

The graph is entered with arguments temperature and pressure to find a zone letter; using as arguments this zone letter and apparent altitude (sextant altitude corrected for dip), a correction is taken from the table. This correction is to be applied to the sextant altitude in addition to the corrections for standard conditions (for the Sun, stars and planets from page A2 and for the Moon from pages xxxiv and xxxv).

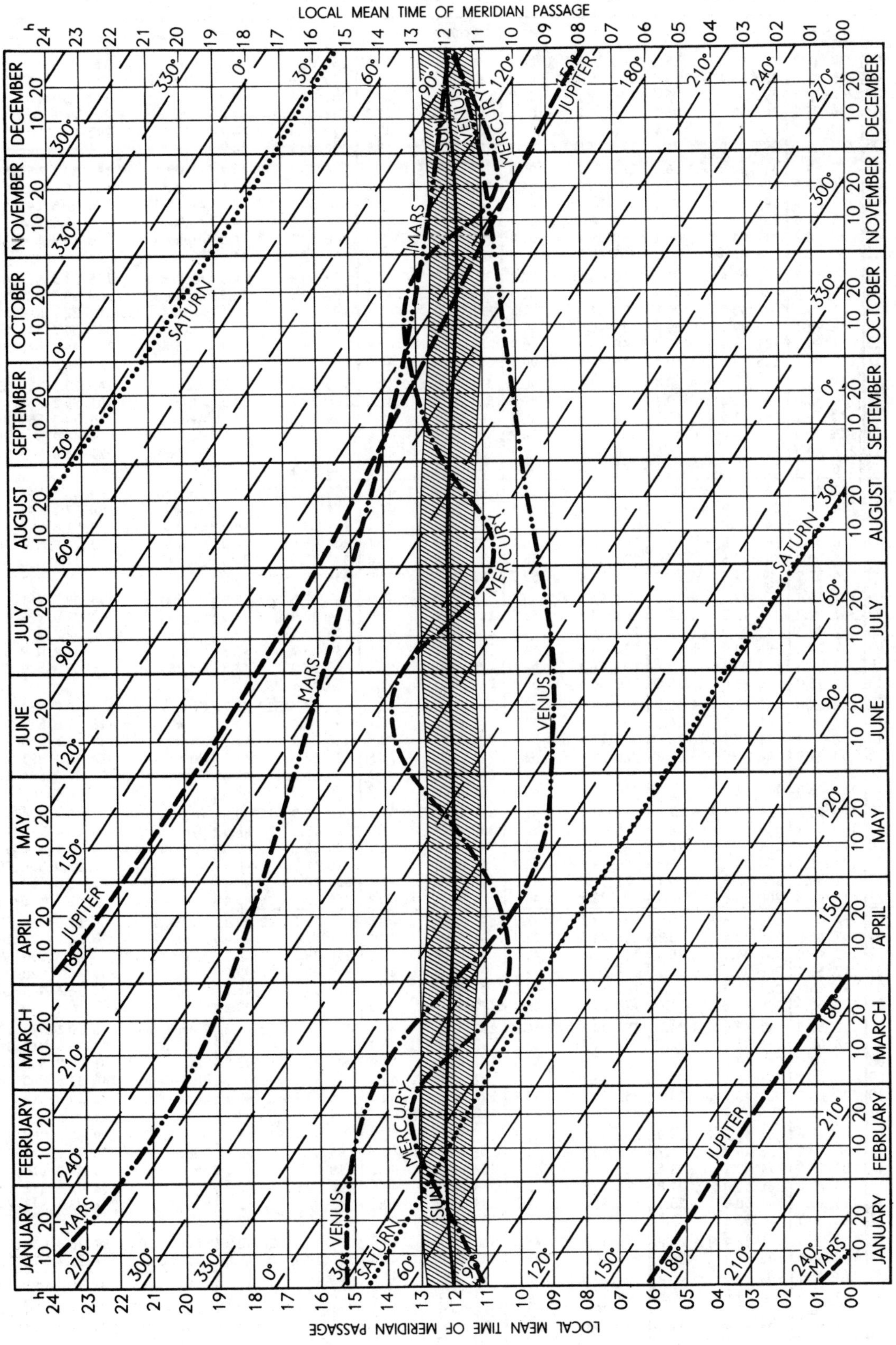
LOCAL MEAN TIME OF MERIDIAN PASSAGE
JANUARY
FEBRUARY
MARCH
APRIL
MAY
JUNE
JULY
AUGUST
SEPTEMBER
OCTOBER
NOVEMBER
DECEMBER
SUN
MERCURY
VENUS
MARS
JUPITER
SATURN
LOCAL MEAN TIME OF MERIDIAN PASSAGE

UT (GMT)	ARIES	VENUS −4.5		MARS +1.2		JUPITER −2.3		SATURN +0.9		STARS		
d h	G.H.A. ° ′	G.H.A. ° ′	Dec. ° ′	G.H.A. ° ′	Dec. ° ′	G.H.A. ° ′	Dec. ° ′	G.H.A. ° ′	Dec. ° ′	Name	S.H.A. ° ′	Dec. ° ′
13 00	230 45.8	220 55.5	N 4 05.3	100 27.8	N20 06.7	45 17.9	S 0 43.3	258 31.9	S12 44.9	Acamar	315 30.1	S40 19.8
01	245 48.2	235 56.4	05.5	115 28.9	06.3	60 20.5	43.3	273 34.3	44.8	Achernar	335 38.4	S57 16.0
02	260 50.7	250 57.3	05.8	130 30.1	05.9	75 23.1	43.2	288 36.6	44.8	Acrux	173 25.7	S63 04.1
03	275 53.2	265 58.2	· · 06.1	145 31.2	· · 05.6	90 25.6	· · 43.2	303 39.0	· · 44.8	Adhara	255 24.6	S28 58.0
04	290 55.6	280 59.1	06.4	160 32.4	05.2	105 28.2	43.1	318 41.4	44.7	Aldebaran	291 07.0	N16 29.7
05	305 58.1	296 00.0	06.6	175 33.6	04.8	120 30.8	43.1	333 43.7	44.7			
06	321 00.6	311 00.9	N 4 06.9	190 34.7	N20 04.5	135 33.4	S 0 43.0	348 46.1	S12 44.7	Alioth	166 33.3	N55 59.8
07	336 03.0	326 01.8	07.2	205 35.9	04.1	150 36.0	43.0	3 48.4	44.6	Alkaid	153 10.2	N49 20.8
T 08	351 05.5	341 02.8	07.5	220 37.0	03.7	165 38.6	43.0	18 50.8	44.6	Al Na'ir	28 02.5	S46 59.3
H 09	6 08.0	356 03.7	· · 07.8	235 38.2	· · 03.4	180 41.2	· · 42.9	33 53.1	· · 44.6	Alnilam	276 01.9	S 1 12.5
U 10	21 10.4	11 04.6	08.1	250 39.4	03.0	195 43.8	42.9	48 55.5	44.5	Alphard	218 10.9	S 8 38.0
R 11	36 12.9	26 05.5	08.3	265 40.5	02.6	210 46.4	42.8	63 57.8	44.5			
S 12	51 15.3	41 06.4	N 4 08.6	280 41.7	N20 02.2	225 49.0	S 0 42.8	79 00.2	S12 44.5	Alphecca	126 23.3	N26 44.1
D 13	66 17.8	56 07.3	08.9	295 42.8	01.9	240 51.6	42.7	94 02.5	44.5	Alpheratz	357 59.3	N29 03.1
A 14	81 20.3	71 08.2	09.2	310 44.0	01.5	255 54.2	42.7	109 04.9	44.4	Altair	62 22.7	N 8 51.0
Y 15	96 22.7	86 09.1	· · 09.5	325 45.1	· · 01.1	270 56.8	· · 42.6	124 07.2	· · 44.4	Ankaa	353 30.7	S42 20.3
16	111 25.2	101 09.9	09.8	340 46.3	00.8	285 59.4	42.6	139 09.6	44.4	Antares	112 44.3	S26 25.1
17	126 27.7	116 10.8	10.1	355 47.5	00.4	301 02.0	42.5	154 12.0	44.3			
18	141 30.1	131 11.7	N 4 10.4	10 48.6	N20 00.0	316 04.6	S 0 42.5	169 14.3	S12 44.3	Arcturus	146 09.1	N19 12.9
19	156 32.6	146 12.6	10.7	25 49.8	19 59.7	331 07.2	42.4	184 16.7	44.3	Atria	107 59.2	S69 00.9
20	171 35.1	161 13.5	11.0	40 50.9	59.3	346 09.8	42.4	199 19.0	44.2	Avior	234 24.4	S59 29.7
21	186 37.5	176 14.4	· · 11.3	55 52.1	· · 58.9	1 12.3	· · 42.4	214 21.4	· · 44.2	Bellatrix	278 48.4	N 6 20.6
22	201 40.0	191 15.3	11.6	70 53.3	58.5	16 14.9	42.3	229 23.7	44.2	Betelgeuse	271 17.8	N 7 24.3
23	216 42.4	206 16.1	11.9	85 54.4	58.2	31 17.5	42.3	244 26.1	44.1			
14 00	231 44.9	221 17.0	N 4 12.2	100 55.6	N19 57.8	46 20.1	S 0 42.2	259 28.5	S12 44.1	Canopus	264 03.2	S52 41.8
01	246 47.4	236 17.9	12.5	115 56.7	57.4	61 22.7	42.2	274 30.8	44.1	Capella	280 57.1	N45 59.5
02	261 49.8	251 18.8	12.8	130 57.9	57.1	76 25.3	42.1	289 33.2	44.0	Deneb	49 41.6	N45 15.2
03	276 52.3	266 19.6	· · 13.1	145 59.0	· · 56.7	91 27.9	· · 42.1	304 35.5	· · 44.0	Denebola	182 48.8	N14 36.4
04	291 54.8	281 20.5	13.4	161 00.2	56.3	106 30.5	42.0	319 37.9	44.0	Diphda	349 11.2	S18 01.3
05	306 57.2	296 21.4	13.7	176 01.4	55.9	121 33.1	42.0	334 40.2	44.0			
06	321 59.7	311 22.3	N 4 14.0	191 02.5	N19 55.6	136 35.7	S 0 42.0	349 42.6	S12 43.9	Dubhe	194 09.7	N61 47.3
07	337 02.2	326 23.1	14.3	206 03.7	55.2	151 38.3	41.9	4 44.9	43.9	Elnath	278 32.0	N28 36.1
08	352 04.6	341 24.0	14.6	221 04.8	54.8	166 40.8	41.9	19 47.3	43.9	Eltanin	90 52.7	N51 29.2
F 09	7 07.1	356 24.9	· · 14.9	236 06.0	· · 54.4	181 43.4	· · 41.8	34 49.7	· · 43.8	Enif	34 01.9	N 9 50.6
R 10	22 09.6	11 25.7	15.2	251 07.1	54.1	196 46.0	41.8	49 52.0	43.8	Fomalhaut	15 40.6	S29 39.3
I 11	37 12.0	26 26.6	15.6	266 08.3	53.7	211 48.6	41.7	64 54.4	43.8			
D 12	52 14.5	41 27.4	N 4 15.9	281 09.5	N19 53.3	226 51.2	S 0 41.7	79 56.7	S12 43.7	Gacrux	172 17.3	S57 04.9
A 13	67 16.9	56 28.3	16.2	296 10.6	53.0	241 53.8	41.7	94 59.1	43.7	Gienah	176 07.6	S17 30.6
Y 14	82 19.4	71 29.1	16.5	311 11.8	52.6	256 56.4	41.6	110 01.5	43.7	Hadar	149 08.7	S60 20.7
15	97 21.9	86 30.0	· · 16.8	326 12.9	· · 52.2	271 59.0	· · 41.6	125 03.8	· · 43.7	Hamal	328 18.1	N23 25.8
16	112 24.3	101 30.8	17.1	341 14.1	51.8	287 01.6	41.5	140 06.2	43.6	Kaus Aust.	84 03.4	S34 23.2
17	127 26.8	116 31.7	17.5	356 15.2	51.5	302 04.1	41.5	155 08.5	43.6			
18	142 29.3	131 32.5	N 4 17.8	11 16.4	N19 51.1	317 06.7	S 0 41.4	170 10.9	S12 43.6	Kochab	137 18.2	N74 11.0
19	157 31.7	146 33.4	18.1	26 17.6	50.7	332 09.3	41.4	185 13.2	43.5	Markab	13 53.4	N15 10.1
20	172 34.2	161 34.2	18.4	41 18.7	50.3	347 11.9	41.4	200 15.6	43.5	Menkar	314 31.1	N 4 03.8
21	187 36.7	176 35.1	· · 18.8	56 19.9	· · 50.0	2 14.5	· · 41.3	215 18.0	· · 43.5	Menkent	148 25.0	S36 20.5
22	202 39.1	191 35.9	19.1	71 21.0	49.6	17 17.1	41.3	230 20.3	43.4	Miaplacidus	221 43.0	S69 41.8
23	217 41.6	206 36.8	19.4	86 22.2	49.2	32 19.7	41.2	245 22.7	43.4			
15 00	232 44.1	221 37.6	N 4 19.7	101 23.3	N19 48.8	47 22.2	S 0 41.2	260 25.0	S12 43.4	Mirfak	309 02.4	N49 50.2
01	247 46.5	236 38.4	20.1	116 24.5	48.4	62 24.8	41.1	275 27.4	43.4	Nunki	76 16.7	S26 18.2
02	262 49.0	251 39.3	20.4	131 25.6	48.1	77 27.4	41.1	290 29.8	43.3	Peacock	53 42.6	S56 45.1
03	277 51.4	266 40.1	· · 20.7	146 26.8	· · 47.7	92 30.0	· · 41.1	305 32.1	· · 43.3	Pollux	243 46.2	N28 02.5
04	292 53.9	281 40.9	21.1	161 27.9	47.3	107 32.6	41.0	320 34.5	43.3	Procyon	245 15.6	N 5 14.4
05	307 56.4	296 41.8	21.4	176 29.1	46.9	122 35.2	41.0	335 36.8	43.2			
06	322 58.8	311 42.6	N 4 21.7	191 30.3	N19 46.6	137 37.7	S 0 40.9	350 39.2	S12 43.2	Rasalhague	96 20.1	N12 33.8
07	338 01.3	326 43.4	22.1	206 31.4	46.2	152 40.3	40.9	5 41.6	43.2	Regulus	207 59.4	N11 59.9
S 08	353 03.8	341 44.2	22.4	221 32.6	45.8	167 42.9	40.9	20 43.9	43.2	Rigel	281 26.8	S 8 12.7
A 09	8 06.2	356 45.1	· · 22.7	236 33.7	· · 45.4	182 45.5	· · 40.8	35 46.3	· · 43.1	Rigil Kent.	140 11.6	S60 48.6
T 10	23 08.7	11 45.9	23.1	251 34.9	45.1	197 48.1	40.8	50 48.7	43.1	Sabik	102 29.5	S15 43.0
U 11	38 11.2	26 46.7	23.4	266 36.0	44.7	212 50.7	40.7	65 51.0	43.1			
R 12	53 13.6	41 47.5	N 4 23.8	281 37.2	N19 44.3	227 53.2	S 0 40.7	80 53.4	S12 43.0	Schedar	349 58.2	N56 29.9
D 13	68 16.1	56 48.3	24.1	296 38.3	43.9	242 55.8	40.7	95 55.7	43.0	Shaula	96 41.9	S37 05.9
A 14	83 18.5	71 49.2	24.5	311 39.5	43.5	257 58.4	40.6	110 58.1	43.0	Sirius	258 47.2	S16 42.6
Y 15	98 21.0	86 50.0	· · 24.8	326 40.6	· · 43.2	273 01.0	· · 40.6	126 00.5	· · 42.9	Spica	158 46.8	S11 07.8
16	113 23.5	101 50.8	25.1	341 41.8	42.8	288 03.6	40.5	141 02.8	42.9	Suhail	223 03.6	S43 24.7
17	128 25.9	116 51.6	25.5	356 42.9	42.4	303 06.1	40.5	156 05.2	42.9			
18	143 28.4	131 52.4	N 4 25.8	11 44.1	N19 42.0	318 08.7	S 0 40.5	171 07.6	S12 42.9	Vega	80 48.8	N38 46.5
19	158 30.9	146 53.2	26.2	26 45.2	41.6	333 11.3	40.4	186 09.9	42.8	Zuben'ubi	137 21.7	S16 01.0
20	173 33.3	161 54.0	26.5	41 46.4	41.3	348 13.9	40.4	201 12.3	42.8		S.H.A.	Mer. Pass.
21	188 35.8	176 54.8	· · 26.9	56 47.6	· · 40.9	3 16.5	· · 40.3	216 14.6	· · 42.8		° ′	h m
22	203 38.3	191 55.6	27.2	71 48.7	40.5	18 19.0	40.3	231 17.0	42.8	Venus	349 32.1	9 14
23	218 40.7	206 56.4	27.6	86 49.9	40.1	33 21.6	40.3	246 19.4	42.7	Mars	229 10.7	17 15
Mer. Pass.	h m 8 31.6	*v* 0.9	*d* 0.3	*v* 1.2	*d* 0.4	*v* 2.6	*d* 0.0	*v* 2.4	*d* 0.0	Jupiter	174 35.2	20 51
										Saturn	27 43.5	6 41

UT (GMT) d	h	SUN G.H.A.	SUN Dec.	MOON G.H.A.	v	MOON Dec.	d	H.P.
		° ′	° ′	° ′	′	° ′	′	′
13	00	180 55.6	N18 19.9	273 13.2	14.4	S11 21.8	10.1	54.9
	01	195 55.6	20.5	287 46.6	14.4	11 11.7	10.1	54.8
	02	210 55.6	21.1	302 20.0	14.5	11 01.6	10.1	54.8
	03	225 55.6	. . 21.8	316 53.5	14.5	10 51.5	10.2	54.8
	04	240 55.6	22.4	331 27.0	14.6	10 41.3	10.3	54.8
	05	255 55.6	23.0	346 00.6	14.7	10 31.0	10.2	54.8
	06	270 55.6	N18 23.6	0 34.3	14.7	S10 20.8	10.3	54.7
	07	285 55.6	24.2	15 08.0	14.7	10 10.5	10.4	54.7
T	08	300 55.6	24.8	29 41.7	14.8	10 00.1	10.3	54.7
H	09	315 55.6	. . 25.5	44 15.5	14.9	9 49.8	10.4	54.7
U	10	330 55.6	26.1	58 49.4	14.9	9 39.4	10.5	54.7
R	11	345 55.6	26.7	73 23.3	14.9	9 28.9	10.4	54.7
S	12	0 55.6	N18 27.3	87 57.2	15.0	S 9 18.5	10.5	54.6
D	13	15 55.6	27.9	102 31.2	15.0	9 08.0	10.6	54.6
A	14	30 55.7	28.5	117 05.2	15.1	8 57.4	10.5	54.6
Y	15	45 55.7	. . 29.1	131 39.3	15.1	8 46.9	10.6	54.6
	16	60 55.7	29.7	146 13.4	15.2	8 36.3	10.6	54.6
	17	75 55.7	30.4	160 47.6	15.2	8 25.7	10.7	54.6
	18	90 55.7	N18 31.0	175 21.8	15.2	S 8 15.0	10.7	54.5
	19	105 55.7	31.6	189 56.0	15.3	8 04.3	10.7	54.5
	20	120 55.7	32.2	204 30.3	15.3	7 53.6	10.7	54.5
	21	135 55.7	. . 32.8	219 04.6	15.4	7 42.9	10.7	54.5
	22	150 55.7	33.4	233 39.0	15.4	7 32.2	10.8	54.5
	23	165 55.7	34.0	248 13.4	15.4	7 21.4	10.8	54.5
14	00	180 55.7	N18 34.6	262 47.8	15.5	S 7 10.6	10.8	54.5
	01	195 55.7	35.2	277 22.3	15.5	6 59.8	10.9	54.4
	02	210 55.7	35.8	291 56.8	15.5	6 48.9	10.9	54.4
	03	225 55.7	. . 36.4	306 31.3	15.6	6 38.0	10.8	54.4
	04	240 55.7	37.0	321 05.9	15.6	6 27.2	11.0	54.4
	05	255 55.7	37.6	335 40.5	15.7	6 16.2	10.9	54.4
	06	270 55.7	N18 38.2	350 15.2	15.6	S 6 05.3	10.9	54.4
	07	285 55.7	38.8	4 49.8	15.7	5 54.4	11.0	54.4
	08	300 55.7	39.4	19 24.5	15.7	5 43.4	11.0	54.4
F	09	315 55.7	. . 40.0	33 59.2	15.8	5 32.4	11.0	54.3
R	10	330 55.7	40.6	48 34.0	15.8	5 21.4	11.0	54.3
I	11	345 55.7	41.2	63 08.8	15.8	5 10.4	11.0	54.3
D	12	0 55.7	N18 41.8	77 43.6	15.8	S 4 59.4	11.1	54.3
A	13	15 55.7	42.4	92 18.4	15.8	4 48.3	11.1	54.3
Y	14	30 55.7	43.0	106 53.2	15.9	4 37.2	11.0	54.3
	15	45 55.7	. . 43.6	121 28.1	15.9	4 26.2	11.1	54.3
	16	60 55.7	44.2	136 03.0	15.9	4 15.1	11.1	54.3
	17	75 55.7	44.8	150 37.9	16.0	4 04.0	11.2	54.3
	18	90 55.7	N18 45.4	165 12.9	15.9	S 3 52.8	11.1	54.3
	19	105 55.6	46.0	179 47.8	16.0	3 41.7	11.1	54.3
	20	120 55.6	46.6	194 22.8	16.0	3 30.6	11.2	54.2
	21	135 55.6	. . 47.2	208 57.8	16.0	3 19.4	11.1	54.2
	22	150 55.6	47.8	223 32.8	16.0	3 08.3	11.2	54.2
	23	165 55.6	48.4	238 07.8	16.0	2 57.1	11.2	54.2
15	00	180 55.6	N18 49.0	252 42.8	16.1	S 2 45.9	11.2	54.2
	01	195 55.6	49.6	267 17.9	16.0	2 34.7	11.2	54.2
	02	210 55.6	50.2	281 52.9	16.1	2 23.5	11.2	54.2
	03	225 55.6	. . 50.8	296 28.0	16.1	2 12.3	11.2	54.2
	04	240 55.6	51.4	311 03.1	16.1	2 01.1	11.2	54.2
	05	255 55.6	52.0	325 38.2	16.1	1 49.9	11.3	54.2
	06	270 55.6	N18 52.5	340 13.3	16.1	S 1 38.6	11.2	54.2
	07	285 55.6	53.1	354 48.4	16.1	1 27.4	11.2	54.2
S	08	300 55.6	53.7	9 23.5	16.2	1 16.2	11.3	54.2
A	09	315 55.6	. . 54.3	23 58.7	16.1	1 04.9	11.2	54.2
T	10	330 55.6	54.9	38 33.8	16.1	0 53.7	11.2	54.2
U	11	345 55.6	55.5	53 08.9	16.2	0 42.5	11.3	54.2
R	12	0 55.5	N18 56.1	67 44.1	16.1	S 0 31.2	11.2	54.2
D	13	15 55.5	56.7	82 19.2	16.2	0 20.0	11.3	54.2
A	14	30 55.5	57.2	96 54.4	16.1	S 0 08.7	11.2	54.2
Y	15	45 55.5	. . 57.8	111 29.5	16.1	N 0 02.5	11.3	54.2
	16	60 55.5	58.4	126 04.6	16.2	0 13.8	11.2	54.2
	17	75 55.5	59.0	140 39.8	16.1	0 25.0	11.3	54.2
	18	90 55.5	N18 59.6	155 14.9	16.2	N 0 36.3	11.2	54.2
	19	105 55.5	19 00.2	169 50.1	16.1	0 47.5	11.3	54.2
	20	120 55.5	00.7	184 25.2	16.1	0 58.8	11.2	54.2
	21	135 55.5	. . 01.3	199 00.3	16.2	1 10.0	11.2	54.2
	22	150 55.5	01.9	213 35.5	16.1	1 21.2	11.3	54.2
	23	165 55.4	02.5	228 10.6	16.1	1 32.5	11.2	54.2
		S.D. 15.8	*d* 0.6	S.D. 14.9		14.8		14.8

Lat.	Twilight Naut.	Twilight Civil	Sunrise	Moonrise 13	14	15	16
°	h m	h m	h m	h m	h m	h m	h m
N 72	▭	▭	▭	02 19	02 00	01 43	01 28
N 70	////	////	00 54	02 02	01 50	01 40	01 30
68	////	////	01 53	01 47	01 42	01 37	01 32
66	////	////	02 26	01 36	01 35	01 35	01 34
64	////	01 15	02 50	01 26	01 30	01 33	01 36
62	////	01 56	03 09	01 17	01 25	01 31	01 37
60	////	02 22	03 25	01 10	01 21	01 30	01 38
N 58	01 07	02 43	03 38	01 04	01 17	01 28	01 39
56	01 44	03 00	03 49	00 58	01 13	01 27	01 40
54	02 09	03 14	03 59	00 53	01 10	01 26	01 41
52	02 29	03 26	04 08	00 48	01 07	01 25	01 42
50	02 45	03 36	04 16	00 44	01 05	01 24	01 43
45	03 15	03 58	04 32	00 35	00 59	01 22	01 44
N 40	03 38	04 15	04 46	00 27	00 55	01 20	01 45
35	03 55	04 29	04 57	00 20	00 51	01 19	01 47
30	04 10	04 41	05 07	00 15	00 47	01 18	01 48
20	04 33	05 01	05 24	00 04	00 41	01 15	01 49
N 10	04 51	05 17	05 39	24 35	00 35	01 13	01 51
0	05 06	05 31	05 53	24 30	00 30	01 11	01 52
S 10	05 19	05 44	06 06	24 25	00 25	01 10	01 54
20	05 31	05 58	06 21	24 19	00 19	01 08	01 56
30	05 43	06 12	06 37	24 13	00 13	01 05	01 57
35	05 49	06 20	06 47	24 09	00 09	01 04	01 59
40	05 55	06 28	06 57	24 05	00 05	01 03	02 00
45	06 02	06 38	07 10	24 00	00 00	01 01	02 01
S 50	06 10	06 50	07 25	23 55	24 59	00 59	02 03
52	06 14	06 55	07 32	23 52	24 58	00 58	02 04
54	06 17	07 01	07 40	23 49	24 57	00 57	02 05
56	06 21	07 07	07 49	23 46	24 56	00 56	02 06
58	06 25	07 14	07 59	23 42	24 55	00 55	02 07
S 60	06 30	07 22	08 11	23 38	24 53	00 53	02 08

Lat.	Sunset	Twilight Civil	Twilight Naut.	Moonset 13	14	15	16
°	h m	h m	h m	h m	h m	h m	h m
N 72	▭	▭	▭	09 57	11 44	13 27	15 10
N 70	23 11	////	////	10 13	11 52	13 28	15 03
68	22 05	////	////	10 26	11 58	13 28	14 58
66	21 30	////	////	10 36	12 03	13 28	14 54
64	21 05	22 44	////	10 45	12 07	13 29	14 50
62	20 45	22 01	////	10 52	12 11	13 29	14 47
60	20 29	21 33	////	10 58	12 14	13 29	14 44
N 58	20 16	21 12	22 51	11 04	12 17	13 29	14 41
56	20 05	20 55	22 12	11 09	12 20	13 29	14 39
54	19 55	20 40	21 46	11 13	12 22	13 30	14 37
52	19 46	20 28	21 26	11 17	12 24	13 30	14 35
50	19 38	20 17	21 10	11 21	12 26	13 30	14 34
45	19 21	19 55	20 39	11 29	12 30	13 30	14 30
N 40	19 07	19 38	20 16	11 35	12 33	13 30	14 27
35	18 56	19 24	19 58	11 41	12 36	13 30	14 25
30	18 46	19 12	19 43	11 45	12 38	13 31	14 23
20	18 28	18 52	19 20	11 54	12 43	13 31	14 19
N 10	18 14	18 36	19 02	12 01	12 47	13 31	14 15
0	18 00	18 22	18 47	12 08	12 50	13 31	14 12
S 10	17 46	18 08	18 34	12 15	12 54	13 31	14 09
20	17 32	17 55	18 22	12 22	12 57	13 32	14 06
30	17 15	17 41	18 09	12 30	13 02	13 32	14 02
35	17 06	17 33	18 03	12 35	13 04	13 32	14 00
40	16 55	17 24	17 57	12 40	13 07	13 32	13 57
45	16 42	17 14	17 50	12 46	13 10	13 32	13 54
S 50	16 27	17 02	17 42	12 53	13 13	13 32	13 51
52	16 20	16 57	17 38	12 57	13 15	13 32	13 49
54	16 12	16 51	17 35	13 00	13 17	13 32	13 48
56	16 03	16 45	17 31	13 04	13 19	13 32	13 46
58	15 53	16 38	17 27	13 09	13 21	13 32	13 44
S 60	15 41	16 30	17 22	13 14	13 24	13 32	13 41

Day	SUN Eqn. of Time 00 h	SUN Eqn. of Time 12 h	SUN Mer. Pass.	MOON Mer. Pass. Upper	MOON Mer. Pass. Lower	Age	Phase
	m s	m s	h m	h m	h m	d	
13	03 42	03 43	11 56	05 58	18 19	22	◑
14	03 43	03 43	11 56	06 40	19 01	23	
15	03 43	03 42	11 56	07 21	19 42	24	

UT (GMT) d	h	ARIES G.H.A.	VENUS −4.1 G.H.A.	VENUS Dec.	MARS +1.6 G.H.A.	MARS Dec.	JUPITER −2.0 G.H.A.	JUPITER Dec.	SATURN +0.6 G.H.A.	SATURN Dec.	STARS Name	S.H.A.	Dec.
		° ′	° ′	° ′	° ′	° ′	° ′	° ′	° ′	° ′		° ′	° ′
6	00	283 59.3	225 34.7	N17 28.8	124 30.8	N 9 43.7	97 30.1	S 1 24.1	311 34.9	S12 52.5	Acamar	315 29.8	S40 19.6
	01	299 01.7	240 34.4	29.5	139 31.9	43.1	112 32.4	24.2	326 37.4	52.5	Achernar	335 37.9	S57 15.8
	02	314 04.2	255 34.2	30.1	154 33.0	42.5	127 34.6	24.4	341 40.0	52.6	Acrux	173 26.1	S63 04.2
	03	329 06.7	270 33.9	· · 30.7	169 34.0	· · 42.0	142 36.9	· · 24.5	356 42.6	· · 52.6	Adhara	255 24.6	S28 57.8
	04	344 09.1	285 33.6	31.3	184 35.1	41.4	157 39.1	24.6	11 45.1	52.7	Aldebaran	291 06.8	N16 29.8
	05	359 11.6	300 33.3	31.9	199 36.2	40.8	172 41.4	24.7	26 47.7	52.7			
	06	14 14.0	315 33.0	N17 32.5	214 37.3	N 9 40.2	187 43.6	S 1 24.8	41 50.2	S12 52.8	Alioth	166 33.7	N55 59.9
	07	29 16.5	330 32.7	33.1	229 38.4	39.6	202 45.8	24.9	56 52.8	52.8	Alkaid	153 10.5	N49 20.9
T	08	44 19.0	345 32.4	33.7	244 39.5	39.1	217 48.1	25.0	71 55.3	52.8	Al Na'ir	28 02.0	S46 59.2
U	09	59 21.4	0 32.1	· · 34.3	259 40.6	· · 38.5	232 50.3	· · 25.1	86 57.9	· · 52.9	Alnilam	276 01.8	S 1 12.4
E	10	74 23.9	15 31.9	34.9	274 41.6	37.9	247 52.6	25.2	102 00.5	52.9	Alphard	218 11.0	S 8 37.9
S	11	89 26.4	30 31.6	35.5	289 42.7	37.3	262 54.8	25.3	117 03.0	53.0			
D	12	104 28.8	45 31.3	N17 36.1	304 43.8	N 9 36.8	277 57.1	S 1 25.4	132 05.6	S12 53.0	Alphecca	126 23.4	N26 44.3
A	13	119 31.3	60 31.0	36.7	319 44.9	36.2	292 59.3	25.5	147 08.1	53.1	Alpheratz	357 58.8	N29 03.3
Y	14	134 33.8	75 30.7	37.3	334 46.0	35.6	308 01.6	25.6	162 10.7	53.1	Altair	62 22.4	N 8 51.2
	15	149 36.2	90 30.4	· · 37.9	349 47.1	· · 35.0	323 03.8	· · 25.7	177 13.3	· · 53.1	Ankaa	353 30.3	S42 20.1
	16	164 38.7	105 30.1	38.5	4 48.2	34.4	338 06.0	25.8	192 15.8	53.2	Antares	112 44.2	S26 25.1
	17	179 41.2	120 29.8	39.1	19 49.3	33.8	353 08.3	25.9	207 18.4	53.2			
	18	194 43.6	135 29.5	N17 39.7	34 50.3	N 9 33.3	8 10.5	S 1 26.0	222 20.9	S12 53.3	Arcturus	146 09.2	N19 13.0
	19	209 46.1	150 29.2	40.3	49 51.4	32.7	23 12.8	26.1	237 23.5	53.3	Atria	107 59.0	S69 01.1
	20	224 48.5	165 28.9	40.9	64 52.5	32.1	38 15.0	26.2	252 26.1	53.4	Avior	234 24.7	S59 29.5
	21	239 51.0	180 28.6	· · 41.5	79 53.6	· · 31.5	53 17.2	· · 26.4	267 28.6	· · 53.4	Bellatrix	278 48.3	N 6 20.6
	22	254 53.5	195 28.3	42.1	94 54.7	30.9	68 19.5	26.5	282 31.2	53.4	Betelgeuse	271 17.7	N 7 24.3
	23	269 55.9	210 28.0	42.7	109 55.8	30.4	83 21.7	26.6	297 33.7	53.5			
7	00	284 58.4	225 27.7	N17 43.3	124 56.9	N 9 29.8	98 24.0	S 1 26.7	312 36.3	S12 53.5	Canopus	264 03.3	S52 41.5
	01	300 00.9	240 27.4	43.9	139 57.9	29.2	113 26.2	26.8	327 38.9	53.6	Capella	280 56.9	N45 59.4
	02	315 03.3	255 27.1	44.5	154 59.0	28.6	128 28.5	26.9	342 41.4	53.6	Deneb	49 41.2	N45 15.5
	03	330 05.8	270 26.8	· · 45.1	170 00.1	· · 28.0	143 30.7	· · 27.0	357 44.0	· · 53.7	Denebola	182 48.9	N14 36.5
	04	345 08.3	285 26.5	45.7	185 01.2	27.5	158 32.9	27.1	12 46.5	53.7	Diphda	349 10.8	S18 01.1
	05	0 10.7	300 26.2	46.3	200 02.3	26.9	173 35.2	27.2	27 49.1	53.8			
	06	15 13.2	315 25.9	N17 46.9	215 03.4	N 9 26.3	188 37.4	S 1 27.3	42 51.7	S12 53.8	Dubhe	194 10.1	N61 47.3
W	07	30 15.6	330 25.6	47.5	230 04.4	25.7	203 39.7	27.4	57 54.2	53.8	Elnath	278 31.8	N28 36.1
E	08	45 18.1	345 25.3	48.1	245 05.5	25.1	218 41.9	27.5	72 56.8	53.9	Eltanin	90 52.5	N51 29.5
D	09	60 20.6	0 25.0	· · 48.7	260 06.6	· · 24.5	233 44.1	· · 27.6	87 59.4	· · 53.9	Enif	34 01.5	N 9 50.8
N	10	75 23.0	15 24.7	49.3	275 07.7	24.0	248 46.4	27.7	103 01.9	54.0	Fomalhaut	15 40.1	S29 39.1
E	11	90 25.5	30 24.4	49.9	290 08.8	23.4	263 48.6	27.8	118 04.5	54.0			
S	12	105 28.0	45 24.1	N17 50.5	305 09.9	N 9 22.8	278 50.8	S 1 28.0	133 07.0	S12 54.1	Gacrux	172 17.6	S57 05.0
D	13	120 30.4	60 23.8	51.1	320 11.0	22.2	293 53.1	28.1	148 09.6	54.1	Gienah	176 07.7	S17 30.5
A	14	135 32.9	75 23.5	51.7	335 12.0	21.6	308 55.3	28.2	163 12.2	54.2	Hadar	149 08.9	S60 20.8
Y	15	150 35.4	90 23.2	· · 52.2	350 13.1	· · 21.0	323 57.6	· · 28.3	178 14.7	· · 54.2	Hamal	328 17.7	N23 25.9
	16	165 37.8	105 22.9	52.8	5 14.2	20.5	338 59.8	28.4	193 17.3	54.2	Kaus Aust.	84 03.1	S34 23.2
	17	180 40.3	120 22.5	53.4	20 15.3	19.9	354 02.0	28.5	208 19.9	54.3			
	18	195 42.8	135 22.2	N17 54.0	35 16.4	N 9 19.3	9 04.3	S 1 28.6	223 22.4	S12 54.3	Kochab	137 18.8	N74 11.2
	19	210 45.2	150 21.9	54.6	50 17.5	18.7	24 06.5	28.7	238 25.0	54.4	Markab	13 53.0	N15 10.3
	20	225 47.7	165 21.6	55.2	65 18.6	18.1	39 08.7	28.8	253 27.6	54.4	Menkar	314 30.8	N 4 03.9
	21	240 50.1	180 21.3	· · 55.8	80 19.6	· · 17.5	54 11.0	· · 28.9	268 30.1	· · 54.5	Menkent	148 25.0	S36 20.5
	22	255 52.6	195 21.0	56.4	95 20.7	17.0	69 13.2	29.0	283 32.7	54.5	Miaplacidus	221 43.6	S69 41.7
	23	270 55.1	210 20.7	57.0	110 21.8	16.4	84 15.4	29.1	298 35.2	54.6			
8	00	285 57.5	225 20.4	N17 57.5	125 22.9	N 9 15.8	99 17.7	S 1 29.3	313 37.8	S12 54.6	Mirfak	309 02.0	N49 50.2
	01	301 00.0	240 20.0	58.1	140 24.0	15.2	114 19.9	29.4	328 40.4	54.6	Nunki	76 16.4	S26 18.2
	02	316 02.5	255 19.7	58.7	155 25.1	14.6	129 22.2	29.5	343 42.9	54.7	Peacock	53 42.0	S56 45.1
	03	331 04.9	270 19.4	· · 59.3	170 26.1	· · 14.0	144 24.4	· · 29.6	358 45.5	· · 54.7	Pollux	243 46.3	N28 02.5
	04	346 07.4	285 19.1	17 59.9	185 27.2	13.5	159 26.6	29.7	13 48.1	54.8	Procyon	245 15.6	N 5 14.5
	05	1 09.9	300 18.8	18 00.5	200 28.3	12.9	174 28.9	29.8	28 50.6	54.8			
	06	16 12.3	315 18.5	N18 01.0	215 29.4	N 9 12.3	189 31.1	S 1 29.9	43 53.2	S12 54.9	Rasalhague	96 19.9	N12 34.0
	07	31 14.8	330 18.1	01.6	230 30.5	11.7	204 33.3	30.0	58 55.8	54.9	Regulus	207 59.6	N11 59.9
T	08	46 17.3	345 17.8	02.2	245 31.6	11.1	219 35.6	30.1	73 58.3	55.0	Rigel	281 26.7	S 8 12.5
H	09	61 19.7	0 17.5	· · 02.8	260 32.7	· · 10.5	234 37.8	· · 30.2	89 00.9	· · 55.0	Rigil Kent.	140 11.8	S60 48.8
U	10	76 22.2	15 17.2	03.4	275 33.7	09.9	249 40.0	30.3	104 03.5	55.1	Sabik	102 29.3	S15 43.0
R	11	91 24.6	30 16.8	03.9	290 34.8	09.4	264 42.3	30.5	119 06.0	55.1			
S	12	106 27.1	45 16.5	N18 04.5	305 35.9	N 9 08.8	279 44.5	S 1 30.6	134 08.6	S12 55.1	Schedar	349 57.5	N56 30.0
D	13	121 29.6	60 16.2	05.1	320 37.0	08.2	294 46.7	30.7	149 11.2	55.2	Shaula	96 41.7	S37 06.0
A	14	136 32.0	75 15.9	05.7	335 38.1	07.6	309 49.0	30.8	164 13.7	55.2	Sirius	258 47.2	S16 42.5
Y	15	151 34.5	90 15.6	· · 06.3	350 39.2	· · 07.0	324 51.2	· · 30.9	179 16.3	· · 55.3	Spica	158 46.9	S11 07.8
	16	166 37.0	105 15.2	06.8	5 40.2	06.4	339 53.4	31.0	194 18.9	55.3	Suhail	223 03.8	S43 24.6
	17	181 39.4	120 14.9	07.4	20 41.3	05.8	354 55.6	31.1	209 21.4	55.4			
	18	196 41.9	135 14.6	N18 08.0	35 42.4	N 9 05.3	9 57.9	S 1 31.2	224 24.0	S12 55.4	Vega	80 48.6	N38 46.8
	19	211 44.4	150 14.3	08.6	50 43.5	04.7	25 00.1	31.3	239 26.6	55.5	Zuben'ubi	137 21.8	S16 01.0
	20	226 46.8	165 13.9	09.1	65 44.6	04.1	40 02.3	31.4	254 29.1	55.5			
	21	241 49.3	180 13.6	· · 09.7	80 45.7	· · 03.5	55 04.6	· · 31.6	269 31.7	· · 55.6			
	22	256 51.7	195 13.3	10.3	95 46.8	02.9	70 06.8	31.7	284 34.3	55.6			
	23	271 54.2	210 12.9	10.9	110 47.8	02.3	85 09.0	31.8	299 36.8	55.6			
Mer. Pass.		h m 4 59.3	v −0.3	d 0.6	v 1.1	d 0.6	v 2.2	d 0.1	v 2.6	d 0.0			

	S.H.A.	Mer. Pass.
	° ′	h m
Venus	300 29.3	8 58
Mars	199 58.5	15 39
Jupiter	173 25.6	17 24
Saturn	27 37.9	3 09

UT (GMT) d	h	SUN G.H.A. ° ′	SUN Dec. ° ′	MOON G.H.A. ° ′	MOON *v* ′	MOON Dec. ° ′	MOON *d* ′	MOON H.P. ′
6	00	178 50.5	N22 42.5	334 56.4	12.7	S14 09.8	9.4	55.6
	01	193 50.4	42.3	349 28.1	12.7	14 00.4	9.5	55.6
	02	208 50.3	42.0	3 59.8	12.8	13 50.9	9.6	55.6
	03	223 50.2	. . 41.8	18 31.6	12.8	13 41.3	9.6	55.5
	04	238 50.1	41.5	33 03.4	12.9	13 31.7	9.6	55.5
	05	253 50.0	41.3	47 35.3	13.0	13 22.1	9.8	55.5
	06	268 49.9	N22 41.0	62 07.3	13.1	S13 12.3	9.7	55.5
	07	283 49.8	40.8	76 39.4	13.1	13 02.6	9.8	55.4
T	08	298 49.7	40.5	91 11.5	13.2	12 52.8	9.9	55.4
U	09	313 49.6	. . 40.3	105 43.7	13.2	12 42.9	9.9	55.4
E	10	328 49.5	40.0	120 15.9	13.4	12 33.0	10.0	55.4
S	11	343 49.4	39.8	134 48.3	13.3	12 23.0	10.0	55.4
D	12	358 49.3	N22 39.5	149 20.6	13.5	S12 13.0	10.0	55.3
A	13	13 49.1	39.3	163 53.1	13.5	12 03.0	10.1	55.3
Y	14	28 49.0	39.0	178 25.6	13.5	11 52.9	10.1	55.3
	15	43 48.9	. . 38.8	192 58.1	13.7	11 42.8	10.2	55.3
	16	58 48.8	38.5	207 30.8	13.6	11 32.6	10.3	55.3
	17	73 48.7	38.2	222 03.4	13.8	11 22.3	10.2	55.2
	18	88 48.6	N22 38.0	236 36.2	13.8	S11 12.1	10.3	55.2
	19	103 48.5	37.7	251 09.0	13.8	11 01.8	10.4	55.2
	20	118 48.4	37.5	265 41.8	14.0	10 51.4	10.4	55.2
	21	133 48.3	. . 37.2	280 14.8	13.9	10 41.0	10.4	55.2
	22	148 48.2	36.9	294 47.7	14.1	10 30.6	10.4	55.1
	23	163 48.1	36.7	309 20.8	14.1	10 20.2	10.5	55.1
7	00	178 48.0	N22 36.4	323 53.9	14.1	S10 09.7	10.6	55.1
	01	193 47.9	36.1	338 27.0	14.2	9 59.1	10.5	55.1
	02	208 47.8	35.9	353 00.2	14.2	9 48.6	10.6	55.1
	03	223 47.7	. . 35.6	7 33.4	14.3	9 38.0	10.7	55.0
	04	238 47.6	35.3	22 06.7	14.4	9 27.3	10.6	55.0
	05	253 47.5	35.1	36 40.1	14.4	9 16.7	10.7	55.0
	06	268 47.4	N22 34.8	51 13.5	14.5	S 9 06.0	10.7	55.0
W	07	283 47.3	34.5	65 47.0	14.5	8 55.3	10.8	55.0
E	08	298 47.2	34.3	80 20.5	14.5	8 44.5	10.8	54.9
D	09	313 47.1	. . 34.0	94 54.0	14.6	8 33.7	10.8	54.9
N	10	328 47.0	33.7	109 27.6	14.7	8 22.9	10.8	54.9
E	11	343 46.9	33.5	124 01.3	14.7	8 12.1	10.9	54.9
S	12	358 46.8	N22 33.2	138 35.0	14.7	S 8 01.2	10.9	54.9
D	13	13 46.7	32.9	153 08.7	14.8	7 50.3	10.9	54.9
A	14	28 46.6	32.6	167 42.5	14.8	7 39.4	10.9	54.8
Y	15	43 46.5	. . 32.4	182 16.3	14.9	7 28.5	11.0	54.8
	16	58 46.4	32.1	196 50.2	14.9	7 17.5	10.9	54.8
	17	73 46.3	31.8	211 24.1	15.0	7 06.6	11.0	54.8
	18	88 46.2	N22 31.5	225 58.1	15.0	S 6 55.6	11.1	54.8
	19	103 46.1	31.3	240 32.1	15.0	6 44.5	11.0	54.8
	20	118 46.0	31.0	255 06.1	15.1	6 33.5	11.1	54.7
	21	133 45.9	. . 30.7	269 40.2	15.1	6 22.4	11.1	54.7
	22	148 45.8	30.4	284 14.3	15.2	6 11.3	11.1	54.7
	23	163 45.8	30.2	298 48.5	15.2	6 00.2	11.1	54.7
8	00	178 45.7	N22 29.9	313 22.7	15.2	S 5 49.1	11.1	54.7
	01	193 45.6	29.6	327 56.9	15.3	5 38.0	11.1	54.7
	02	208 45.5	29.3	342 31.2	15.3	5 26.9	11.2	54.7
	03	223 45.4	. . 29.0	357 05.5	15.3	5 15.7	11.2	54.6
	04	238 45.3	28.7	11 39.8	15.4	5 04.5	11.2	54.6
	05	253 45.2	28.5	26 14.2	15.4	4 53.3	11.2	54.6
	06	268 45.1	N22 28.2	40 48.6	15.4	S 4 42.1	11.2	54.6
	07	283 45.0	27.9	55 23.0	15.5	4 30.9	11.2	54.6
T	08	298 44.9	27.6	69 57.5	15.5	4 19.7	11.2	54.6
H	09	313 44.8	. . 27.3	84 32.0	15.5	4 08.5	11.3	54.6
U	10	328 44.7	27.0	99 06.5	15.6	3 57.2	11.3	54.5
R	11	343 44.6	26.8	113 41.1	15.6	3 45.9	11.2	54.5
S	12	358 44.5	N22 26.5	128 15.7	15.6	S 3 34.7	11.3	54.5
D	13	13 44.4	26.2	142 50.3	15.6	3 23.4	11.3	54.5
A	14	28 44.3	25.9	157 24.9	15.6	3 12.1	11.2	54.5
Y	15	43 44.2	. . 25.6	171 59.5	15.7	3 00.9	11.3	54.5
	16	58 44.1	25.3	186 34.2	15.7	2 49.6	11.3	54.5
	17	73 44.0	25.0	201 08.9	15.7	2 38.3	11.3	54.5
	18	88 43.9	N22 24.7	215 43.6	15.8	S 2 27.0	11.3	54.5
	19	103 43.8	24.4	230 18.4	15.8	2 15.7	11.4	54.4
	20	118 43.7	24.1	244 53.2	15.7	2 04.3	11.3	54.4
	21	133 43.7	. . 23.8	259 27.9	15.8	1 53.0	11.3	54.4
	22	148 43.6	23.5	274 02.7	15.9	1 41.7	11.3	54.4
	23	163 43.5	23.2	288 37.6	15.8	1 30.4	11.3	54.4
		S.D. 15.8	*d* 0.3	S.D.	15.1	15.0		14.9

Lat. °	Twilight Naut. h m	Twilight Civil h m	Sunrise h m	Moonrise 6 h m	Moonrise 7 h m	Moonrise 8 h m	Moonrise 9 h m
N 72	▭	▭	▭	22 41	22 22	22 06	21 50
N 70	▭	▭	▭	22 25	22 13	22 04	21 54
68	▭	▭	▭	22 11	22 06	22 02	21 57
66	////	////	00 57	22 00	22 00	22 00	21 59
64	////	////	01 54	21 51	21 55	21 59	22 02
62	////	////	02 26	21 43	21 51	21 57	22 04
60	////	01 19	02 50	21 36	21 47	21 56	22 05
N 58	////	01 58	03 08	21 30	21 44	21 55	22 07
56	////	02 25	03 24	21 25	21 41	21 54	22 08
54	01 13	02 45	03 37	21 20	21 38	21 54	22 09
52	01 49	03 02	03 49	21 16	21 35	21 53	22 10
50	02 14	03 16	03 59	21 12	21 33	21 52	22 11
45	02 56	03 44	04 21	21 03	21 28	21 51	22 13
N 40	03 25	04 06	04 38	20 56	21 24	21 50	22 15
35	03 47	04 23	04 52	20 50	21 20	21 49	22 17
30	04 05	04 38	05 05	20 44	21 17	21 48	22 18
20	04 33	05 02	05 26	20 35	21 11	21 46	22 21
N 10	04 55	05 21	05 44	20 26	21 06	21 45	22 23
0	05 13	05 39	06 01	20 18	21 02	21 44	22 25
S 10	05 29	05 55	06 18	20 10	20 57	21 42	22 27
20	05 45	06 12	06 36	20 02	20 52	21 41	22 29
30	06 00	06 30	06 56	19 52	20 46	21 40	22 32
35	06 09	06 40	07 08	19 46	20 43	21 39	22 33
40	06 17	06 51	07 22	19 40	20 40	21 38	22 35
45	06 27	07 04	07 38	19 32	20 35	21 37	22 37
S 50	06 38	07 19	07 57	19 23	20 30	21 35	22 40
52	06 43	07 26	08 07	19 19	20 28	21 35	22 41
54	06 48	07 34	08 17	19 15	20 25	21 34	22 42
56	06 54	07 43	08 29	19 09	20 22	21 33	22 43
58	07 00	07 52	08 43	19 04	20 19	21 32	22 45
S 60	07 07	08 03	08 59	18 57	20 15	21 31	22 46

Lat. °	Sunset h m	Twilight Civil h m	Twilight Naut. h m	Moonset 6 h m	Moonset 7 h m	Moonset 8 h m	Moonset 9 h m
N 72	▭	▭	▭	04 38	06 38	08 27	10 10
N 70	▭	▭	▭	05 05	06 53	08 33	10 10
68	▭	▭	▭	05 26	07 05	08 38	10 09
66	23 08	////	////	05 42	07 14	08 42	10 09
64	22 14	////	////	05 55	07 22	08 46	10 08
62	21 42	////	////	06 06	07 29	08 49	10 08
60	21 19	22 48	////	06 15	07 35	08 52	10 07
N 58	21 00	22 10	////	06 24	07 40	08 54	10 07
56	20 45	21 44	////	06 31	07 44	08 56	10 07
54	20 32	21 24	22 54	06 37	07 49	08 58	10 07
52	20 20	21 07	22 19	06 43	07 52	09 00	10 06
50	20 10	20 53	21 55	06 48	07 56	09 02	10 06
45	19 49	20 25	21 13	06 59	08 03	09 05	10 06
N 40	19 31	20 04	20 45	07 08	08 09	09 08	10 06
35	19 17	19 46	20 23	07 16	08 14	09 10	10 05
30	19 05	19 32	20 05	07 23	08 18	09 12	10 05
20	18 44	19 08	19 37	07 35	08 26	09 16	10 04
N 10	18 25	18 48	19 15	07 45	08 33	09 19	10 04
0	18 09	18 31	18 57	07 55	08 39	09 22	10 04
S 10	17 52	18 14	18 41	08 04	08 46	09 25	10 03
20	17 34	17 58	18 25	08 14	08 52	09 28	10 03
30	17 14	17 40	18 10	08 26	09 00	09 32	10 02
35	17 02	17 30	18 01	08 32	09 04	09 34	10 02
40	16 48	17 19	17 52	08 40	09 09	09 36	10 02
45	16 32	17 06	17 43	08 48	09 15	09 39	10 01
S 50	16 13	16 51	17 32	08 59	09 21	09 42	10 01
52	16 03	16 44	17 27	09 04	09 25	09 43	10 00
54	15 53	16 36	17 22	09 09	09 28	09 45	10 00
56	15 41	16 27	17 16	09 15	09 32	09 46	10 00
58	15 27	16 18	17 10	09 21	09 36	09 48	10 00
S 60	15 11	16 07	17 03	09 28	09 40	09 50	09 59

Day	SUN Eqn. of Time 00ʰ m s	SUN Eqn. of Time 12ʰ m s	SUN Mer. Pass. h m	MOON Mer. Pass. Upper h m	MOON Mer. Pass. Lower h m	MOON Age d	MOON Phase
6	04 38	04 43	12 05	01 43	14 06	16	◑
7	04 48	04 52	12 05	02 29	14 51	17	
8	04 57	05 02	12 05	03 12	15 33	18	

1993 SEPTEMBER 10, 11, 12 (FRI., SAT., SUN.)

UT (GMT)	ARIES	VENUS −4.0		MARS +1.6		JUPITER −1.7		SATURN +0.4	
d h	G.H.A. ° ′	G.H.A. ° ′	Dec. ° ′	G.H.A. ° ′	Dec. ° ′	G.H.A. ° ′	Dec. ° ′	G.H.A. ° ′	Dec. ° ′
10 00	349 02.4	210 30.5	N16 25.9	151 46.1	S 7 04.0	153 01.7	S 5 37.4	20 48.6	S14 29.7
01	4 04.9	225 29.9	25.1	166 47.1	04.6	168 03.7	37.6	35 51.2	29.7
02	19 07.3	240 29.3	24.3	181 48.0	05.3	183 05.7	37.8	50 53.9	29.8
03	34 09.8	255 28.8	· · 23.6	196 48.9	· · 05.9	198 07.7	· · 38.0	65 56.5	· · 29.8
04	49 12.3	270 28.2	22.8	211 49.9	06.6	213 09.7	38.2	80 59.1	29.9
05	64 14.7	285 27.6	22.0	226 50.8	07.2	228 11.7	38.4	96 01.7	30.0
06	79 17.2	300 27.0	N16 21.2	241 51.8	S 7 07.9	243 13.7	S 5 38.6	111 04.4	S14 30.0
07	94 19.7	315 26.5	20.4	256 52.7	08.5	258 15.7	38.8	126 07.0	30.1
08	109 22.1	330 25.9	19.6	271 53.6	09.2	273 17.7	39.0	141 09.6	30.1
F 09	124 24.6	345 25.3	· · 18.8	286 54.6	· · 09.8	288 19.7	· · 39.2	156 12.2	· · 30.2
R 10	139 27.1	0 24.8	18.0	301 55.5	10.5	303 21.7	39.4	171 14.9	30.2
I 11	154 29.5	15 24.2	17.2	316 56.5	11.1	318 23.7	39.6	186 17.5	30.3
D 12	169 32.0	30 23.6	N16 16.5	331 57.4	S 7 11.8	333 25.7	S 5 39.8	201 20.1	S14 30.4
A 13	184 34.4	45 23.1	15.7	346 58.4	12.4	348 27.7	40.0	216 22.7	30.4
Y 14	199 36.9	60 22.5	14.9	1 59.3	13.0	3 29.7	40.2	231 25.4	30.5
15	214 39.4	75 21.9	· · 14.1	17 00.2	· · 13.7	18 31.6	· · 40.4	246 28.0	· · 30.5
16	229 41.8	90 21.4	13.3	32 01.2	14.3	33 33.6	40.6	261 30.6	30.6
17	244 44.3	105 20.8	12.5	47 02.1	15.0	48 35.6	40.8	276 33.2	30.6
18	259 46.8	120 20.3	N16 11.7	62 03.1	S 7 15.6	63 37.6	S 5 41.0	291 35.9	S14 30.7
19	274 49.2	135 19.7	10.9	77 04.0	16.3	78 39.6	41.2	306 38.5	30.8
20	289 51.7	150 19.1	10.1	92 04.9	16.9	93 41.6	41.4	321 41.1	30.8
21	304 54.2	165 18.6	· · 09.3	107 05.9	· · 17.6	108 43.6	· · 41.6	336 43.7	· · 30.9
22	319 56.6	180 18.0	08.5	122 06.8	18.2	123 45.6	41.8	351 46.4	30.9
23	334 59.1	195 17.4	07.7	137 07.8	18.9	138 47.6	42.0	6 49.0	31.0
11 00	350 01.6	210 16.9	N16 06.9	152 08.7	S 7 19.5	153 49.6	S 5 42.2	21 51.6	S14 31.0
01	5 04.0	225 16.3	06.1	167 09.6	20.2	168 51.6	42.3	36 54.2	31.1
02	20 06.5	240 15.7	05.3	182 10.6	20.8	183 53.6	42.5	51 56.9	31.1
03	35 08.9	255 15.2	· · 04.5	197 11.5	· · 21.5	198 55.6	· · 42.7	66 59.5	· · 31.2
04	50 11.4	270 14.6	03.7	212 12.4	22.1	213 57.6	42.9	82 02.1	31.3
05	65 13.9	285 14.1	02.9	227 13.4	22.8	228 59.6	43.1	97 04.7	31.3
06	80 16.3	300 13.5	N16 02.1	242 14.3	S 7 23.4	244 01.6	S 5 43.3	112 07.4	S14 31.4
07	95 18.8	315 12.9	01.2	257 15.3	24.1	259 03.6	43.5	127 10.0	31.4
S 08	110 21.3	330 12.4	16 00.4	272 16.2	24.7	274 05.6	43.7	142 12.6	31.5
A 09	125 23.7	345 11.8	15 59.6	287 17.1	· · 25.4	289 07.6	· · 43.9	157 15.2	· · 31.5
T 10	140 26.2	0 11.3	58.8	302 18.1	26.0	304 09.6	44.1	172 17.9	31.6
U 11	155 28.7	15 10.7	58.0	317 19.0	26.7	319 11.6	44.3	187 20.5	31.7
R 12	170 31.1	30 10.1	N15 57.2	332 19.9	S 7 27.3	334 13.6	S 5 44.5	202 23.1	S14 31.7
D 13	185 33.6	45 09.6	56.4	347 20.9	28.0	349 15.6	44.7	217 25.7	31.8
A 14	200 36.0	60 09.0	55.6	2 21.8	28.6	4 17.6	44.9	232 28.4	31.8
Y 15	215 38.5	75 08.5	· · 54.8	17 22.8	· · 29.2	19 19.6	· · 45.1	247 31.0	· · 31.9
16	230 41.0	90 07.9	53.9	32 23.7	29.9	34 21.6	45.3	262 33.6	31.9
17	245 43.4	105 07.4	53.1	47 24.6	30.5	49 23.6	45.5	277 36.2	32.0
18	260 45.9	120 06.8	N15 52.3	62 25.6	S 7 31.2	64 25.6	S 5 45.7	292 38.9	S14 32.0
19	275 48.4	135 06.2	51.5	77 26.5	31.8	79 27.5	45.9	307 41.5	32.1
20	290 50.8	150 05.7	50.7	92 27.4	32.5	94 29.5	46.1	322 44.1	32.2
21	305 53.3	165 05.1	· · 49.9	107 28.4	· · 33.1	109 31.5	· · 46.3	337 46.7	· · 32.2
22	320 55.8	180 04.6	49.0	122 29.3	33.8	124 33.5	46.5	352 49.3	32.3
23	335 58.2	195 04.0	48.2	137 30.2	34.4	139 35.5	46.7	7 52.0	32.3
12 00	351 00.7	210 03.5	N15 47.4	152 31.2	S 7 35.1	154 37.5	S 5 46.9	22 54.6	S14 32.4
01	6 03.2	225 02.9	46.6	167 32.1	35.7	169 39.5	47.1	37 57.2	32.4
02	21 05.6	240 02.3	45.8	182 33.0	36.4	184 41.5	47.3	52 59.8	32.5
03	36 08.1	255 01.8	· · 44.9	197 34.0	· · 37.0	199 43.5	· · 47.5	68 02.5	· · 32.5
04	51 10.5	270 01.2	44.1	212 34.9	37.7	214 45.5	47.7	83 05.1	32.6
05	66 13.0	285 00.7	43.3	227 35.8	38.3	229 47.5	47.9	98 07.7	32.7
06	81 15.5	300 00.1	N15 42.5	242 36.8	S 7 38.9	244 49.5	S 5 48.1	113 10.3	S14 32.7
07	96 17.9	314 59.6	41.6	257 37.7	39.6	259 51.5	48.3	128 13.0	32.8
08	111 20.4	329 59.0	40.8	272 38.6	40.2	274 53.5	48.5	143 15.6	32.8
S 09	126 22.9	344 58.5	· · 40.0	287 39.6	· · 40.9	289 55.5	· · 48.7	158 18.2	· · 32.9
U 10	141 25.3	359 57.9	39.1	302 40.5	41.5	304 57.5	48.8	173 20.8	32.9
N 11	156 27.8	14 57.4	38.3	317 41.4	42.2	319 59.5	49.0	188 23.4	33.0
D 12	171 30.3	29 56.8	N15 37.5	332 42.4	S 7 42.8	335 01.5	S 5 49.2	203 26.1	S14 33.0
A 13	186 32.7	44 56.3	36.7	347 43.3	43.5	350 03.4	49.4	218 28.7	33.1
Y 14	201 35.2	59 55.7	35.8	2 44.2	44.1	5 05.4	49.6	233 31.3	33.1
15	216 37.7	74 55.2	· · 35.0	17 45.2	· · 44.8	20 07.4	· · 49.8	248 33.9	· · 33.2
16	231 40.1	89 54.6	34.2	32 46.1	45.4	35 09.4	50.0	263 36.5	33.3
17	246 42.6	104 54.1	33.3	47 47.0	46.1	50 11.4	50.2	278 39.2	33.3
18	261 45.0	119 53.5	N15 32.5	62 48.0	S 7 46.7	65 13.4	S 5 50.4	293 41.8	S14 33.4
19	276 47.5	134 53.0	31.7	77 48.9	47.3	80 15.4	50.6	308 44.4	33.4
20	291 50.0	149 52.4	30.8	92 49.8	48.0	95 17.4	50.8	323 47.0	33.5
21	306 52.4	164 51.9	· · 30.0	107 50.8	· · 48.6	110 19.4	· · 51.0	338 49.7	· · 33.5
22	321 54.9	179 51.3	29.1	122 51.7	49.3	125 21.4	51.2	353 52.3	33.6
23	336 57.4	194 50.8	28.3	137 52.6	49.9	140 23.4	51.4	8 54.9	33.6
Mer. Pass.	h m 0 39.8	v −0.6	d 0.8	v 0.9	d 0.6	v 2.0	d 0.2	v 2.6	d 0.1

STARS

Name	S.H.A. ° ′	Dec. ° ′
Acamar	315 29.3	S40 19.5
Achernar	335 37.2	S57 15.8
Acrux	173 26.5	S63 04.0
Adhara	255 24.3	S28 57.6
Aldebaran	291 06.3	N16 29.9
Alioth	166 34.1	N55 59.7
Alkaid	153 10.8	N49 20.8
Al Na'ir	28 01.7	S46 59.4
Alnilam	276 01.4	S 1 12.2
Alphard	218 10.9	S 8 37.8
Alphecca	126 23.6	N26 44.4
Alpheratz	357 58.5	N29 03.5
Altair	62 22.4	N 8 51.3
Ankaa	353 29.8	S42 20.1
Antares	112 44.5	S26 25.1
Arcturus	146 09.4	N19 13.0
Atria	107 59.6	S69 01.2
Avior	234 24.5	S59 29.2
Bellatrix	278 47.9	N 6 20.7
Betelgeuse	271 17.3	N 7 24.4
Canopus	264 02.8	S52 41.3
Capella	280 56.3	N45 59.4
Deneb	49 41.2	N45 15.8
Denebola	182 49.0	N14 36.5
Diphda	349 10.4	S18 01.0
Dubhe	194 10.3	N61 47.0
Elnath	278 31.3	N28 36.1
Eltanin	90 52.9	N51 29.7
Enif	34 01.3	N 9 51.0
Fomalhaut	15 39.8	S29 39.1
Gacrux	172 18.0	S57 04.8
Gienah	176 07.8	S17 30.4
Hadar	149 09.4	S60 20.8
Hamal	328 17.2	N23 26.1
Kaus Aust.	84 03.3	S34 23.3
Kochab	137 20.0	N74 11.1
Markab	13 52.7	N15 10.5
Menkar	314 30.3	N 4 04.1
Menkent	148 25.3	S36 20.4
Miaplacidus	221 43.6	S69 41.4
Mirfak	309 01.3	N49 50.3
Nunki	76 16.4	S26 18.2
Peacock	53 41.9	S56 45.3
Pollux	243 46.0	N28 02.4
Procyon	245 15.3	N 5 14.5
Rasalhague	96 20.1	N12 34.1
Regulus	207 59.5	N11 59.9
Rigel	281 26.2	S 8 12.4
Rigil Kent.	140 12.3	S60 48.7
Sabik	102 29.5	S15 43.0
Schedar	349 56.9	N56 30.2
Shaula	96 41.9	S37 06.0
Sirius	258 46.8	S16 42.3
Spica	158 47.1	S11 07.7
Suhail	223 03.7	S43 24.3
Vega	80 48.8	N38 47.0
Zuben'ubi	137 22.0	S16 00.9

	S.H.A. ° ′	Mer. Pass. h m
Venus	220 15.3	9 59
Mars	162 07.1	13 51
Jupiter	163 48.1	13 43
Saturn	31 50.1	22 29

UT (GMT) d	h	SUN G.H.A. ° ′	SUN Dec. ° ′	MOON G.H.A. ° ′	MOON v ′	MOON Dec. ° ′	MOON d ′	MOON H.P. ′
10	00	180 43.3	N 5 01.0	263 27.2	8.8	N21 42.3	1.8	56.6
	01	195 43.6	5 00.0	277 55.0	8.7	21 40.5	1.8	56.7
	02	210 43.8	4 59.1	292 22.7	8.7	21 38.7	2.0	56.7
	03	225 44.0	. . 58.1	306 50.4	8.7	21 36.7	2.2	56.7
	04	240 44.2	57.2	321 18.1	8.7	21 34.5	2.2	56.8
	05	255 44.4	56.2	335 45.8	8.6	21 32.3	2.4	56.8
	06	270 44.7	N 4 55.3	350 13.4	8.6	N21 29.9	2.5	56.9
	07	285 44.9	54.3	4 41.0	8.6	21 27.4	2.6	56.9
	08	300 45.1	53.4	19 08.6	8.5	21 24.8	2.8	56.9
F	09	315 45.3	. . 52.4	33 36.1	8.6	21 22.0	2.8	57.0
R	10	330 45.5	51.5	48 03.7	8.5	21 19.2	3.0	57.0
I	11	345 45.7	50.5	62 31.2	8.5	21 16.2	3.1	57.0
D	12	0 46.0	N 4 49.6	76 58.7	8.4	N21 13.1	3.3	57.1
A	13	15 46.2	48.6	91 26.1	8.4	21 09.8	3.3	57.1
Y	14	30 46.4	47.7	105 53.5	8.5	21 06.5	3.5	57.2
	15	45 46.6	. . 46.7	120 21.0	8.4	21 03.0	3.6	57.2
	16	60 46.8	45.8	134 48.4	8.3	20 59.4	3.8	57.2
	17	75 47.1	44.8	149 15.7	8.4	20 55.6	3.8	57.3
	18	90 47.3	N 4 43.9	163 43.1	8.3	N20 51.8	4.0	57.3
	19	105 47.5	42.9	178 10.4	8.3	20 47.8	4.1	57.3
	20	120 47.7	42.0	192 37.7	8.3	20 43.7	4.3	57.4
	21	135 47.9	. . 41.0	207 05.0	8.3	20 39.4	4.3	57.4
	22	150 48.1	40.1	221 32.3	8.3	20 35.1	4.5	57.5
	23	165 48.4	39.1	235 59.6	8.3	20 30.6	4.6	57.5
11	00	180 48.6	N 4 38.2	250 26.9	8.2	N20 26.0	4.8	57.5
	01	195 48.8	37.2	264 54.1	8.2	20 21.2	4.8	57.6
	02	210 49.0	36.3	279 21.3	8.3	20 16.4	5.0	57.6
	03	225 49.2	. . 35.3	293 48.6	8.2	20 11.4	5.1	57.7
	04	240 49.5	34.4	308 15.8	8.1	20 06.3	5.2	57.7
	05	255 49.7	33.4	322 42.9	8.2	20 01.1	5.4	57.7
	06	270 49.9	N 4 32.5	337 10.1	8.2	N19 55.7	5.5	57.8
	07	285 50.1	31.5	351 37.3	8.2	19 50.2	5.6	57.8
S	08	300 50.3	30.6	6 04.5	8.1	19 44.6	5.7	57.9
A	09	315 50.6	. . 29.6	20 31.6	8.2	19 38.9	5.9	57.9
T	10	330 50.8	28.7	34 58.8	8.1	19 33.0	5.9	57.9
U	11	345 51.0	27.7	49 25.9	8.1	19 27.1	6.1	58.0
R	12	0 51.2	N 4 26.8	63 53.0	8.1	N19 21.0	6.2	58.0
D	13	15 51.4	25.8	78 20.1	8.2	19 14.8	6.4	58.1
A	14	30 51.6	24.9	92 47.3	8.1	19 08.4	6.4	58.1
Y	15	45 51.9	. . 23.9	107 14.4	8.1	19 02.0	6.6	58.1
	16	60 52.1	23.0	121 41.5	8.1	18 55.4	6.7	58.2
	17	75 52.3	22.0	136 08.6	8.1	18 48.7	6.8	58.2
	18	90 52.5	N 4 21.1	150 35.7	8.1	N18 41.9	7.0	58.3
	19	105 52.7	20.1	165 02.8	8.1	18 34.9	7.0	58.3
	20	120 53.0	19.2	179 29.9	8.1	18 27.9	7.2	58.3
	21	135 53.2	. . 18.2	193 57.0	8.0	18 20.7	7.3	58.4
	22	150 53.4	17.3	208 24.0	8.1	18 13.4	7.5	58.4
	23	165 53.6	16.3	222 51.1	8.1	18 05.9	7.5	58.5
12	00	180 53.8	N 4 15.4	237 18.2	8.1	N17 58.4	7.6	58.5
	01	195 54.1	14.4	251 45.3	8.1	17 50.8	7.8	58.5
	02	210 54.3	13.5	266 12.4	8.1	17 43.0	7.9	58.6
	03	225 54.5	. . 12.5	280 39.5	8.1	17 35.1	8.0	58.6
	04	240 54.7	11.5	295 06.6	8.1	17 27.1	8.1	58.7
	05	255 54.9	10.6	309 33.7	8.0	17 19.0	8.3	58.7
	06	270 55.2	N 4 09.6	324 00.7	8.1	N17 10.7	8.3	58.7
	07	285 55.4	08.7	338 27.8	8.1	17 02.4	8.5	58.8
	08	300 55.6	07.7	352 54.9	8.1	16 53.9	8.6	58.8
S	09	315 55.8	. . 06.8	7 22.0	8.1	16 45.3	8.7	58.9
U	10	330 56.0	05.8	21 49.1	8.1	16 36.6	8.8	58.9
N	11	345 56.3	04.9	36 16.2	8.1	16 27.8	8.9	58.9
D	12	0 56.5	N 4 03.9	50 43.3	8.1	N16 18.9	9.0	59.0
A	13	15 56.7	03.0	65 10.4	8.2	16 09.9	9.1	59.0
Y	14	30 56.9	02.0	79 37.6	8.1	16 00.8	9.3	59.1
	15	45 57.1	. . 01.0	94 04.7	8.1	15 51.5	9.3	59.1
	16	60 57.4	4 00.1	108 31.8	8.1	15 42.2	9.5	59.1
	17	75 57.6	3 59.1	122 58.9	8.2	15 32.7	9.5	59.2
	18	90 57.8	N 3 58.2	137 26.1	8.1	N15 23.2	9.7	59.2
	19	105 58.0	57.2	151 53.2	8.2	15 13.5	9.8	59.3
	20	120 58.2	56.3	166 20.4	8.1	15 03.7	9.9	59.3
	21	135 58.5	. . 55.3	180 47.5	8.2	14 53.8	9.9	59.3
	22	150 58.7	54.4	195 14.7	8.1	14 43.9	10.1	59.4
	23	165 58.9	53.4	209 41.8	8.2	14 33.8	10.2	59.4
		S.D. 15.9	*d* 1.0	S.D.		15.6	15.8	16.1

Lat. °	Twilight Naut. h m	Twilight Civil h m	Sunrise h m	Moonrise 10 h m	Moonrise 11 h m	Moonrise 12 h m	Moonrise 13 h m
N 72	01 27	03 33	04 49	▭	▭	23 36	26 00
N 70	02 12	03 50	04 56	▭	21 50	24 07	00 07
68	02 40	04 03	05 02	20 38	22 33	24 30	00 30
66	03 01	04 13	05 07	21 24	23 02	24 48	00 48
64	03 17	04 22	05 11	21 54	23 24	25 02	01 02
62	03 31	04 29	05 15	22 17	23 41	25 14	01 14
60	03 42	04 35	05 18	22 35	23 55	25 24	01 24
N 58	03 51	04 41	05 21	22 50	24 07	00 07	01 33
56	03 59	04 45	05 24	23 03	24 18	00 18	01 41
54	04 06	04 50	05 26	23 14	24 27	00 27	01 48
52	04 12	04 53	05 28	23 24	24 35	00 35	01 54
50	04 17	04 57	05 30	23 33	24 43	00 43	01 59
45	04 29	05 04	05 34	23 51	24 59	00 59	02 11
N 40	04 37	05 10	05 37	24 06	00 06	01 12	02 21
35	04 44	05 14	05 40	24 19	00 19	01 23	02 30
30	04 50	05 18	05 42	24 30	00 30	01 32	02 37
20	04 59	05 24	05 46	24 50	00 50	01 49	02 50
N 10	05 04	05 29	05 50	00 11	01 06	02 03	03 01
0	05 09	05 33	05 53	00 28	01 22	02 16	03 11
S 10	05 11	05 35	05 56	00 44	01 37	02 30	03 21
20	05 12	05 38	06 00	01 02	01 54	02 44	03 32
30	05 12	05 39	06 03	01 23	02 13	03 00	03 45
35	05 11	05 40	06 05	01 35	02 24	03 10	03 52
40	05 09	05 41	06 08	01 49	02 37	03 20	04 00
45	05 07	05 41	06 10	02 05	02 52	03 33	04 10
S 50	05 04	05 41	06 13	02 25	03 10	03 48	04 21
52	05 02	05 41	06 15	02 35	03 19	03 55	04 26
54	05 00	05 41	06 16	02 45	03 28	04 03	04 32
56	04 58	05 41	06 18	02 58	03 39	04 12	04 39
58	04 55	05 40	06 20	03 12	03 52	04 22	04 46
S 60	04 52	05 40	06 22	03 28	04 06	04 34	04 54

Lat. °	Sunset h m	Twilight Civil h m	Twilight Naut. h m	Moonset 10 h m	Moonset 11 h m	Moonset 12 h m	Moonset 13 h m
N 72	19 02	20 16	22 16	▭	▭	18 37	18 08
N 70	18 55	20 00	21 35	▭	18 26	18 04	17 51
68	18 49	19 48	21 08	17 43	17 42	17 40	17 37
66	18 44	19 38	20 48	16 57	17 13	17 21	17 25
64	18 40	19 29	20 33	16 27	16 51	17 05	17 15
62	18 37	19 22	20 20	16 04	16 33	16 53	17 07
60	18 33	19 16	20 09	15 46	16 18	16 42	17 00
N 58	18 31	19 11	20 00	15 30	16 05	16 32	16 53
56	18 28	19 06	19 53	15 17	15 54	16 24	16 48
54	18 26	19 02	19 46	15 06	15 44	16 16	16 43
52	18 24	18 59	19 40	14 56	15 36	16 09	16 38
50	18 23	18 55	19 34	14 46	15 28	16 03	16 34
45	18 19	18 48	19 23	14 27	15 11	15 50	16 25
N 40	18 16	18 43	19 15	14 12	14 57	15 39	16 17
35	18 13	18 38	19 08	13 58	14 46	15 30	16 10
30	18 10	18 34	19 03	13 47	14 36	15 21	16 05
20	18 06	18 29	18 54	13 27	14 18	15 07	15 54
N 10	18 03	18 24	18 49	13 10	14 02	14 54	15 45
0	18 00	18 21	18 45	12 53	13 48	14 42	15 37
S 10	17 57	18 18	18 42	12 37	13 33	14 30	15 28
20	17 54	18 16	18 41	12 20	13 18	14 18	15 19
30	17 50	18 14	18 42	12 00	12 59	14 03	15 08
35	17 48	18 14	18 43	11 48	12 49	13 54	15 02
40	17 46	18 13	18 45	11 34	12 37	13 44	14 55
45	17 44	18 13	18 47	11 18	12 22	13 33	14 47
S 50	17 41	18 13	18 51	10 58	12 05	13 18	14 37
52	17 40	18 13	18 52	10 49	11 56	13 12	14 32
54	17 38	18 13	18 54	10 38	11 47	13 04	14 27
56	17 37	18 14	18 57	10 26	11 37	12 56	14 22
58	17 35	18 14	19 00	10 12	11 25	12 47	14 15
S 60	17 33	18 14	19 03	09 56	11 11	12 36	14 08

Day	SUN Eqn. of Time 00ʰ m s	SUN Eqn. of Time 12ʰ m s	SUN Mer. Pass. h m	MOON Mer. Pass. Upper h m	MOON Mer. Pass. Lower h m	MOON Age d	MOON Phase
10	02 53	03 03	11 57	06 41	19 08	24	◐
11	03 14	03 24	11 57	07 35	20 02	25	
12	03 35	03 45	11 56	08 29	20 57	26	

UT (GMT) d h	ARIES G.H.A.	VENUS −3.9 G.H.A.	VENUS Dec.	MARS +1.2 G.H.A.	MARS Dec.	JUPITER −1.8 G.H.A.	JUPITER Dec.	SATURN +0.8 G.H.A.	SATURN Dec.
	° ′	° ′	° ′	° ′	° ′	° ′	° ′	° ′	° ′
18 00	86 37.2	188 41.9	S22 46.9	178 17.2	S24 08.9	231 09.7	S12 56.6	118 05.7	S14 16.8
01	101 39.6	203 41.0	47.3	193 17.7	09.0	246 11.7	56.8	133 07.9	16.8
02	116 42.1	218 40.1	47.6	208 18.1	09.0	261 13.8	56.9	148 10.2	16.7
03	131 44.5	233 39.1	· · 48.0	223 18.5	· · 09.0	276 15.8	· · 57.0	163 12.5	· · 16.6
04	146 47.0	248 38.2	48.3	238 18.9	09.1	291 17.8	57.2	178 14.8	16.6
05	161 49.5	263 37.3	48.7	253 19.3	09.1	306 19.9	57.3	193 17.0	16.5
06	176 51.9	278 36.3	S22 49.1	268 19.7	S24 09.2	321 21.9	S12 57.5	208 19.3	S14 16.4
07	191 54.4	293 35.4	49.4	283 20.1	09.2	336 24.0	57.6	223 21.6	16.4
S 08	206 56.9	308 34.5	49.8	298 20.5	09.2	351 26.0	57.7	238 23.8	16.3
A 09	221 59.3	323 33.6	· · 50.1	313 20.9	· · 09.3	6 28.0	· · 57.9	253 26.1	· · 16.2
T 10	237 01.8	338 32.6	50.5	328 21.3	09.3	21 30.1	58.0	268 28.4	16.1
U 11	252 04.3	353 31.7	50.8	343 21.7	09.3	36 32.1	58.2	283 30.7	16.1
R 12	267 06.7	8 30.8	S22 51.2	358 22.1	S24 09.4	51 34.1	S12 58.3	298 32.9	S14 16.0
D 13	282 09.2	23 29.8	51.5	13 22.5	09.4	66 36.2	58.4	313 35.2	15.9
A 14	297 11.7	38 28.9	51.9	28 22.9	09.5	81 38.2	58.6	328 37.5	15.9
Y 15	312 14.1	53 28.0	· · 52.2	43 23.3	· · 09.5	96 40.2	· · 58.7	343 39.7	· · 15.8
16	327 16.6	68 27.0	52.5	58 23.7	09.5	111 42.3	58.8	358 42.0	15.7
17	342 19.0	83 26.1	52.9	73 24.1	09.6	126 44.3	59.0	13 44.3	15.6
18	357 21.5	98 25.2	S22 53.2	88 24.5	S24 09.6	141 46.3	S12 59.1	28 46.5	S14 15.6
19	12 24.0	113 24.2	53.6	103 24.9	09.6	156 48.4	59.3	43 48.8	15.5
20	27 26.4	128 23.3	53.9	118 25.4	09.7	171 50.4	59.4	58 51.1	15.4
21	42 28.9	143 22.4	· · 54.2	133 25.8	· · 09.7	186 52.5	· · 59.5	73 53.3	· · 15.4
22	57 31.4	158 21.4	54.6	148 26.2	09.7	201 54.5	59.7	88 55.6	15.3
23	72 33.8	173 20.5	54.9	163 26.6	09.8	216 56.5	59.8	103 57.9	15.2
19 00	87 36.3	188 19.6	S22 55.2	178 27.0	S24 09.8	231 58.6	S12 59.9	119 00.2	S14 15.1
01	102 38.8	203 18.6	55.6	193 27.4	09.8	247 00.6	13 00.1	134 02.4	15.1
02	117 41.2	218 17.7	55.9	208 27.8	09.9	262 02.6	00.2	149 04.7	15.0
03	132 43.7	233 16.8	· · 56.2	223 28.2	· · 09.9	277 04.7	· · 00.4	164 07.0	· · 14.9
04	147 46.2	248 15.8	56.6	238 28.6	09.9	292 06.7	00.5	179 09.2	14.9
05	162 48.6	263 14.9	56.9	253 29.0	09.9	307 08.8	00.6	194 11.5	14.8
06	177 51.1	278 14.0	S22 57.2	268 29.4	S24 10.0	322 10.8	S13 00.8	209 13.8	S14 14.7
07	192 53.5	293 13.0	57.5	283 29.8	10.0	337 12.8	00.9	224 16.0	14.6
08	207 56.0	308 12.1	57.9	298 30.2	10.0	352 14.9	01.0	239 18.3	14.6
S 09	222 58.5	323 11.1	· · 58.2	313 30.6	· · 10.1	7 16.9	· · 01.2	254 20.6	· · 14.5
U 10	238 00.9	338 10.2	58.5	328 31.0	10.1	22 18.9	01.3	269 22.8	14.4
N 11	253 03.4	353 09.3	58.8	343 31.4	10.1	37 21.0	01.5	284 25.1	14.4
D 12	268 05.9	8 08.3	S22 59.2	358 31.8	S24 10.1	52 23.0	S13 01.6	299 27.4	S14 14.3
A 13	283 08.3	23 07.4	59.5	13 32.2	10.2	67 25.1	01.7	314 29.6	14.2
Y 14	298 10.8	38 06.5	22 59.8	28 32.6	10.2	82 27.1	01.9	329 31.9	14.1
15	313 13.3	53 05.5	23 00.1	43 33.0	· · 10.2	97 29.1	· · 02.0	344 34.2	· · 14.1
16	328 15.7	68 04.6	00.4	58 33.4	10.2	112 31.2	02.1	359 36.4	14.0
17	343 18.2	83 03.6	00.7	73 33.8	10.3	127 33.2	02.3	14 38.7	13.9
18	358 20.7	98 02.7	S23 01.1	88 34.2	S24 10.3	142 35.3	S13 02.4	29 41.0	S14 13.9
19	13 23.1	113 01.8	01.4	103 34.6	10.3	157 37.3	02.5	44 43.2	13.8
20	28 25.6	128 00.8	01.7	118 35.0	10.3	172 39.3	02.7	59 45.5	13.7
21	43 28.0	142 59.9	· · 02.0	133 35.4	· · 10.3	187 41.4	· · 02.8	74 47.8	· · 13.6
22	58 30.5	157 58.9	02.3	148 35.8	10.4	202 43.4	03.0	89 50.0	13.6
23	73 33.0	172 58.0	02.6	163 36.2	10.4	217 45.5	03.1	104 52.3	13.5
20 00	88 35.4	187 57.1	S23 02.9	178 36.6	S24 10.4	232 47.5	S13 03.2	119 54.6	S14 13.4
01	103 37.9	202 56.1	03.2	193 37.0	10.4	247 49.5	03.4	134 56.8	13.4
02	118 40.4	217 55.2	03.5	208 37.4	10.4	262 51.6	03.5	149 59.1	13.3
03	133 42.8	232 54.2	· · 03.8	223 37.8	· · 10.5	277 53.6	· · 03.6	165 01.4	· · 13.2
04	148 45.3	247 53.3	04.1	238 38.2	10.5	292 55.7	03.8	180 03.6	13.1
05	163 47.8	262 52.4	04.4	253 38.6	10.5	307 57.7	03.9	195 05.9	13.1
06	178 50.2	277 51.4	S23 04.7	268 39.0	S24 10.5	322 59.7	S13 04.0	210 08.1	S14 13.0
07	193 52.7	292 50.5	05.0	283 39.4	10.5	338 01.8	04.2	225 10.4	12.9
08	208 55.1	307 49.5	05.3	298 39.8	10.6	353 03.8	04.3	240 12.7	12.8
M 09	223 57.6	322 48.6	· · 05.6	313 40.2	· · 10.6	8 05.9	· · 04.4	255 14.9	· · 12.8
O 10	239 00.1	337 47.6	05.9	328 40.6	10.6	23 07.9	04.6	270 17.2	12.7
N 11	254 02.5	352 46.7	06.2	343 41.0	10.6	38 10.0	04.7	285 19.5	12.6
D 12	269 05.0	7 45.8	S23 06.5	358 41.4	S24 10.6	53 12.0	S13 04.9	300 21.7	S14 12.6
A 13	284 07.5	22 44.8	06.7	13 41.8	10.6	68 14.0	05.0	315 24.0	12.5
Y 14	299 09.9	37 43.9	07.0	28 42.2	10.6	83 16.1	05.1	330 26.3	12.4
15	314 12.4	52 42.9	· · 07.3	43 42.6	· · 10.7	98 18.1	· · 05.3	345 28.5	· · 12.3
16	329 14.9	67 42.0	07.6	58 43.0	10.7	113 20.2	05.4	0 30.8	12.3
17	344 17.3	82 41.0	07.9	73 43.4	10.7	128 22.2	05.5	15 33.1	12.2
18	359 19.8	97 40.1	S23 08.2	88 43.8	S24 10.7	143 24.2	S13 05.7	30 35.3	S14 12.1
19	14 22.3	112 39.1	08.5	103 44.2	10.7	158 26.3	05.8	45 37.6	12.0
20	29 24.7	127 38.2	08.7	118 44.6	10.7	173 28.3	05.9	60 39.8	12.0
21	44 27.2	142 37.2	· · 09.0	133 45.0	· · 10.7	188 30.4	· · 06.1	75 42.1	· · 11.9
22	59 29.6	157 36.3	09.3	148 45.4	10.7	203 32.4	06.2	90 44.4	11.8
23	74 32.1	172 35.4	09.6	163 45.8	10.7	218 34.5	06.3	105 46.6	11.8
Mer. Pass.	h m 18 06.6	v −0.9	d 0.3	v 0.4	d 0.0	v 2.0	d 0.1	v 2.3	d 0.1

STARS

Name	S.H.A.	Dec.
	° ′	° ′
Acamar	315 29.1	S40 19.9
Achernar	335 37.3	S57 16.2
Acrux	173 25.7	S63 03.7
Adhara	255 23.6	S28 57.9
Aldebaran	291 05.8	N16 29.8
Alioth	166 33.6	N55 59.2
Alkaid	153 10.6	N49 20.3
Al Na'ir	28 02.2	S46 59.5
Alnilam	276 00.8	S 1 12.4
Alphard	218 10.2	S 8 38.0
Alphecca	126 23.7	N26 44.0
Alpheratz	357 58.6	N29 03.7
Altair	62 22.7	N 8 51.3
Ankaa	353 30.0	S42 20.5
Antares	112 44.5	S26 25.0
Arcturus	146 09.2	N19 12.7
Atria	108 00.0	S69 00.9
Avior	234 23.4	S59 29.3
Bellatrix	278 47.3	N 6 20.6
Betelgeuse	271 16.7	N 7 24.3
Canopus	264 02.0	S52 41.6
Capella	280 55.5	N45 59.5
Deneb	49 41.8	N45 15.8
Denebola	182 48.5	N14 36.2
Diphda	349 10.4	S18 01.2
Dubhe	194 09.3	N61 46.6
Elnath	278 30.7	N28 36.1
Eltanin	90 53.5	N51 29.5
Enif	34 01.6	N 9 51.0
Fomalhaut	15 40.1	S29 39.3
Gacrux	172 17.3	S57 04.6
Gienah	176 07.4	S17 30.5
Hadar	149 09.0	S60 20.4
Hamal	328 17.0	N23 26.2
Kaus Aust.	84 03.5	S34 23.2
Kochab	137 20.3	N74 10.6
Markab	13 52.9	N15 10.6
Menkar	314 30.0	N 4 04.0
Menkent	148 25.0	S36 20.3
Miaplacidus	221 42.1	S69 41.4
Mirfak	309 00.8	N49 50.6
Nunki	76 16.7	S26 18.2
Peacock	53 42.6	S56 45.3
Pollux	243 45.2	N28 02.3
Procyon	245 14.7	N 5 14.3
Rasalhague	96 20.3	N12 33.9
Regulus	207 58.8	N11 59.6
Rigel	281 25.7	S 8 12.6
Rigil Kent.	140 12.1	S60 48.4
Sabik	102 29.6	S15 43.0
Schedar	349 57.1	N56 30.6
Shaula	96 42.1	S37 05.9
Sirius	258 46.2	S16 42.5
Spica	158 46.8	S11 07.8
Suhail	223 02.8	S43 24.4
Vega	80 49.3	N38 46.8
Zuben'ubi	137 21.8	S16 01.0

	S.H.A.	Mer. Pass.
	° ′	h m
Venus	100 43.3	11 27
Mars	90 50.7	12 06
Jupiter	144 22.3	8 31
Saturn	31 23.9	16 02

UT (GMT) d h	SUN G.H.A. ° ′	SUN Dec. ° ′	MOON G.H.A. ° ′	v ′	Dec. ° ′	d ′	H.P. ′
18 00	180 53.4	S23 22.8	122 46.2	13.6	S 8 53.1	10.6	55.8
01	195 53.1	22.9	137 18.8	13.6	8 42.5	10.6	55.8
02	210 52.8	23.0	151 51.4	13.7	8 31.9	10.7	55.7
03	225 52.5	. . 23.0	166 24.1	13.8	8 21.2	10.7	55.7
04	240 52.2	23.1	180 56.9	13.8	8 10.5	10.6	55.7
05	255 51.8	23.2	195 29.7	13.9	7 59.9	10.8	55.7
06	270 51.5	S23 23.3	210 02.6	13.9	S 7 49.1	10.7	55.6
07	285 51.2	23.3	224 35.5	14.0	7 38.4	10.8	55.6
S 08	300 50.9	23.4	239 08.5	14.1	7 27.6	10.7	55.6
A 09	315 50.6	. . 23.5	253 41.6	14.1	7 16.9	10.8	55.5
T 10	330 50.3	23.5	268 14.7	14.1	7 06.1	10.9	55.5
U 11	345 50.0	23.6	282 47.8	14.2	6 55.2	10.8	55.5
R 12	0 49.7	S23 23.7	297 21.0	14.3	S 6 44.4	10.8	55.5
D 13	15 49.4	23.7	311 54.3	14.3	6 33.6	10.9	55.4
A 14	30 49.1	23.8	326 27.6	14.3	6 22.7	10.9	55.4
Y 15	45 48.8	. . 23.9	341 00.9	14.4	6 11.8	10.9	55.4
16	60 48.5	23.9	355 34.3	14.5	6 00.9	10.9	55.4
17	75 48.1	24.0	10 07.8	14.5	5 50.0	11.0	55.3
18	90 47.8	S23 24.0	24 41.3	14.5	S 5 39.0	10.9	55.3
19	105 47.5	24.1	39 14.8	14.6	5 28.1	11.0	55.3
20	120 47.2	24.2	53 48.4	14.7	5 17.1	10.9	55.3
21	135 46.9	. . 24.2	68 22.1	14.7	5 06.2	11.0	55.2
22	150 46.6	24.3	82 55.8	14.7	4 55.2	11.0	55.2
23	165 46.3	24.3	97 29.5	14.7	4 44.2	11.0	55.2
19 00	180 46.0	S23 24.4	112 03.2	14.8	S 4 33.2	11.0	55.2
01	195 45.7	24.5	126 37.0	14.9	4 22.2	11.1	55.1
02	210 45.4	24.5	141 10.9	14.9	4 11.1	11.0	55.1
03	225 45.1	. . 24.6	155 44.8	14.9	4 00.1	11.0	55.1
04	240 44.8	24.6	170 18.7	14.9	3 49.1	11.0	55.1
05	255 44.4	24.7	184 52.6	15.0	3 38.1	11.1	55.1
06	270 44.1	S23 24.7	199 26.6	15.1	S 3 27.0	11.0	55.0
07	285 43.8	24.8	214 00.7	15.0	3 16.0	11.1	55.0
08	300 43.5	24.8	228 34.7	15.1	3 04.9	11.1	55.0
S 09	315 43.2	. . 24.9	243 08.8	15.2	2 53.8	11.0	55.0
U 10	330 42.9	24.9	257 43.0	15.1	2 42.8	11.1	54.9
N 11	345 42.6	25.0	272 17.1	15.2	2 31.7	11.0	54.9
D 12	0 42.3	S23 25.0	286 51.3	15.2	S 2 20.7	11.1	54.9
A 13	15 42.0	25.1	301 25.5	15.3	2 09.6	11.1	54.9
Y 14	30 41.7	25.1	315 59.8	15.3	1 58.5	11.0	54.9
15	45 41.3	. . 25.1	330 34.1	15.3	1 47.5	11.1	54.8
16	60 41.0	25.2	345 08.4	15.3	1 36.4	11.0	54.8
17	75 40.7	25.2	359 42.7	15.4	1 25.4	11.1	54.8
18	90 40.4	S23 25.3	14 17.1	15.3	S 1 14.3	11.1	54.8
19	105 40.1	25.3	28 51.4	15.4	1 03.2	11.0	54.8
20	120 39.8	25.4	43 25.8	15.5	0 52.2	11.1	54.8
21	135 39.5	. . 25.4	58 00.3	15.4	0 41.1	11.0	54.7
22	150 39.2	25.4	72 34.7	15.5	0 30.1	11.0	54.7
23	165 38.9	25.5	87 09.2	15.5	0 19.1	11.1	54.7
20 00	180 38.6	S23 25.5	101 43.7	15.5	S 0 08.0	11.0	54.7
01	195 38.2	25.5	116 18.2	15.5	N 0 03.0	11.0	54.7
02	210 37.9	25.6	130 52.7	15.5	0 14.0	11.0	54.6
03	225 37.6	. . 25.6	145 27.2	15.6	0 25.0	11.0	54.6
04	240 37.3	25.6	160 01.8	15.6	0 36.0	11.0	54.6
05	255 37.0	25.7	174 36.4	15.5	0 47.0	11.0	54.6
06	270 36.7	S23 25.7	189 10.9	15.6	N 0 58.0	11.0	54.6
07	285 36.4	25.7	203 45.5	15.6	1 09.0	10.9	54.6
08	300 36.1	25.8	218 20.1	15.7	1 19.9	11.0	54.6
M 09	315 35.8	. . 25.8	232 54.8	15.6	1 30.9	10.9	54.5
O 10	330 35.5	25.8	247 29.4	15.6	1 41.8	10.9	54.5
N 11	345 35.1	25.9	262 04.0	15.7	1 52.7	10.9	54.5
D 12	0 34.8	S23 25.9	276 38.7	15.6	N 2 03.6	10.9	54.5
A 13	15 34.5	25.9	291 13.3	15.7	2 14.5	10.9	54.5
Y 14	30 34.2	25.9	305 48.0	15.7	2 25.4	10.9	54.5
15	45 33.9	. . 26.0	320 22.7	15.7	2 36.3	10.8	54.5
16	60 33.6	26.0	334 57.4	15.6	2 47.1	10.9	54.4
17	75 33.3	26.0	349 32.0	15.7	2 58.0	10.8	54.4
18	90 33.0	S23 26.0	4 06.7	15.7	N 3 08.8	10.8	54.4
19	105 32.7	26.1	18 41.4	15.7	3 19.6	10.8	54.4
20	120 32.4	26.1	33 16.1	15.7	3 30.4	10.8	54.4
21	135 32.0	. . 26.1	47 50.8	15.7	3 41.2	10.7	54.4
22	150 31.7	26.1	62 25.5	15.6	3 51.9	10.7	54.4
23	165 31.4	26.1	77 00.1	15.7	4 02.6	10.8	54.4
	S.D. 16.3	d 0.0	S.D. 15.1		15.0		14.8

Lat. °	Twilight Naut. h m	Twilight Civil h m	Sunrise h m	Moonrise 18 h m	19 h m	20 h m	21 h m
N 72	08 24	10 56	▬	11 38	11 23	11 10	10 56
N 70	08 04	09 53	▬	11 28	11 20	11 12	11 04
68	07 48	09 18	▬	11 20	11 17	11 14	11 11
66	07 35	08 52	10 33	11 13	11 15	11 16	11 17
64	07 24	08 33	09 51	11 07	11 13	11 17	11 22
62	07 15	08 17	09 22	11 03	11 11	11 18	11 26
60	07 06	08 03	09 01	10 58	11 09	11 20	11 30
N 58	06 59	07 51	08 43	10 54	11 08	11 21	11 33
56	06 52	07 41	08 29	10 51	11 07	11 21	11 36
54	06 46	07 32	08 16	10 48	11 06	11 22	11 39
52	06 40	07 24	08 05	10 45	11 05	11 23	11 42
50	06 35	07 16	07 55	10 42	11 04	11 24	11 44
45	06 23	07 00	07 34	10 37	11 02	11 25	11 49
N 40	06 13	06 47	07 17	10 32	11 00	11 26	11 53
35	06 03	06 35	07 03	10 28	10 58	11 27	11 57
30	05 54	06 24	06 51	10 25	10 57	11 28	12 00
20	05 38	06 05	06 29	10 18	10 55	11 30	12 05
N 10	05 22	05 48	06 11	10 13	10 53	11 32	12 10
0	05 05	05 31	05 53	10 08	10 51	11 33	12 15
S 10	04 46	05 13	05 36	10 03	10 49	11 34	12 19
20	04 23	04 52	05 17	09 57	10 47	11 36	12 24
30	03 54	04 27	04 55	09 51	10 45	11 38	12 30
35	03 35	04 12	04 42	09 47	10 44	11 39	12 33
40	03 12	03 54	04 27	09 43	10 42	11 40	12 37
45	02 41	03 31	04 09	09 38	10 40	11 41	12 42
S 50	01 56	03 01	03 46	09 32	10 38	11 43	12 47
52	01 28	02 46	03 35	09 30	10 38	11 44	12 49
54	00 41	02 28	03 23	09 27	10 37	11 45	12 52
56	////	02 06	03 09	09 23	10 35	11 46	12 55
58	////	01 36	02 52	09 20	10 34	11 47	12 58
S 60	////	00 46	02 31	09 16	10 33	11 48	13 02

Lat. °	Sunset h m	Twilight Civil h m	Twilight Naut. h m	Moonset 18 h m	19 h m	20 h m	21 h m
N 72	▬	12 59	15 30	21 23	23 07	24 48	00 48
N 70	▬	14 01	15 50	21 30	23 07	24 42	00 42
68	▬	14 37	16 06	21 37	23 08	24 37	00 37
66	13 21	15 02	16 19	21 42	23 08	24 33	00 33
64	14 04	15 22	16 30	21 46	23 08	24 29	00 29
62	14 32	15 38	16 40	21 50	23 09	24 26	00 26
60	14 54	15 51	16 48	21 53	23 09	24 23	00 23
N 58	15 11	16 03	16 56	21 56	23 09	24 21	00 21
56	15 26	16 13	17 02	21 58	23 09	24 19	00 19
54	15 38	16 22	17 09	22 00	23 09	24 17	00 17
52	15 50	16 31	17 14	22 02	23 10	24 15	00 15
50	15 59	16 38	17 20	22 04	23 10	24 14	00 14
45	16 20	16 54	17 31	22 08	23 10	24 11	00 11
N 40	16 37	17 08	17 42	22 12	23 10	24 08	00 08
35	16 51	17 19	17 51	22 15	23 10	24 05	00 05
30	17 04	17 30	18 00	22 17	23 11	24 03	00 03
20	17 25	17 49	18 16	22 21	23 11	23 59	24 48
N 10	17 43	18 06	18 33	22 25	23 11	23 56	24 41
0	18 01	18 23	18 50	22 29	23 11	23 53	24 35
S 10	18 19	18 42	19 09	22 32	23 12	23 50	24 29
20	18 38	19 02	19 31	22 36	23 12	23 47	24 22
30	19 00	19 27	20 00	22 40	23 12	23 43	24 15
35	19 13	19 42	20 19	22 42	23 12	23 41	24 10
40	19 28	20 01	20 42	22 45	23 12	23 39	24 06
45	19 46	20 23	21 13	22 48	23 12	23 36	24 00
S 50	20 08	20 53	21 59	22 52	23 12	23 33	23 53
52	20 19	21 08	22 27	22 54	23 13	23 31	23 50
54	20 32	21 26	23 14	22 55	23 13	23 29	23 47
56	20 46	21 48	////	22 57	23 13	23 28	23 43
58	21 03	22 18	////	23 00	23 13	23 26	23 39
S 60	21 23	23 10	////	23 02	23 13	23 23	23 34

Day	SUN Eqn. of Time 00ʰ m s	SUN Eqn. of Time 12ʰ m s	SUN Mer. Pass. h m	MOON Mer. Pass. Upper h m	MOON Mer. Pass. Lower h m	MOON Age d	MOON Phase
18	03 34	03 19	11 57	16 18	03 56	05	◐
19	03 05	02 50	11 57	17 01	04 40	06	
20	02 35	02 20	11 58	17 43	05 22	07	

Mag.	Name and Number		SHA							Declination						
				JULY	AUG.	SEPT.	OCT.	NOV.	DEC.		JULY	AUG.	SEPT.	OCT.	NOV.	DEC.
			°	′	′	′	′	′	′	°	′	′	′	′	′	′
3·4	γ Cephei		**5**	12·7	12·1	11·8	12·0	12·4	13·0	N **77**	35·6	35·8	36·0	36·1	36·3	36·4
2·6	*Markab*	57	**13**	53·0	52·8	52·7	52·7	52·8	52·9	N **15**	10·3	10·4	10·5	10·6	10·6	10·6
2·6	*Scheat*		**14**	07·6	07·4	07·4	07·4	07·5	07·6	N **28**	02·9	03·0	03·2	03·3	03·3	03·3
1·3	*Fomalhaut*	56	**15**	40·1	39·9	39·8	39·9	40·0	40·1	S **29**	39·1	39·1	39·1	39·2	39·3	39·3
2·2	β Gruis		**19**	25·2	25·0	24·9	25·0	25·1	25·3	S **46**	54·8	54·8	54·9	55·0	55·1	55·1
2·9	α Tucanæ		**25**	28·3	28·0	28·0	28·1	28·4	28·6	S **60**	17·2	17·2	17·4	17·5	17·6	17·5
2·2	*Al Na'ir*	55	**28**	01·9	01·8	01·7	01·8	02·0	02·2	S **46**	59·2	59·3	59·4	59·5	59·5	59·5
3·0	δ Capricorni		**33**	19·2	19·1	19·0	19·1	19·2	19·3	S **16**	09·2	09·1	09·2	09·2	09·2	09·2
2·5	*Enif*	54	**34**	01·4	01·3	01·3	01·4	01·5	01·6	N **9**	50·9	51·0	51·0	51·1	51·1	51·0
3·1	β Aquarii		**37**	11·2	11·1	11·1	11·1	11·3	11·3	S **5**	35·8	35·7	35·7	35·7	35·7	35·8
2·6	*Alderamin*		**40**	22·8	22·7	22·8	23·0	23·4	23·7	N **62**	33·5	33·7	33·9	34·0	34·0	34·0
2·6	ε Cygni		**48**	30·2	30·1	30·2	30·3	30·5	30·6	N **33**	56·8	57·0	57·1	57·2	57·2	57·1
1·3	*Deneb*	53	**49**	41·2	41·1	41·2	41·4	41·6	41·7	N **45**	15·5	15·7	15·8	15·9	15·9	15·8
3·2	α Indi		**50**	42·5	42·4	42·5	42·6	42·8	42·9	S **47**	18·6	18·7	18·8	18·8	18·9	18·8
2·1	*Peacock*	52	**53**	42·0	41·9	42·0	42·2	42·4	42·6	S **56**	45·2	45·3	45·4	45·4	45·4	45·4
2·3	γ Cygni		**54**	29·4	29·4	29·5	29·7	29·8	30·0	N **40**	14·2	14·4	14·5	14·6	14·6	14·5
0·9	*Altair*	51	**62**	22·4	22·4	22·4	22·6	22·7	22·7	N **8**	51·2	51·3	51·4	51·4	51·3	51·3
2·8	γ Aquilæ		**63**	30·1	30·1	30·2	30·3	30·4	30·5	N **10**	36·0	36·1	36·1	36·1	36·1	36·1
3·0	δ Cygni		**63**	47·7	47·8	47·9	48·1	48·3	48·4	N **45**	07·0	07·2	07·3	07·3	07·3	07·2
3·2	*Albireo*		**67**	22·5	22·5	22·6	22·8	22·9	23·0	N **27**	56·9	57·0	57·1	57·1	57·1	57·0
3·0	π Sagittarii		**72**	38·7	38·6	38·7	38·9	39·0	39·0	S **21**	01·9	01·9	01·9	01·9	01·9	01·9
3·0	ζ Aquilæ		**73**	42·7	42·7	42·8	43·0	43·1	43·1	N **13**	51·4	51·4	51·5	51·5	51·5	51·4
2·7	ζ Sagittarii		**74**	26·3	26·3	26·4	26·5	26·7	26·7	S **29**	53·3	53·3	53·3	53·3	53·3	53·3
2·1	*Nunki*	50	**76**	16·4	16·3	16·4	16·6	16·7	16·7	S **26**	18·2	18·2	18·2	18·2	18·2	18·2
0·1	*Vega*	49	**80**	48·6	48·7	48·8	49·0	49·2	49·3	N **38**	46·8	46·9	47·0	47·0	47·0	46·8
2·9	λ Sagittarii		**83**	05·8	05·8	05·9	06·0	06·1	06·1	S **25**	25·5	25·5	25·5	25·5	25·5	25·5
2·0	*Kaus Australis*	48	**84**	03·1	03·1	03·3	03·4	03·5	03·5	S **34**	23·2	23·3	23·3	23·3	23·3	23·2
2·8	δ Sagittarii		**84**	50·6	50·6	50·7	50·9	51·0	51·0	S **29**	49·8	49·8	49·9	49·9	49·8	49·8
3·1	γ Sagittarii		**88**	38·4	38·5	38·6	38·7	38·8	38·8	S **30**	25·4	25·5	25·5	25·5	25·4	25·4
2·4	*Eltanin*	47	**90**	52·6	52·7	52·9	53·2	53·4	53·5	N **51**	29·6	29·7	29·7	29·7	29·6	29·5
2·9	β Ophiuchi		**94**	12·2	12·2	12·3	12·5	12·5	12·5	N **4**	34·3	34·3	34·4	34·4	34·3	34·3
2·5	κ Scorpii		**94**	28·7	28·7	28·9	29·0	29·1	29·1	S **39**	01·6	01·7	01·7	01·7	01·6	01·6
2·0	θ Scorpii		**95**	46·4	46·5	46·6	46·8	46·9	46·9	S **42**	59·7	59·7	59·7	59·7	59·7	59·6
2·1	*Rasalhague*	46	**96**	19·9	20·0	20·1	20·3	20·4	20·3	N **12**	34·0	34·1	34·1	34·1	34·0	33·9
1·7	*Shaula*	45	**96**	41·7	41·8	41·9	42·1	42·2	42·1	S **37**	06·0	06·0	06·0	06·0	06·0	05·9
3·0	α Aræ		**97**	09·0	09·1	09·3	09·5	09·6	09·6	S **49**	52·3	52·4	52·4	52·4	52·3	52·2
2·8	υ Scorpii		**97**	24·4	24·5	24·6	24·8	24·9	24·8	S **37**	17·5	17·5	17·5	17·5	17·5	17·4
3·0	β Draconis		**97**	25·1	25·3	25·5	25·8	26·0	26·0	N **52**	18·6	18·7	18·7	18·7	18·6	18·4
2·8	β Aræ		**98**	47·6	47·7	48·0	48·2	48·3	48·3	S **55**	31·5	31·6	31·6	31·6	31·5	31·4
Var.‡	α Herculis		**101**	24·2	24·3	24·4	24·5	24·6	24·6	N **14**	24·0	24·0	24·1	24·0	24·0	23·9
2·6	*Sabik*	44	**102**	29·3	29·4	29·5	29·6	29·7	29·6	S **15**	43·0	43·0	43·0	43·0	43·0	43·0
3·1	ζ Aræ		**105**	27·8	28·0	28·2	28·4	28·5	28·4	S **55**	58·9	59·0	59·0	59·0	58·9	58·8
2·4	ε Scorpii		**107**	33·2	33·3	33·4	33·6	33·6	33·5	S **34**	17·0	17·0	17·0	17·0	16·9	16·9
1·9	*Atria*	43	**107**	59·0	59·3	59·7	60·0	60·2	60·1	S **69**	01·1	01·2	01·2	01·2	01·0	00·9
3·0	ζ Herculis		**109**	43·9	44·0	44·2	44·3	44·4	44·4	N **31**	37·0	37·1	37·1	37·1	37·0	36·8
2·7	ζ Ophiuchi		**110**	47·4	47·5	47·6	47·7	47·8	47·7	S **10**	33·2	33·2	33·2	33·2	33·2	33·2
2·9	τ Scorpii		**111**	07·2	07·3	07·4	07·6	07·6	07·5	S **28**	12·2	12·2	12·2	12·2	12·2	12·1
2·8	β Herculis		**112**	30·4	30·5	30·6	30·8	30·8	30·8	N **21**	30·3	30·4	30·4	30·4	30·3	30·2
1·2	*Antares*	42	**112**	44·3	44·3	44·5	44·6	44·6	44·5	S **26**	25·1	25·1	25·1	25·1	25·1	25·0
2·9	η Draconis		**114**	00·9	01·2	01·6	01·9	02·1	02·1	N **61**	32·0	32·0	32·0	31·9	31·8	31·6
3·0	δ Ophiuchi		**116**	29·4	29·5	29·6	29·7	29·8	29·7	S **3**	40·6	40·6	40·6	40·6	40·6	40·7
2·8	β Scorpii		**118**	43·6	43·6	43·8	43·9	43·9	43·8	S **19**	47·3	47·3	47·3	47·3	47·3	47·3
2·5	*Dschubba*		**120**	00·2	00·3	00·4	00·5	00·5	00·4	S **22**	36·3	36·3	36·2	36·2	36·2	36·2
3·0	π Scorpii		**120**	22·5	22·6	22·8	22·9	22·9	22·8	S **26**	05·8	05·8	05·8	05·8	05·7	05·7
3·0	β Trianguli Aust.		**121**	20·5	20·7	21·0	21·3	21·3	21·1	S **63**	24·9	24·9	24·9	24·8	24·7	24·6
2·8	α Serpentis		**124**	00·3	00·4	00·5	00·6	00·6	00·5	N **6**	26·8	26·8	26·9	26·8	26·8	26·7
3·0	γ Lupi		**126**	18·8	18·9	19·1	19·2	19·2	19·0	S **41**	08·9	08·9	08·9	08·8	08·7	08·7
2·3	*Alphecca*	41	**126**	23·4	23·5	23·7	23·8	23·8	23·7	N **26**	44·3	44·4	44·4	44·3	44·2	44·0

‡ 3·0 — 3·7

CONVERSION OF ARC TO TIME

0°–59°		60°–119°		120°–179°		180°–239°		240°–299°		300°–359°			0'·00	0'·25	0'·50	0'·75
°	h m	°	h m	°	h m	°	h m	°	h m	°	h m	'	m s	m s	m s	m s
0	0 00	**60**	4 00	**120**	8 00	**180**	12 00	**240**	16 00	**300**	20 00	**0**	0 00	0 01	0 02	0 03
1	0 04	**61**	4 04	**121**	8 04	**181**	12 04	**241**	16 04	**301**	20 04	**1**	0 04	0 05	0 06	0 07
2	0 08	**62**	4 08	**122**	8 08	**182**	12 08	**242**	16 08	**302**	20 08	**2**	0 08	0 09	0 10	0 11
3	0 12	**63**	4 12	**123**	8 12	**183**	12 12	**243**	16 12	**303**	20 12	**3**	0 12	0 13	0 14	0 15
4	0 16	**64**	4 16	**124**	8 16	**184**	12 16	**244**	16 16	**304**	20 16	**4**	0 16	0 17	0 18	0 19
5	0 20	**65**	4 20	**125**	8 20	**185**	12 20	**245**	16 20	**305**	20 20	**5**	0 20	0 21	0 22	0 23
6	0 24	**66**	4 24	**126**	8 24	**186**	12 24	**246**	16 24	**306**	20 24	**6**	0 24	0 25	0 26	0 27
7	0 28	**67**	4 28	**127**	8 28	**187**	12 28	**247**	16 28	**307**	20 28	**7**	0 28	0 29	0 30	0 31
8	0 32	**68**	4 32	**128**	8 32	**188**	12 32	**248**	16 32	**308**	20 32	**8**	0 32	0 33	0 34	0 35
9	0 36	**69**	4 36	**129**	8 36	**189**	12 36	**249**	16 36	**309**	20 36	**9**	0 36	0 37	0 38	0 39
10	0 40	**70**	4 40	**130**	8 40	**190**	12 40	**250**	16 40	**310**	20 40	**10**	0 40	0 41	0 42	0 43
11	0 44	**71**	4 44	**131**	8 44	**191**	12 44	**251**	16 44	**311**	20 44	**11**	0 44	0 45	0 46	0 47
12	0 48	**72**	4 48	**132**	8 48	**192**	12 48	**252**	16 48	**312**	20 48	**12**	0 48	0 49	0 50	0 51
13	0 52	**73**	4 52	**133**	8 52	**193**	12 52	**253**	16 52	**313**	20 52	**13**	0 52	0 53	0 54	0 55
14	0 56	**74**	4 56	**134**	8 56	**194**	12 56	**254**	16 56	**314**	20 56	**14**	0 56	0 57	0 58	0 59
15	1 00	**75**	5 00	**135**	9 00	**195**	13 00	**255**	17 00	**315**	21 00	**15**	1 00	1 01	1 02	1 03
16	1 04	**76**	5 04	**136**	9 04	**196**	13 04	**256**	17 04	**316**	21 04	**16**	1 04	1 05	1 06	1 07
17	1 08	**77**	5 08	**137**	9 08	**197**	13 08	**257**	17 08	**317**	21 08	**17**	1 08	1 09	1 10	1 11
18	1 12	**78**	5 12	**138**	9 12	**198**	13 12	**258**	17 12	**318**	21 12	**18**	1 12	1 13	1 14	1 15
19	1 16	**79**	5 16	**139**	9 16	**199**	13 16	**259**	17 16	**319**	21 16	**19**	1 16	1 17	1 18	1 19
20	1 20	**80**	5 20	**140**	9 20	**200**	13 20	**260**	17 20	**320**	21 20	**20**	1 20	1 21	1 22	1 23
21	1 24	**81**	5 24	**141**	9 24	**201**	13 24	**261**	17 24	**321**	21 24	**21**	1 24	1 25	1 26	1 27
22	1 28	**82**	5 28	**142**	9 28	**202**	13 28	**262**	17 28	**322**	21 28	**22**	1 28	1 29	1 30	1 31
23	1 32	**83**	5 32	**143**	9 32	**203**	13 32	**263**	17 32	**323**	21 32	**23**	1 32	1 33	1 34	1 35
24	1 36	**84**	5 36	**144**	9 36	**204**	13 36	**264**	17 36	**324**	21 36	**24**	1 36	1 37	1 38	1 39
25	1 40	**85**	5 40	**145**	9 40	**205**	13 40	**265**	17 40	**325**	21 40	**25**	1 40	1 41	1 42	1 43
26	1 44	**86**	5 44	**146**	9 44	**206**	13 44	**266**	17 44	**326**	21 44	**26**	1 44	1 45	1 46	1 47
27	1 48	**87**	5 48	**147**	9 48	**207**	13 48	**267**	17 48	**327**	21 48	**27**	1 48	1 49	1 50	1 51
28	1 52	**88**	5 52	**148**	9 52	**208**	13 52	**268**	17 52	**328**	21 52	**28**	1 52	1 53	1 54	1 55
29	1 56	**89**	5 56	**149**	9 56	**209**	13 56	**269**	17 56	**329**	21 56	**29**	1 56	1 57	1 58	1 59
30	2 00	**90**	6 00	**150**	10 00	**210**	14 00	**270**	18 00	**330**	22 00	**30**	2 00	2 01	2 02	2 03
31	2 04	**91**	6 04	**151**	10 04	**211**	14 04	**271**	18 04	**331**	22 04	**31**	2 04	2 05	2 06	2 07
32	2 08	**92**	6 08	**152**	10 08	**212**	14 08	**272**	18 08	**332**	22 08	**32**	2 08	2 09	2 10	2 11
33	2 12	**93**	6 12	**153**	10 12	**213**	14 12	**273**	18 12	**333**	22 12	**33**	2 12	2 13	2 14	2 15
34	2 16	**94**	6 16	**154**	10 16	**214**	14 16	**274**	18 16	**334**	22 16	**34**	2 16	2 17	2 18	2 19
35	2 20	**95**	6 20	**155**	10 20	**215**	14 20	**275**	18 20	**335**	22 20	**35**	2 20	2 21	2 22	2 23
36	2 24	**96**	6 24	**156**	10 24	**216**	14 24	**276**	18 24	**336**	22 24	**36**	2 24	2 25	2 26	2 27
37	2 28	**97**	6 28	**157**	10 28	**217**	14 28	**277**	18 28	**337**	22 28	**37**	2 28	2 29	2 30	2 31
38	2 32	**98**	6 32	**158**	10 32	**218**	14 32	**278**	18 32	**338**	22 32	**38**	2 32	2 33	2 34	2 35
39	2 36	**99**	6 36	**159**	10 36	**219**	14 36	**279**	18 36	**339**	22 36	**39**	2 36	2 37	2 38	2 39
40	2 40	**100**	6 40	**160**	10 40	**220**	14 40	**280**	18 40	**340**	22 40	**40**	2 40	2 41	2 42	2 43
41	2 44	**101**	6 44	**161**	10 44	**221**	14 44	**281**	18 44	**341**	22 44	**41**	2 44	2 45	2 46	2 47
42	2 48	**102**	6 48	**162**	10 48	**222**	14 48	**282**	18 48	**342**	22 48	**42**	2 48	2 49	2 50	2 51
43	2 52	**103**	6 52	**163**	10 52	**223**	14 52	**283**	18 52	**343**	22 52	**43**	2 52	2 53	2 54	2 55
44	2 56	**104**	6 56	**164**	10 56	**224**	14 56	**284**	18 56	**344**	22 56	**44**	2 56	2 57	2 58	2 59
45	3 00	**105**	7 00	**165**	11 00	**225**	15 00	**285**	19 00	**345**	23 00	**45**	3 00	3 01	3 02	3 03
46	3 04	**106**	7 04	**166**	11 04	**226**	15 04	**286**	19 04	**346**	23 04	**46**	3 04	3 05	3 06	3 07
47	3 08	**107**	7 08	**167**	11 08	**227**	15 08	**287**	19 08	**347**	23 08	**47**	3 08	3 09	3 10	3 11
48	3 12	**108**	7 12	**168**	11 12	**228**	15 12	**288**	19 12	**348**	23 12	**48**	3 12	3 13	3 14	3 15
49	3 16	**109**	7 16	**169**	11 16	**229**	15 16	**289**	19 16	**349**	23 16	**49**	3 16	3 17	3 18	3 19
50	3 20	**110**	7 20	**170**	11 20	**230**	15 20	**290**	19 20	**350**	23 20	**50**	3 20	3 21	3 22	3 23
51	3 24	**111**	7 24	**171**	11 24	**231**	15 24	**291**	19 24	**351**	23 24	**51**	3 24	3 25	3 26	3 27
52	3 28	**112**	7 28	**172**	11 28	**232**	15 28	**292**	19 28	**352**	23 28	**52**	3 28	3 29	3 30	3 31
53	3 32	**113**	7 32	**173**	11 32	**233**	15 32	**293**	19 32	**353**	23 32	**53**	3 32	3 33	3 34	3 35
54	3 36	**114**	7 36	**174**	11 36	**234**	15 36	**294**	19 36	**354**	23 36	**54**	3 36	3 37	3 38	3 39
55	3 40	**115**	7 40	**175**	11 40	**235**	15 40	**295**	19 40	**355**	23 40	**55**	3 40	3 41	3 42	3 43
56	3 44	**116**	7 44	**176**	11 44	**236**	15 44	**296**	19 44	**356**	23 44	**56**	3 44	3 45	3 46	3 47
57	3 48	**117**	7 48	**177**	11 48	**237**	15 48	**297**	19 48	**357**	23 48	**57**	3 48	3 49	3 50	3 51
58	3 52	**118**	7 52	**178**	11 52	**238**	15 52	**298**	19 52	**358**	23 52	**58**	3 52	3 53	3 54	3 55
59	3 56	**119**	7 56	**179**	11 56	**239**	15 56	**299**	19 56	**359**	23 56	**59**	3 56	3 57	3 58	3 59

The above table is for converting expressions in arc to their equivalent in time; its main use in this Almanac is for the conversion of longitude for application to L.M.T. (*added* if *west*, *subtracted* if *east*) to give G.M.T. or vice versa, particularly in the case of sunrise, sunset, etc.

0m	SUN PLANETS	ARIES	MOON	v or d	Corrⁿ	v or d	Corrⁿ	v or d	Corrⁿ
s	° ′	° ′	° ′	′	′	′	′	′	′
00	0 00·0	0 00·0	0 00·0	0·0	0·0	6·0	0·1	12·0	0·1
01	0 00·3	0 00·3	0 00·2	0·1	0·0	6·1	0·1	12·1	0·1
02	0 00·5	0 00·5	0 00·5	0·2	0·0	6·2	0·1	12·2	0·1
03	0 00·8	0 00·8	0 00·7	0·3	0·0	6·3	0·1	12·3	0·1
04	0 01·0	0 01·0	0 01·0	0·4	0·0	6·4	0·1	12·4	0·1
05	0 01·3	0 01·3	0 01·2	0·5	0·0	6·5	0·1	12·5	0·1
06	0 01·5	0 01·5	0 01·4	0·6	0·0	6·6	0·1	12·6	0·1
07	0 01·8	0 01·8	0 01·7	0·7	0·0	6·7	0·1	12·7	0·1
08	0 02·0	0 02·0	0 01·9	0·8	0·0	6·8	0·1	12·8	0·1
09	0 02·3	0 02·3	0 02·1	0·9	0·0	6·9	0·1	12·9	0·1
10	0 02·5	0 02·5	0 02·4	1·0	0·0	7·0	0·1	13·0	0·1
11	0 02·8	0 02·8	0 02·6	1·1	0·0	7·1	0·1	13·1	0·1
12	0 03·0	0 03·0	0 02·9	1·2	0·0	7·2	0·1	13·2	0·1
13	0 03·3	0 03·3	0 03·1	1·3	0·0	7·3	0·1	13·3	0·1
14	0 03·5	0 03·5	0 03·3	1·4	0·0	7·4	0·1	13·4	0·1
15	0 03·8	0 03·8	0 03·6	1·5	0·0	7·5	0·1	13·5	0·1
16	0 04·0	0 04·0	0 03·8	1·6	0·0	7·6	0·1	13·6	0·1
17	0 04·3	0 04·3	0 04·1	1·7	0·0	7·7	0·1	13·7	0·1
18	0 04·5	0 04·5	0 04·3	1·8	0·0	7·8	0·1	13·8	0·1
19	0 04·8	0 04·8	0 04·5	1·9	0·0	7·9	0·1	13·9	0·1
20	0 05·0	0 05·0	0 04·8	2·0	0·0	8·0	0·1	14·0	0·1
21	0 05·3	0 05·3	0 05·0	2·1	0·0	8·1	0·1	14·1	0·1
22	0 05·5	0 05·5	0 05·2	2·2	0·0	8·2	0·1	14·2	0·1
23	0 05·8	0 05·8	0 05·5	2·3	0·0	8·3	0·1	14·3	0·1
24	0 06·0	0 06·0	0 05·7	2·4	0·0	8·4	0·1	14·4	0·1
25	0 06·3	0 06·3	0 06·0	2·5	0·0	8·5	0·1	14·5	0·1
26	0 06·5	0 06·5	0 06·2	2·6	0·0	8·6	0·1	14·6	0·1
27	0 06·8	0 06·8	0 06·4	2·7	0·0	8·7	0·1	14·7	0·1
28	0 07·0	0 07·0	0 06·7	2·8	0·0	8·8	0·1	14·8	0·1
29	0 07·3	0 07·3	0 06·9	2·9	0·0	8·9	0·1	14·9	0·1
30	0 07·5	0 07·5	0 07·2	3·0	0·0	9·0	0·1	15·0	0·1
31	0 07·8	0 07·8	0 07·4	3·1	0·0	9·1	0·1	15·1	0·1
32	0 08·0	0 08·0	0 07·6	3·2	0·0	9·2	0·1	15·2	0·1
33	0 08·3	0 08·3	0 07·9	3·3	0·0	9·3	0·1	15·3	0·1
34	0 08·5	0 08·5	0 08·1	3·4	0·0	9·4	0·1	15·4	0·1
35	0 08·8	0 08·8	0 08·4	3·5	0·0	9·5	0·1	15·5	0·1
36	0 09·0	0 09·0	0 08·6	3·6	0·0	9·6	0·1	15·6	0·1
37	0 09·3	0 09·3	0 08·8	3·7	0·0	9·7	0·1	15·7	0·1
38	0 09·5	0 09·5	0 09·1	3·8	0·0	9·8	0·1	15·8	0·1
39	0 09·8	0 09·8	0 09·3	3·9	0·0	9·9	0·1	15·9	0·1
40	0 10·0	0 10·0	0 09·5	4·0	0·0	10·0	0·1	16·0	0·1
41	0 10·3	0 10·3	0 09·8	4·1	0·0	10·1	0·1	16·1	0·1
42	0 10·5	0 10·5	0 10·0	4·2	0·0	10·2	0·1	16·2	0·1
43	0 10·8	0 10·8	0 10·3	4·3	0·0	10·3	0·1	16·3	0·1
44	0 11·0	0 11·0	0 10·5	4·4	0·0	10·4	0·1	16·4	0·1
45	0 11·3	0 11·3	0 10·7	4·5	0·0	10·5	0·1	16·5	0·1
46	0 11·5	0 11·5	0 11·0	4·6	0·0	10·6	0·1	16·6	0·1
47	0 11·8	0 11·8	0 11·2	4·7	0·0	10·7	0·1	16·7	0·1
48	0 12·0	0 12·0	0 11·5	4·8	0·0	10·8	0·1	16·8	0·1
49	0 12·3	0 12·3	0 11·7	4·9	0·0	10·9	0·1	16·9	0·1
50	0 12·5	0 12·5	0 11·9	5·0	0·0	11·0	0·1	17·0	0·1
51	0 12·8	0 12·8	0 12·2	5·1	0·0	11·1	0·1	17·1	0·1
52	0 13·0	0 13·0	0 12·4	5·2	0·0	11·2	0·1	17·2	0·1
53	0 13·3	0 13·3	0 12·6	5·3	0·0	11·3	0·1	17·3	0·1
54	0 13·5	0 13·5	0 12·9	5·4	0·0	11·4	0·1	17·4	0·1
55	0 13·8	0 13·8	0 13·1	5·5	0·0	11·5	0·1	17·5	0·1
56	0 14·0	0 14·0	0 13·4	5·6	0·0	11·6	0·1	17·6	0·1
57	0 14·3	0 14·3	0 13·6	5·7	0·0	11·7	0·1	17·7	0·1
58	0 14·5	0 14·5	0 13·8	5·8	0·0	11·8	0·1	17·8	0·1
59	0 14·8	0 14·8	0 14·1	5·9	0·0	11·9	0·1	17·9	0·1
60	0 15·0	0 15·0	0 14·3	6·0	0·1	12·0	0·1	18·0	0·2

1m	SUN PLANETS	ARIES	MOON	v or d	Corrⁿ	v or d	Corrⁿ	v or d	Corrⁿ
s	° ′	° ′	° ′	′	′	′	′	′	′
00	0 15·0	0 15·0	0 14·3	0·0	0·0	6·0	0·2	12·0	0·3
01	0 15·3	0 15·3	0 14·6	0·1	0·0	6·1	0·2	12·1	0·3
02	0 15·5	0 15·5	0 14·8	0·2	0·0	6·2	0·2	12·2	0·3
03	0 15·8	0 15·8	0 15·0	0·3	0·0	6·3	0·2	12·3	0·3
04	0 16·0	0 16·0	0 15·3	0·4	0·0	6·4	0·2	12·4	0·3
05	0 16·3	0 16·3	0 15·5	0·5	0·0	6·5	0·2	12·5	0·3
06	0 16·5	0 16·5	0 15·7	0·6	0·0	6·6	0·2	12·6	0·3
07	0 16·8	0 16·8	0 16·0	0·7	0·0	6·7	0·2	12·7	0·3
08	0 17·0	0 17·0	0 16·2	0·8	0·0	6·8	0·2	12·8	0·3
09	0 17·3	0 17·3	0 16·5	0·9	0·0	6·9	0·2	12·9	0·3
10	0 17·5	0 17·5	0 16·7	1·0	0·0	7·0	0·2	13·0	0·3
11	0 17·8	0 17·8	0 16·9	1·1	0·0	7·1	0·2	13·1	0·3
12	0 18·0	0 18·0	0 17·2	1·2	0·0	7·2	0·2	13·2	0·3
13	0 18·3	0 18·3	0 17·4	1·3	0·0	7·3	0·2	13·3	0·3
14	0 18·5	0 18·6	0 17·7	1·4	0·0	7·4	0·2	13·4	0·3
15	0 18·8	0 18·8	0 17·9	1·5	0·0	7·5	0·2	13·5	0·3
16	0 19·0	0 19·1	0 18·1	1·6	0·0	7·6	0·2	13·6	0·3
17	0 19·3	0 19·3	0 18·4	1·7	0·0	7·7	0·2	13·7	0·3
18	0 19·5	0 19·6	0 18·6	1·8	0·0	7·8	0·2	13·8	0·3
19	0 19·8	0 19·8	0 18·9	1·9	0·0	7·9	0·2	13·9	0·3
20	0 20·0	0 20·1	0 19·1	2·0	0·1	8·0	0·2	14·0	0·4
21	0 20·3	0 20·3	0 19·3	2·1	0·1	8·1	0·2	14·1	0·4
22	0 20·5	0 20·6	0 19·6	2·2	0·1	8·2	0·2	14·2	0·4
23	0 20·8	0 20·8	0 19·8	2·3	0·1	8·3	0·2	14·3	0·4
24	0 21·0	0 21·1	0 20·0	2·4	0·1	8·4	0·2	14·4	0·4
25	0 21·3	0 21·3	0 20·3	2·5	0·1	8·5	0·2	14·5	0·4
26	0 21·5	0 21·6	0 20·5	2·6	0·1	8·6	0·2	14·6	0·4
27	0 21·8	0 21·8	0 20·8	2·7	0·1	8·7	0·2	14·7	0·4
28	0 22·0	0 22·1	0 21·0	2·8	0·1	8·8	0·2	14·8	0·4
29	0 22·3	0 22·3	0 21·2	2·9	0·1	8·9	0·2	14·9	0·4
30	0 22·5	0 22·6	0 21·5	3·0	0·1	9·0	0·2	15·0	0·4
31	0 22·8	0 22·8	0 21·7	3·1	0·1	9·1	0·2	15·1	0·4
32	0 23·0	0 23·1	0 22·0	3·2	0·1	9·2	0·2	15·2	0·4
33	0 23·3	0 23·3	0 22·2	3·3	0·1	9·3	0·2	15·3	0·4
34	0 23·5	0 23·6	0 22·4	3·4	0·1	9·4	0·2	15·4	0·4
35	0 23·8	0 23·8	0 22·7	3·5	0·1	9·5	0·2	15·5	0·4
36	0 24·0	0 24·1	0 22·9	3·6	0·1	9·6	0·2	15·6	0·4
37	0 24·3	0 24·3	0 23·1	3·7	0·1	9·7	0·2	15·7	0·4
38	0 24·5	0 24·6	0 23·4	3·8	0·1	9·8	0·2	15·8	0·4
39	0 24·8	0 24·8	0 23·6	3·9	0·1	9·9	0·2	15·9	0·4
40	0 25·0	0 25·1	0 23·9	4·0	0·1	10·0	0·3	16·0	0·4
41	0 25·3	0 25·3	0 24·1	4·1	0·1	10·1	0·3	16·1	0·4
42	0 25·5	0 25·6	0 24·3	4·2	0·1	10·2	0·3	16·2	0·4
43	0 25·8	0 25·8	0 24·6	4·3	0·1	10·3	0·3	16·3	0·4
44	0 26·0	0 26·1	0 24·8	4·4	0·1	10·4	0·3	16·4	0·4
45	0 26·3	0 26·3	0 25·1	4·5	0·1	10·5	0·3	16·5	0·4
46	0 26·5	0 26·6	0 25·3	4·6	0·1	10·6	0·3	16·6	0·4
47	0 26·8	0 26·8	0 25·5	4·7	0·1	10·7	0·3	16·7	0·4
48	0 27·0	0 27·1	0 25·8	4·8	0·1	10·8	0·3	16·8	0·4
49	0 27·3	0 27·3	0 26·0	4·9	0·1	10·9	0·3	16·9	0·4
50	0 27·5	0 27·6	0 26·2	5·0	0·1	11·0	0·3	17·0	0·4
51	0 27·8	0 27·8	0 26·5	5·1	0·1	11·1	0·3	17·1	0·4
52	0 28·0	0 28·1	0 26·7	5·2	0·1	11·2	0·3	17·2	0·4
53	0 28·3	0 28·3	0 27·0	5·3	0·1	11·3	0·3	17·3	0·4
54	0 28·5	0 28·6	0 27·2	5·4	0·1	11·4	0·3	17·4	0·4
55	0 28·8	0 28·8	0 27·4	5·5	0·1	11·5	0·3	17·5	0·4
56	0 29·0	0 29·1	0 27·7	5·6	0·1	11·6	0·3	17·6	0·4
57	0 29·3	0 29·3	0 27·9	5·7	0·1	11·7	0·3	17·7	0·4
58	0 29·5	0 29·6	0 28·2	5·8	0·1	11·8	0·3	17·8	0·4
59	0 29·8	0 29·8	0 28·4	5·9	0·1	11·9	0·3	17·9	0·4
60	0 30·0	0 30·1	0 28·6	6·0	0·2	12·0	0·3	18·0	0·5

2^m s	SUN PLANETS ° ′	ARIES ° ′	MOON ° ′	v or d ′	Corr^n ′	v or d ′	Corr^n ′	v or d ′	Corr^n ′
00	0 30.0	0 30.1	0 28.6	0.0	0.0	6.0	0.3	12.0	0.5
01	0 30.3	0 30.3	0 28.9	0.1	0.0	6.1	0.3	12.1	0.5
02	0 30.5	0 30.6	0 29.1	0.2	0.0	6.2	0.3	12.2	0.5
03	0 30.8	0 30.8	0 29.3	0.3	0.0	6.3	0.3	12.3	0.5
04	0 31.0	0 31.1	0 29.6	0.4	0.0	6.4	0.3	12.4	0.5
05	0 31.3	0 31.3	0 29.8	0.5	0.0	6.5	0.3	12.5	0.5
06	0 31.5	0 31.6	0 30.1	0.6	0.0	6.6	0.3	12.6	0.5
07	0 31.8	0 31.8	0 30.3	0.7	0.0	6.7	0.3	12.7	0.5
08	0 32.0	0 32.1	0 30.5	0.8	0.0	6.8	0.3	12.8	0.5
09	0 32.3	0 32.3	0 30.8	0.9	0.0	6.9	0.3	12.9	0.5
10	0 32.5	0 32.6	0 31.0	1.0	0.0	7.0	0.3	13.0	0.5
11	0 32.8	0 32.8	0 31.3	1.1	0.0	7.1	0.3	13.1	0.5
12	0 33.0	0 33.1	0 31.5	1.2	0.1	7.2	0.3	13.2	0.6
13	0 33.3	0 33.3	0 31.7	1.3	0.1	7.3	0.3	13.3	0.6
14	0 33.5	0 33.6	0 32.0	1.4	0.1	7.4	0.3	13.4	0.6
15	0 33.8	0 33.8	0 32.2	1.5	0.1	7.5	0.3	13.5	0.6
16	0 34.0	0 34.1	0 32.5	1.6	0.1	7.6	0.3	13.6	0.6
17	0 34.3	0 34.3	0 32.7	1.7	0.1	7.7	0.3	13.7	0.6
18	0 34.5	0 34.6	0 32.9	1.8	0.1	7.8	0.3	13.8	0.6
19	0 34.8	0 34.8	0 33.2	1.9	0.1	7.9	0.3	13.9	0.6
20	0 35.0	0 35.1	0 33.4	2.0	0.1	8.0	0.3	14.0	0.6
21	0 35.3	0 35.3	0 33.6	2.1	0.1	8.1	0.3	14.1	0.6
22	0 35.5	0 35.6	0 33.9	2.2	0.1	8.2	0.3	14.2	0.6
23	0 35.8	0 35.8	0 34.1	2.3	0.1	8.3	0.3	14.3	0.6
24	0 36.0	0 36.1	0 34.4	2.4	0.1	8.4	0.4	14.4	0.6
25	0 36.3	0 36.3	0 34.6	2.5	0.1	8.5	0.4	14.5	0.6
26	0 36.5	0 36.6	0 34.8	2.6	0.1	8.6	0.4	14.6	0.6
27	0 36.8	0 36.9	0 35.1	2.7	0.1	8.7	0.4	14.7	0.6
28	0 37.0	0 37.1	0 35.3	2.8	0.1	8.8	0.4	14.8	0.6
29	0 37.3	0 37.4	0 35.6	2.9	0.1	8.9	0.4	14.9	0.6
30	0 37.5	0 37.6	0 35.8	3.0	0.1	9.0	0.4	15.0	0.6
31	0 37.8	0 37.9	0 36.0	3.1	0.1	9.1	0.4	15.1	0.6
32	0 38.0	0 38.1	0 36.3	3.2	0.1	9.2	0.4	15.2	0.6
33	0 38.3	0 38.4	0 36.5	3.3	0.1	9.3	0.4	15.3	0.6
34	0 38.5	0 38.6	0 36.7	3.4	0.1	9.4	0.4	15.4	0.6
35	0 38.8	0 38.9	0 37.0	3.5	0.1	9.5	0.4	15.5	0.6
36	0 39.0	0 39.1	0 37.2	3.6	0.2	9.6	0.4	15.6	0.7
37	0 39.3	0 39.4	0 37.5	3.7	0.2	9.7	0.4	15.7	0.7
38	0 39.5	0 39.6	0 37.7	3.8	0.2	9.8	0.4	15.8	0.7
39	0 39.8	0 39.9	0 37.9	3.9	0.2	9.9	0.4	15.9	0.7
40	0 40.0	0 40.1	0 38.2	4.0	0.2	10.0	0.4	16.0	0.7
41	0 40.3	0 40.4	0 38.4	4.1	0.2	10.1	0.4	16.1	0.7
42	0 40.5	0 40.6	0 38.7	4.2	0.2	10.2	0.4	16.2	0.7
43	0 40.8	0 40.9	0 38.9	4.3	0.2	10.3	0.4	16.3	0.7
44	0 41.0	0 41.1	0 39.1	4.4	0.2	10.4	0.4	16.4	0.7
45	0 41.3	0 41.4	0 39.4	4.5	0.2	10.5	0.4	16.5	0.7
46	0 41.5	0 41.6	0 39.6	4.6	0.2	10.6	0.4	16.6	0.7
47	0 41.8	0 41.9	0 39.8	4.7	0.2	10.7	0.4	16.7	0.7
48	0 42.0	0 42.1	0 40.1	4.8	0.2	10.8	0.5	16.8	0.7
49	0 42.3	0 42.4	0 40.3	4.9	0.2	10.9	0.5	16.9	0.7
50	0 42.5	0 42.6	0 40.6	5.0	0.2	11.0	0.5	17.0	0.7
51	0 42.8	0 42.9	0 40.8	5.1	0.2	11.1	0.5	17.1	0.7
52	0 43.0	0 43.1	0 41.0	5.2	0.2	11.2	0.5	17.2	0.7
53	0 43.3	0 43.4	0 41.3	5.3	0.2	11.3	0.5	17.3	0.7
54	0 43.5	0 43.6	0 41.5	5.4	0.2	11.4	0.5	17.4	0.7
55	0 43.8	0 43.9	0 41.8	5.5	0.2	11.5	0.5	17.5	0.7
56	0 44.0	0 44.1	0 42.0	5.6	0.2	11.6	0.5	17.6	0.7
57	0 44.3	0 44.4	0 42.2	5.7	0.2	11.7	0.5	17.7	0.7
58	0 44.5	0 44.6	0 42.5	5.8	0.2	11.8	0.5	17.8	0.7
59	0 44.8	0 44.9	0 42.7	5.9	0.2	11.9	0.5	17.9	0.7
60	0 45.0	0 45.1	0 43.0	6.0	0.3	12.0	0.5	18.0	0.8

3^m s	SUN PLANETS ° ′	ARIES ° ′	MOON ° ′	v or d ′	Corr^n ′	v or d ′	Corr^n ′	v or d ′	Corr^n ′
00	0 45.0	0 45.1	0 43.0	0.0	0.0	6.0	0.4	12.0	0.7
01	0 45.3	0 45.4	0 43.2	0.1	0.0	6.1	0.4	12.1	0.7
02	0 45.5	0 45.6	0 43.4	0.2	0.0	6.2	0.4	12.2	0.7
03	0 45.8	0 45.9	0 43.7	0.3	0.0	6.3	0.4	12.3	0.7
04	0 46.0	0 46.1	0 43.9	0.4	0.0	6.4	0.4	12.4	0.7
05	0 46.3	0 46.4	0 44.1	0.5	0.0	6.5	0.4	12.5	0.7
06	0 46.5	0 46.6	0 44.4	0.6	0.0	6.6	0.4	12.6	0.7
07	0 46.8	0 46.9	0 44.6	0.7	0.0	6.7	0.4	12.7	0.7
08	0 47.0	0 47.1	0 44.9	0.8	0.0	6.8	0.4	12.8	0.7
09	0 47.3	0 47.4	0 45.1	0.9	0.1	6.9	0.4	12.9	0.8
10	0 47.5	0 47.6	0 45.3	1.0	0.1	7.0	0.4	13.0	0.8
11	0 47.8	0 47.9	0 45.6	1.1	0.1	7.1	0.4	13.1	0.8
12	0 48.0	0 48.1	0 45.8	1.2	0.1	7.2	0.4	13.2	0.8
13	0 48.3	0 48.4	0 46.1	1.3	0.1	7.3	0.4	13.3	0.8
14	0 48.5	0 48.6	0 46.3	1.4	0.1	7.4	0.4	13.4	0.8
15	0 48.8	0 48.9	0 46.5	1.5	0.1	7.5	0.4	13.5	0.8
16	0 49.0	0 49.1	0 46.8	1.6	0.1	7.6	0.4	13.6	0.8
17	0 49.3	0 49.4	0 47.0	1.7	0.1	7.7	0.4	13.7	0.8
18	0 49.5	0 49.6	0 47.2	1.8	0.1	7.8	0.5	13.8	0.8
19	0 49.8	0 49.9	0 47.5	1.9	0.1	7.9	0.5	13.9	0.8
20	0 50.0	0 50.1	0 47.7	2.0	0.1	8.0	0.5	14.0	0.8
21	0 50.3	0 50.4	0 48.0	2.1	0.1	8.1	0.5	14.1	0.8
22	0 50.5	0 50.6	0 48.2	2.2	0.1	8.2	0.5	14.2	0.8
23	0 50.8	0 50.9	0 48.4	2.3	0.1	8.3	0.5	14.3	0.8
24	0 51.0	0 51.1	0 48.7	2.4	0.1	8.4	0.5	14.4	0.8
25	0 51.3	0 51.4	0 48.9	2.5	0.1	8.5	0.5	14.5	0.8
26	0 51.5	0 51.6	0 49.2	2.6	0.2	8.6	0.5	14.6	0.9
27	0 51.8	0 51.9	0 49.4	2.7	0.2	8.7	0.5	14.7	0.9
28	0 52.0	0 52.1	0 49.6	2.8	0.2	8.8	0.5	14.8	0.9
29	0 52.3	0 52.4	0 49.9	2.9	0.2	8.9	0.5	14.9	0.9
30	0 52.5	0 52.6	0 50.1	3.0	0.2	9.0	0.5	15.0	0.9
31	0 52.8	0 52.9	0 50.3	3.1	0.2	9.1	0.5	15.1	0.9
32	0 53.0	0 53.1	0 50.6	3.2	0.2	9.2	0.5	15.2	0.9
33	0 53.3	0 53.4	0 50.8	3.3	0.2	9.3	0.5	15.3	0.9
34	0 53.5	0 53.6	0 51.1	3.4	0.2	9.4	0.5	15.4	0.9
35	0 53.8	0 53.9	0 51.3	3.5	0.2	9.5	0.6	15.5	0.9
36	0 54.0	0 54.1	0 51.5	3.6	0.2	9.6	0.6	15.6	0.9
37	0 54.3	0 54.4	0 51.8	3.7	0.2	9.7	0.6	15.7	0.9
38	0 54.5	0 54.6	0 52.0	3.8	0.2	9.8	0.6	15.8	0.9
39	0 54.8	0 54.9	0 52.3	3.9	0.2	9.9	0.6	15.9	0.9
40	0 55.0	0 55.2	0 52.5	4.0	0.2	10.0	0.6	16.0	0.9
41	0 55.3	0 55.4	0 52.7	4.1	0.2	10.1	0.6	16.1	0.9
42	0 55.5	0 55.7	0 53.0	4.2	0.2	10.2	0.6	16.2	0.9
43	0 55.8	0 55.9	0 53.2	4.3	0.3	10.3	0.6	16.3	1.0
44	0 56.0	0 56.2	0 53.4	4.4	0.3	10.4	0.6	16.4	1.0
45	0 56.3	0 56.4	0 53.7	4.5	0.3	10.5	0.6	16.5	1.0
46	0 56.5	0 56.7	0 53.9	4.6	0.3	10.6	0.6	16.6	1.0
47	0 56.8	0 56.9	0 54.2	4.7	0.3	10.7	0.6	16.7	1.0
48	0 57.0	0 57.2	0 54.4	4.8	0.3	10.8	0.6	16.8	1.0
49	0 57.3	0 57.4	0 54.6	4.9	0.3	10.9	0.6	16.9	1.0
50	0 57.5	0 57.7	0 54.9	5.0	0.3	11.0	0.6	17.0	1.0
51	0 57.8	0 57.9	0 55.1	5.1	0.3	11.1	0.6	17.1	1.0
52	0 58.0	0 58.2	0 55.4	5.2	0.3	11.2	0.7	17.2	1.0
53	0 58.3	0 58.4	0 55.6	5.3	0.3	11.3	0.7	17.3	1.0
54	0 58.5	0 58.7	0 55.8	5.4	0.3	11.4	0.7	17.4	1.0
55	0 58.8	0 58.9	0 56.1	5.5	0.3	11.5	0.7	17.5	1.0
56	0 59.0	0 59.2	0 56.3	5.6	0.3	11.6	0.7	17.6	1.0
57	0 59.3	0 59.4	0 56.6	5.7	0.3	11.7	0.7	17.7	1.0
58	0 59.5	0 59.7	0 56.8	5.8	0.3	11.8	0.7	17.8	1.0
59	0 59.8	0 59.9	0 57.0	5.9	0.3	11.9	0.7	17.9	1.0
60	1 00.0	1 00.2	0 57.3	6.0	0.4	12.0	0.7	18.0	1.1

4ᵐ	SUN PLANETS	ARIES	MOON	v or d	Corrⁿ	v or d	Corrⁿ	v or d	Corrⁿ
s	° ′	° ′	° ′	′	′	′	′	′	′
00	1 00·0	1 00·2	0 57·3	0·0	0·0	6·0	0·5	12·0	0·9
01	1 00·3	1 00·4	0 57·5	0·1	0·0	6·1	0·5	12·1	0·9
02	1 00·5	1 00·7	0 57·7	0·2	0·0	6·2	0·5	12·2	0·9
03	1 00·8	1 00·9	0 58·0	0·3	0·0	6·3	0·5	12·3	0·9
04	1 01·0	1 01·2	0 58·2	0·4	0·0	6·4	0·5	12·4	0·9
05	1 01·3	1 01·4	0 58·5	0·5	0·0	6·5	0·5	12·5	0·9
06	1 01·5	1 01·7	0 58·7	0·6	0·0	6·6	0·5	12·6	0·9
07	1 01·8	1 01·9	0 58·9	0·7	0·1	6·7	0·5	12·7	1·0
08	1 02·0	1 02·2	0 59·2	0·8	0·1	6·8	0·5	12·8	1·0
09	1 02·3	1 02·4	0 59·4	0·9	0·1	6·9	0·5	12·9	1·0
10	1 02·5	1 02·7	0 59·7	1·0	0·1	7·0	0·5	13·0	1·0
11	1 02·8	1 02·9	0 59·9	1·1	0·1	7·1	0·5	13·1	1·0
12	1 03·0	1 03·2	1 00·1	1·2	0·1	7·2	0·5	13·2	1·0
13	1 03·3	1 03·4	1 00·4	1·3	0·1	7·3	0·5	13·3	1·0
14	1 03·5	1 03·7	1 00·6	1·4	0·1	7·4	0·6	13·4	1·0
15	1 03·8	1 03·9	1 00·8	1·5	0·1	7·5	0·6	13·5	1·0
16	1 04·0	1 04·2	1 01·1	1·6	0·1	7·6	0·6	13·6	1·0
17	1 04·3	1 04·4	1 01·3	1·7	0·1	7·7	0·6	13·7	1·0
18	1 04·5	1 04·7	1 01·6	1·8	0·1	7·8	0·6	13·8	1·0
19	1 04·8	1 04·9	1 01·8	1·9	0·1	7·9	0·6	13·9	1·0
20	1 05·0	1 05·2	1 02·0	2·0	0·2	8·0	0·6	14·0	1·1
21	1 05·3	1 05·4	1 02·3	2·1	0·2	8·1	0·6	14·1	1·1
22	1 05·5	1 05·7	1 02·5	2·2	0·2	8·2	0·6	14·2	1·1
23	1 05·8	1 05·9	1 02·8	2·3	0·2	8·3	0·6	14·3	1·1
24	1 06·0	1 06·2	1 03·0	2·4	0·2	8·4	0·6	14·4	1·1
25	1 06·3	1 06·4	1 03·2	2·5	0·2	8·5	0·6	14·5	1·1
26	1 06·5	1 06·7	1 03·5	2·6	0·2	8·6	0·6	14·6	1·1
27	1 06·8	1 06·9	1 03·7	2·7	0·2	8·7	0·7	14·7	1·1
28	1 07·0	1 07·2	1 03·9	2·8	0·2	8·8	0·7	14·8	1·1
29	1 07·3	1 07·4	1 04·2	2·9	0·2	8·9	0·7	14·9	1·1
30	1 07·5	1 07·7	1 04·4	3·0	0·2	9·0	0·7	15·0	1·1
31	1 07·8	1 07·9	1 04·7	3·1	0·2	9·1	0·7	15·1	1·1
32	1 08·0	1 08·2	1 04·9	3·2	0·2	9·2	0·7	15·2	1·1
33	1 08·3	1 08·4	1 05·1	3·3	0·2	9·3	0·7	15·3	1·1
34	1 08·5	1 08·7	1 05·4	3·4	0·3	9·4	0·7	15·4	1·2
35	1 08·8	1 08·9	1 05·6	3·5	0·3	9·5	0·7	15·5	1·2
36	1 09·0	1 09·2	1 05·9	3·6	0·3	9·6	0·7	15·6	1·2
37	1 09·3	1 09·4	1 06·1	3·7	0·3	9·7	0·7	15·7	1·2
38	1 09·5	1 09·7	1 06·3	3·8	0·3	9·8	0·7	15·8	1·2
39	1 09·8	1 09·9	1 06·6	3·9	0·3	9·9	0·7	15·9	1·2
40	1 10·0	1 10·2	1 06·8	4·0	0·3	10·0	0·8	16·0	1·2
41	1 10·3	1 10·4	1 07·0	4·1	0·3	10·1	0·8	16·1	1·2
42	1 10·5	1 10·7	1 07·3	4·2	0·3	10·2	0·8	16·2	1·2
43	1 10·8	1 10·9	1 07·5	4·3	0·3	10·3	0·8	16·3	1·2
44	1 11·0	1 11·2	1 07·8	4·4	0·3	10·4	0·8	16·4	1·2
45	1 11·3	1 11·4	1 08·0	4·5	0·3	10·5	0·8	16·5	1·2
46	1 11·5	1 11·7	1 08·2	4·6	0·3	10·6	0·8	16·6	1·2
47	1 11·8	1 11·9	1 08·5	4·7	0·4	10·7	0·8	16·7	1·3
48	1 12·0	1 12·2	1 08·7	4·8	0·4	10·8	0·8	16·8	1·3
49	1 12·3	1 12·4	1 09·0	4·9	0·4	10·9	0·8	16·9	1·3
50	1 12·5	1 12·7	1 09·2	5·0	0·4	11·0	0·8	17·0	1·3
51	1 12·8	1 12·9	1 09·4	5·1	0·4	11·1	0·8	17·1	1·3
52	1 13·0	1 13·2	1 09·7	5·2	0·4	11·2	0·8	17·2	1·3
53	1 13·3	1 13·5	1 09·9	5·3	0·4	11·3	0·8	17·3	1·3
54	1 13·5	1 13·7	1 10·2	5·4	0·4	11·4	0·9	17·4	1·3
55	1 13·8	1 14·0	1 10·4	5·5	0·4	11·5	0·9	17·5	1·3
56	1 14·0	1 14·2	1 10·6	5·6	0·4	11·6	0·9	17·6	1·3
57	1 14·3	1 14·5	1 10·9	5·7	0·4	11·7	0·9	17·7	1·3
58	1 14·5	1 14·7	1 11·1	5·8	0·4	11·8	0·9	17·8	1·3
59	1 14·8	1 15·0	1 11·3	5·9	0·4	11·9	0·9	17·9	1·3
60	1 15·0	1 15·2	1 11·6	6·0	0·5	12·0	0·9	18·0	1·4

5ᵐ	SUN PLANETS	ARIES	MOON	v or d	Corrⁿ	v or d	Corrⁿ	v or d	Corrⁿ
s	° ′	° ′	° ′	′	′	′	′	′	′
00	1 15·0	1 15·2	1 11·6	0·0	0·0	6·0	0·6	12·0	1·1
01	1 15·3	1 15·5	1 11·8	0·1	0·0	6·1	0·6	12·1	1·1
02	1 15·5	1 15·7	1 12·1	0·2	0·0	6·2	0·6	12·2	1·1
03	1 15·8	1 16·0	1 12·3	0·3	0·0	6·3	0·6	12·3	1·1
04	1 16·0	1 16·2	1 12·5	0·4	0·0	6·4	0·6	12·4	1·1
05	1 16·3	1 16·5	1 12·8	0·5	0·0	6·5	0·6	12·5	1·1
06	1 16·5	1 16·7	1 13·0	0·6	0·1	6·6	0·6	12·6	1·2
07	1 16·8	1 17·0	1 13·3	0·7	0·1	6·7	0·6	12·7	1·2
08	1 17·0	1 17·2	1 13·5	0·8	0·1	6·8	0·6	12·8	1·2
09	1 17·3	1 17·5	1 13·7	0·9	0·1	6·9	0·6	12·9	1·2
10	1 17·5	1 17·7	1 14·0	1·0	0·1	7·0	0·6	13·0	1·2
11	1 17·8	1 18·0	1 14·2	1·1	0·1	7·1	0·7	13·1	1·2
12	1 18·0	1 18·2	1 14·4	1·2	0·1	7·2	0·7	13·2	1·2
13	1 18·3	1 18·5	1 14·7	1·3	0·1	7·3	0·7	13·3	1·2
14	1 18·5	1 18·7	1 14·9	1·4	0·1	7·4	0·7	13·4	1·2
15	1 18·8	1 19·0	1 15·2	1·5	0·1	7·5	0·7	13·5	1·2
16	1 19·0	1 19·2	1 15·4	1·6	0·1	7·6	0·7	13·6	1·2
17	1 19·3	1 19·5	1 15·6	1·7	0·2	7·7	0·7	13·7	1·3
18	1 19·5	1 19·7	1 15·9	1·8	0·2	7·8	0·7	13·8	1·3
19	1 19·8	1 20·0	1 16·1	1·9	0·2	7·9	0·7	13·9	1·3
20	1 20·0	1 20·2	1 16·4	2·0	0·2	8·0	0·7	14·0	1·3
21	1 20·3	1 20·5	1 16·6	2·1	0·2	8·1	0·7	14·1	1·3
22	1 20·5	1 20·7	1 16·8	2·2	0·2	8·2	0·8	14·2	1·3
23	1 20·8	1 21·0	1 17·1	2·3	0·2	8·3	0·8	14·3	1·3
24	1 21·0	1 21·2	1 17·3	2·4	0·2	8·4	0·8	14·4	1·3
25	1 21·3	1 21·5	1 17·5	2·5	0·2	8·5	0·8	14·5	1·3
26	1 21·5	1 21·7	1 17·8	2·6	0·2	8·6	0·8	14·6	1·3
27	1 21·8	1 22·0	1 18·0	2·7	0·2	8·7	0·8	14·7	1·3
28	1 22·0	1 22·2	1 18·3	2·8	0·3	8·8	0·8	14·8	1·4
29	1 22·3	1 22·5	1 18·5	2·9	0·3	8·9	0·8	14·9	1·4
30	1 22·5	1 22·7	1 18·7	3·0	0·3	9·0	0·8	15·0	1·4
31	1 22·8	1 23·0	1 19·0	3·1	0·3	9·1	0·8	15·1	1·4
32	1 23·0	1 23·2	1 19·2	3·2	0·3	9·2	0·8	15·2	1·4
33	1 23·3	1 23·5	1 19·5	3·3	0·3	9·3	0·9	15·3	1·4
34	1 23·5	1 23·7	1 19·7	3·4	0·3	9·4	0·9	15·4	1·4
35	1 23·8	1 24·0	1 19·9	3·5	0·3	9·5	0·9	15·5	1·4
36	1 24·0	1 24·2	1 20·2	3·6	0·3	9·6	0·9	15·6	1·4
37	1 24·3	1 24·5	1 20·4	3·7	0·3	9·7	0·9	15·7	1·4
38	1 24·5	1 24·7	1 20·7	3·8	0·3	9·8	0·9	15·8	1·4
39	1 24·8	1 25·0	1 20·9	3·9	0·4	9·9	0·9	15·9	1·5
40	1 25·0	1 25·2	1 21·1	4·0	0·4	10·0	0·9	16·0	1·5
41	1 25·3	1 25·5	1 21·4	4·1	0·4	10·1	0·9	16·1	1·5
42	1 25·5	1 25·7	1 21·6	4·2	0·4	10·2	0·9	16·2	1·5
43	1 25·8	1 26·0	1 21·8	4·3	0·4	10·3	0·9	16·3	1·5
44	1 26·0	1 26·2	1 22·1	4·4	0·4	10·4	1·0	16·4	1·5
45	1 26·3	1 26·5	1 22·3	4·5	0·4	10·5	1·0	16·5	1·5
46	1 26·5	1 26·7	1 22·6	4·6	0·4	10·6	1·0	16·6	1·5
47	1 26·8	1 27·0	1 22·8	4·7	0·4	10·7	1·0	16·7	1·5
48	1 27·0	1 27·2	1 23·0	4·8	0·4	10·8	1·0	16·8	1·5
49	1 27·3	1 27·5	1 23·3	4·9	0·4	10·9	1·0	16·9	1·5
50	1 27·5	1 27·7	1 23·5	5·0	0·5	11·0	1·0	17·0	1·6
51	1 27·8	1 28·0	1 23·8	5·1	0·5	11·1	1·0	17·1	1·6
52	1 28·0	1 28·2	1 24·0	5·2	0·5	11·2	1·0	17·2	1·6
53	1 28·3	1 28·5	1 24·2	5·3	0·5	11·3	1·0	17·3	1·6
54	1 28·5	1 28·7	1 24·5	5·4	0·5	11·4	1·0	17·4	1·6
55	1 28·8	1 29·0	1 24·7	5·5	0·5	11·5	1·1	17·5	1·6
56	1 29·0	1 29·2	1 24·9	5·6	0·5	11·6	1·1	17·6	1·6
57	1 29·3	1 29·5	1 25·2	5·7	0·5	11·7	1·1	17·7	1·6
58	1 29·5	1 29·7	1 25·4	5·8	0·5	11·8	1·1	17·8	1·6
59	1 29·8	1 30·0	1 25·7	5·9	0·5	11·9	1·1	17·9	1·6
60	1 30·0	1 30·2	1 25·9	6·0	0·6	12·0	1·1	18·0	1·7

6ᵐ	SUN PLANETS	ARIES	MOON	v or d	Corrⁿ	v or d	Corrⁿ	v or d	Corrⁿ
s	° ′	° ′	° ′	′	′	′	′	′	′
00	1 30·0	1 30·2	1 25·9	0·0	0·0	6·0	0·7	12·0	1·3
01	1 30·3	1 30·5	1 26·1	0·1	0·0	6·1	0·7	12·1	1·3
02	1 30·5	1 30·7	1 26·4	0·2	0·0	6·2	0·7	12·2	1·3
03	1 30·8	1 31·0	1 26·6	0·3	0·0	6·3	0·7	12·3	1·3
04	1 31·0	1 31·2	1 26·9	0·4	0·0	6·4	0·7	12·4	1·3
05	1 31·3	1 31·5	1 27·1	0·5	0·1	6·5	0·7	12·5	1·4
06	1 31·5	1 31·8	1 27·3	0·6	0·1	6·6	0·7	12·6	1·4
07	1 31·8	1 32·0	1 27·6	0·7	0·1	6·7	0·7	12·7	1·4
08	1 32·0	1 32·3	1 27·8	0·8	0·1	6·8	0·7	12·8	1·4
09	1 32·3	1 32·5	1 28·0	0·9	0·1	6·9	0·7	12·9	1·4
10	1 32·5	1 32·8	1 28·3	1·0	0·1	7·0	0·8	13·0	1·4
11	1 32·8	1 33·0	1 28·5	1·1	0·1	7·1	0·8	13·1	1·4
12	1 33·0	1 33·3	1 28·8	1·2	0·1	7·2	0·8	13·2	1·4
13	1 33·3	1 33·5	1 29·0	1·3	0·1	7·3	0·8	13·3	1·4
14	1 33·5	1 33·8	1 29·2	1·4	0·2	7·4	0·8	13·4	1·5
15	1 33·8	1 34·0	1 29·5	1·5	0·2	7·5	0·8	13·5	1·5
16	1 34·0	1 34·3	1 29·7	1·6	0·2	7·6	0·8	13·6	1·5
17	1 34·3	1 34·5	1 30·0	1·7	0·2	7·7	0·8	13·7	1·5
18	1 34·5	1 34·8	1 30·2	1·8	0·2	7·8	0·8	13·8	1·5
19	1 34·8	1 35·0	1 30·4	1·9	0·2	7·9	0·9	13·9	1·5
20	1 35·0	1 35·3	1 30·7	2·0	0·2	8·0	0·9	14·0	1·5
21	1 35·3	1 35·5	1 30·9	2·1	0·2	8·1	0·9	14·1	1·5
22	1 35·5	1 35·8	1 31·1	2·2	0·2	8·2	0·9	14·2	1·5
23	1 35·8	1 36·0	1 31·4	2·3	0·2	8·3	0·9	14·3	1·5
24	1 36·0	1 36·3	1 31·6	2·4	0·3	8·4	0·9	14·4	1·6
25	1 36·3	1 36·5	1 31·9	2·5	0·3	8·5	0·9	14·5	1·6
26	1 36·5	1 36·8	1 32·1	2·6	0·3	8·6	0·9	14·6	1·6
27	1 36·8	1 37·0	1 32·3	2·7	0·3	8·7	0·9	14·7	1·6
28	1 37·0	1 37·3	1 32·6	2·8	0·3	8·8	1·0	14·8	1·6
29	1 37·3	1 37·5	1 32·8	2·9	0·3	8·9	1·0	14·9	1·6
30	1 37·5	1 37·8	1 33·1	3·0	0·3	9·0	1·0	15·0	1·6
31	1 37·8	1 38·0	1 33·3	3·1	0·3	9·1	1·0	15·1	1·6
32	1 38·0	1 38·3	1 33·5	3·2	0·3	9·2	1·0	15·2	1·6
33	1 38·3	1 38·5	1 33·8	3·3	0·4	9·3	1·0	15·3	1·7
34	1 38·5	1 38·8	1 34·0	3·4	0·4	9·4	1·0	15·4	1·7
35	1 38·8	1 39·0	1 34·3	3·5	0·4	9·5	1·0	15·5	1·7
36	1 39·0	1 39·3	1 34·5	3·6	0·4	9·6	1·0	15·6	1·7
37	1 39·3	1 39·5	1 34·7	3·7	0·4	9·7	1·1	15·7	1·7
38	1 39·5	1 39·8	1 35·0	3·8	0·4	9·8	1·1	15·8	1·7
39	1 39·8	1 40·0	1 35·2	3·9	0·4	9·9	1·1	15·9	1·7
40	1 40·0	1 40·3	1 35·4	4·0	0·4	10·0	1·1	16·0	1·7
41	1 40·3	1 40·5	1 35·7	4·1	0·4	10·1	1·1	16·1	1·7
42	1 40·5	1 40·8	1 35·9	4·2	0·5	10·2	1·1	16·2	1·8
43	1 40·8	1 41·0	1 36·2	4·3	0·5	10·3	1·1	16·3	1·8
44	1 41·0	1 41·3	1 36·4	4·4	0·5	10·4	1·1	16·4	1·8
45	1 41·3	1 41·5	1 36·6	4·5	0·5	10·5	1·1	16·5	1·8
46	1 41·5	1 41·8	1 36·9	4·6	0·5	10·6	1·1	16·6	1·8
47	1 41·8	1 42·0	1 37·1	4·7	0·5	10·7	1·2	16·7	1·8
48	1 42·0	1 42·3	1 37·4	4·8	0·5	10·8	1·2	16·8	1·8
49	1 42·3	1 42·5	1 37·6	4·9	0·5	10·9	1·2	16·9	1·8
50	1 42·5	1 42·8	1 37·8	5·0	0·5	11·0	1·2	17·0	1·8
51	1 42·8	1 43·0	1 38·1	5·1	0·6	11·1	1·2	17·1	1·9
52	1 43·0	1 43·3	1 38·3	5·2	0·6	11·2	1·2	17·2	1·9
53	1 43·3	1 43·5	1 38·5	5·3	0·6	11·3	1·2	17·3	1·9
54	1 43·5	1 43·8	1 38·8	5·4	0·6	11·4	1·2	17·4	1·9
55	1 43·8	1 44·0	1 39·0	5·5	0·6	11·5	1·2	17·5	1·9
56	1 44·0	1 44·3	1 39·3	5·6	0·6	11·6	1·3	17·6	1·9
57	1 44·3	1 44·5	1 39·5	5·7	0·6	11·7	1·3	17·7	1·9
58	1 44·5	1 44·8	1 39·7	5·8	0·6	11·8	1·3	17·8	1·9
59	1 44·8	1 45·0	1 40·0	5·9	0·6	11·9	1·3	17·9	1·9
60	1 45·0	1 45·3	1 40·2	6·0	0·7	12·0	1·3	18·0	2·0

7ᵐ	SUN PLANETS	ARIES	MOON	v or d	Corrⁿ	v or d	Corrⁿ	v or d	Corrⁿ
s	° ′	° ′	° ′	′	′	′	′	′	′
00	1 45·0	1 45·3	1 40·2	0·0	0·0	6·0	0·8	12·0	1·5
01	1 45·3	1 45·5	1 40·5	0·1	0·0	6·1	0·8	12·1	1·5
02	1 45·5	1 45·8	1 40·7	0·2	0·0	6·2	0·8	12·2	1·5
03	1 45·8	1 46·0	1 40·9	0·3	0·0	6·3	0·8	12·3	1·5
04	1 46·0	1 46·3	1 41·2	0·4	0·1	6·4	0·8	12·4	1·6
05	1 46·3	1 46·5	1 41·4	0·5	0·1	6·5	0·8	12·5	1·6
06	1 46·5	1 46·8	1 41·6	0·6	0·1	6·6	0·8	12·6	1·6
07	1 46·8	1 47·0	1 41·9	0·7	0·1	6·7	0·8	12·7	1·6
08	1 47·0	1 47·3	1 42·1	0·8	0·1	6·8	0·9	12·8	1·6
09	1 47·3	1 47·5	1 42·4	0·9	0·1	6·9	0·9	12·9	1·6
10	1 47·5	1 47·8	1 42·6	1·0	0·1	7·0	0·9	13·0	1·6
11	1 47·8	1 48·0	1 42·8	1·1	0·1	7·1	0·9	13·1	1·6
12	1 48·0	1 48·3	1 43·1	1·2	0·2	7·2	0·9	13·2	1·7
13	1 48·3	1 48·5	1 43·3	1·3	0·2	7·3	0·9	13·3	1·7
14	1 48·5	1 48·8	1 43·6	1·4	0·2	7·4	0·9	13·4	1·7
15	1 48·8	1 49·0	1 43·8	1·5	0·2	7·5	0·9	13·5	1·7
16	1 49·0	1 49·3	1 44·0	1·6	0·2	7·6	1·0	13·6	1·7
17	1 49·3	1 49·5	1 44·3	1·7	0·2	7·7	1·0	13·7	1·7
18	1 49·5	1 49·8	1 44·5	1·8	0·2	7·8	1·0	13·8	1·7
19	1 49·8	1 50·1	1 44·8	1·9	0·2	7·9	1·0	13·9	1·7
20	1 50·0	1 50·3	1 45·0	2·0	0·3	8·0	1·0	14·0	1·8
21	1 50·3	1 50·6	1 45·2	2·1	0·3	8·1	1·0	14·1	1·8
22	1 50·5	1 50·8	1 45·5	2·2	0·3	8·2	1·0	14·2	1·8
23	1 50·8	1 51·1	1 45·7	2·3	0·3	8·3	1·0	14·3	1·8
24	1 51·0	1 51·3	1 45·9	2·4	0·3	8·4	1·1	14·4	1·8
25	1 51·3	1 51·6	1 46·2	2·5	0·3	8·5	1·1	14·5	1·8
26	1 51·5	1 51·8	1 46·4	2·6	0·3	8·6	1·1	14·6	1·8
27	1 51·8	1 52·1	1 46·7	2·7	0·3	8·7	1·1	14·7	1·8
28	1 52·0	1 52·3	1 46·9	2·8	0·4	8·8	1·1	14·8	1·9
29	1 52·3	1 52·6	1 47·1	2·9	0·4	8·9	1·1	14·9	1·9
30	1 52·5	1 52·8	1 47·4	3·0	0·4	9·0	1·1	15·0	1·9
31	1 52·8	1 53·1	1 47·6	3·1	0·4	9·1	1·1	15·1	1·9
32	1 53·0	1 53·3	1 47·9	3·2	0·4	9·2	1·2	15·2	1·9
33	1 53·3	1 53·6	1 48·1	3·3	0·4	9·3	1·2	15·3	1·9
34	1 53·5	1 53·8	1 48·3	3·4	0·4	9·4	1·2	15·4	1·9
35	1 53·8	1 54·1	1 48·6	3·5	0·4	9·5	1·2	15·5	1·9
36	1 54·0	1 54·3	1 48·8	3·6	0·5	9·6	1·2	15·6	2·0
37	1 54·3	1 54·6	1 49·0	3·7	0·5	9·7	1·2	15·7	2·0
38	1 54·5	1 54·8	1 49·3	3·8	0·5	9·8	1·2	15·8	2·0
39	1 54·8	1 55·1	1 49·5	3·9	0·5	9·9	1·2	15·9	2·0
40	1 55·0	1 55·3	1 49·8	4·0	0·5	10·0	1·3	16·0	2·0
41	1 55·3	1 55·6	1 50·0	4·1	0·5	10·1	1·3	16·1	2·0
42	1 55·5	1 55·8	1 50·2	4·2	0·5	10·2	1·3	16·2	2·0
43	1 55·8	1 56·1	1 50·5	4·3	0·5	10·3	1·3	16·3	2·0
44	1 56·0	1 56·3	1 50·7	4·4	0·6	10·4	1·3	16·4	2·1
45	1 56·3	1 56·6	1 51·0	4·5	0·6	10·5	1·3	16·5	2·1
46	1 56·5	1 56·8	1 51·2	4·6	0·6	10·6	1·3	16·6	2·1
47	1 56·8	1 57·1	1 51·4	4·7	0·6	10·7	1·3	16·7	2·1
48	1 57·0	1 57·3	1 51·7	4·8	0·6	10·8	1·4	16·8	2·1
49	1 57·3	1 57·6	1 51·9	4·9	0·6	10·9	1·4	16·9	2·1
50	1 57·5	1 57·8	1 52·1	5·0	0·6	11·0	1·4	17·0	2·1
51	1 57·8	1 58·1	1 52·4	5·1	0·6	11·1	1·4	17·1	2·1
52	1 58·0	1 58·3	1 52·6	5·2	0·7	11·2	1·4	17·2	2·2
53	1 58·3	1 58·6	1 52·9	5·3	0·7	11·3	1·4	17·3	2·2
54	1 58·5	1 58·8	1 53·1	5·4	0·7	11·4	1·4	17·4	2·2
55	1 58·8	1 59·1	1 53·3	5·5	0·7	11·5	1·4	17·5	2·2
56	1 59·0	1 59·3	1 53·6	5·6	0·7	11·6	1·5	17·6	2·2
57	1 59·3	1 59·6	1 53·8	5·7	0·7	11·7	1·5	17·7	2·2
58	1 59·5	1 59·8	1 54·1	5·8	0·7	11·8	1·5	17·8	2·2
59	1 59·8	2 00·1	1 54·3	5·9	0·7	11·9	1·5	17·9	2·2
60	2 00·0	2 00·3	1 54·5	6·0	0·8	12·0	1·5	18·0	2·3

8^m	SUN PLANETS	ARIES	MOON	v or d	Corr^n	v or d	Corr^n	v or d	Corr^n
s	° ′	° ′	° ′	′	′	′	′	′	′
00	2 00·0	2 00·3	1 54·5	0·0	0·0	6·0	0·9	12·0	1·7
01	2 00·3	2 00·6	1 54·8	0·1	0·0	6·1	0·9	12·1	1·7
02	2 00·5	2 00·8	1 55·0	0·2	0·0	6·2	0·9	12·2	1·7
03	2 00·8	2 01·1	1 55·2	0·3	0·0	6·3	0·9	12·3	1·7
04	2 01·0	2 01·3	1 55·5	0·4	0·1	6·4	0·9	12·4	1·8
05	2 01·3	2 01·6	1 55·7	0·5	0·1	6·5	0·9	12·5	1·8
06	2 01·5	2 01·8	1 56·0	0·6	0·1	6·6	0·9	12·6	1·8
07	2 01·8	2 02·1	1 56·2	0·7	0·1	6·7	0·9	12·7	1·8
08	2 02·0	2 02·3	1 56·4	0·8	0·1	6·8	1·0	12·8	1·8
09	2 02·3	2 02·6	1 56·7	0·9	0·1	6·9	1·0	12·9	1·8
10	2 02·5	2 02·8	1 56·9	1·0	0·1	7·0	1·0	13·0	1·8
11	2 02·8	2 03·1	1 57·2	1·1	0·2	7·1	1·0	13·1	1·9
12	2 03·0	2 03·3	1 57·4	1·2	0·2	7·2	1·0	13·2	1·9
13	2 03·3	2 03·6	1 57·6	1·3	0·2	7·3	1·0	13·3	1·9
14	2 03·5	2 03·8	1 57·9	1·4	0·2	7·4	1·0	13·4	1·9
15	2 03·8	2 04·1	1 58·1	1·5	0·2	7·5	1·1	13·5	1·9
16	2 04·0	2 04·3	1 58·4	1·6	0·2	7·6	1·1	13·6	1·9
17	2 04·3	2 04·6	1 58·6	1·7	0·2	7·7	1·1	13·7	1·9
18	2 04·5	2 04·8	1 58·8	1·8	0·3	7·8	1·1	13·8	2·0
19	2 04·8	2 05·1	1 59·1	1·9	0·3	7·9	1·1	13·9	2·0
20	2 05·0	2 05·3	1 59·3	2·0	0·3	8·0	1·1	14·0	2·0
21	2 05·3	2 05·6	1 59·5	2·1	0·3	8·1	1·1	14·1	2·0
22	2 05·5	2 05·8	1 59·8	2·2	0·3	8·2	1·2	14·2	2·0
23	2 05·8	2 06·1	2 00·0	2·3	0·3	8·3	1·2	14·3	2·0
24	2 06·0	2 06·3	2 00·3	2·4	0·3	8·4	1·2	14·4	2·0
25	2 06·3	2 06·6	2 00·5	2·5	0·4	8·5	1·2	14·5	2·1
26	2 06·5	2 06·8	2 00·7	2·6	0·4	8·6	1·2	14·6	2·1
27	2 06·8	2 07·1	2 01·0	2·7	0·4	8·7	1·2	14·7	2·1
28	2 07·0	2 07·3	2 01·2	2·8	0·4	8·8	1·2	14·8	2·1
29	2 07·3	2 07·6	2 01·5	2·9	0·4	8·9	1·3	14·9	2·1
30	2 07·5	2 07·8	2 01·7	3·0	0·4	9·0	1·3	15·0	2·1
31	2 07·8	2 08·1	2 01·9	3·1	0·4	9·1	1·3	15·1	2·1
32	2 08·0	2 08·4	2 02·2	3·2	0·5	9·2	1·3	15·2	2·2
33	2 08·3	2 08·6	2 02·4	3·3	0·5	9·3	1·3	15·3	2·2
34	2 08·5	2 08·9	2 02·6	3·4	0·5	9·4	1·3	15·4	2·2
35	2 08·8	2 09·1	2 02·9	3·5	0·5	9·5	1·3	15·5	2·2
36	2 09·0	2 09·4	2 03·1	3·6	0·5	9·6	1·4	15·6	2·2
37	2 09·3	2 09·6	2 03·4	3·7	0·5	9·7	1·4	15·7	2·2
38	2 09·5	2 09·9	2 03·6	3·8	0·5	9·8	1·4	15·8	2·2
39	2 09·8	2 10·1	2 03·8	3·9	0·6	9·9	1·4	15·9	2·3
40	2 10·0	2 10·4	2 04·1	4·0	0·6	10·0	1·4	16·0	2·3
41	2 10·3	2 10·6	2 04·3	4·1	0·6	10·1	1·4	16·1	2·3
42	2 10·5	2 10·9	2 04·6	4·2	0·6	10·2	1·4	16·2	2·3
43	2 10·8	2 11·1	2 04·8	4·3	0·6	10·3	1·5	16·3	2·3
44	2 11·0	2 11·4	2 05·0	4·4	0·6	10·4	1·5	16·4	2·3
45	2 11·3	2 11·6	2 05·3	4·5	0·6	10·5	1·5	16·5	2·3
46	2 11·5	2 11·9	2 05·5	4·6	0·7	10·6	1·5	16·6	2·4
47	2 11·8	2 12·1	2 05·7	4·7	0·7	10·7	1·5	16·7	2·4
48	2 12·0	2 12·4	2 06·0	4·8	0·7	10·8	1·5	16·8	2·4
49	2 12·3	2 12·6	2 06·2	4·9	0·7	10·9	1·5	16·9	2·4
50	2 12·5	2 12·9	2 06·5	5·0	0·7	11·0	1·6	17·0	2·4
51	2 12·8	2 13·1	2 06·7	5·1	0·7	11·1	1·6	17·1	2·4
52	2 13·0	2 13·4	2 06·9	5·2	0·7	11·2	1·6	17·2	2·4
53	2 13·3	2 13·6	2 07·2	5·3	0·8	11·3	1·6	17·3	2·5
54	2 13·5	2 13·9	2 07·4	5·4	0·8	11·4	1·6	17·4	2·5
55	2 13·8	2 14·1	2 07·7	5·5	0·8	11·5	1·6	17·5	2·5
56	2 14·0	2 14·4	2 07·9	5·6	0·8	11·6	1·6	17·6	2·5
57	2 14·3	2 14·6	2 08·1	5·7	0·8	11·7	1·7	17·7	2·5
58	2 14·5	2 14·9	2 08·4	5·8	0·8	11·8	1·7	17·8	2·5
59	2 14·8	2 15·1	2 08·6	5·9	0·8	11·9	1·7	17·9	2·5
60	2 15·0	2 15·4	2 08·9	6·0	0·9	12·0	1·7	18·0	2·6

9^m	SUN PLANETS	ARIES	MOON	v or d	Corr^n	v or d	Corr^n	v or d	Corr^n
s	° ′	° ′	° ′	′	′	′	′	′	′
00	2 15·0	2 15·4	2 08·9	0·0	0·0	6·0	1·0	12·0	1·9
01	2 15·3	2 15·6	2 09·1	0·1	0·0	6·1	1·0	12·1	1·9
02	2 15·5	2 15·9	2 09·3	0·2	0·0	6·2	1·0	12·2	1·9
03	2 15·8	2 16·1	2 09·6	0·3	0·0	6·3	1·0	12·3	1·9
04	2 16·0	2 16·4	2 09·8	0·4	0·1	6·4	1·0	12·4	2·0
05	2 16·3	2 16·6	2 10·0	0·5	0·1	6·5	1·0	12·5	2·0
06	2 16·5	2 16·9	2 10·3	0·6	0·1	6·6	1·0	12·6	2·0
07	2 16·8	2 17·1	2 10·5	0·7	0·1	6·7	1·1	12·7	2·0
08	2 17·0	2 17·4	2 10·8	0·8	0·1	6·8	1·1	12·8	2·0
09	2 17·3	2 17·6	2 11·0	0·9	0·1	6·9	1·1	12·9	2·0
10	2 17·5	2 17·9	2 11·2	1·0	0·2	7·0	1·1	13·0	2·1
11	2 17·8	2 18·1	2 11·5	1·1	0·2	7·1	1·1	13·1	2·1
12	2 18·0	2 18·4	2 11·7	1·2	0·2	7·2	1·1	13·2	2·1
13	2 18·3	2 18·6	2 12·0	1·3	0·2	7·3	1·2	13·3	2·1
14	2 18·5	2 18·9	2 12·2	1·4	0·2	7·4	1·2	13·4	2·1
15	2 18·8	2 19·1	2 12·4	1·5	0·2	7·5	1·2	13·5	2·1
16	2 19·0	2 19·4	2 12·7	1·6	0·3	7·6	1·2	13·6	2·2
17	2 19·3	2 19·6	2 12·9	1·7	0·3	7·7	1·2	13·7	2·2
18	2 19·5	2 19·9	2 13·1	1·8	0·3	7·8	1·2	13·8	2·2
19	2 19·8	2 20·1	2 13·4	1·9	0·3	7·9	1·3	13·9	2·2
20	2 20·0	2 20·4	2 13·6	2·0	0·3	8·0	1·3	14·0	2·2
21	2 20·3	2 20·6	2 13·9	2·1	0·3	8·1	1·3	14·1	2·2
22	2 20·5	2 20·9	2 14·1	2·2	0·3	8·2	1·3	14·2	2·2
23	2 20·8	2 21·1	2 14·3	2·3	0·4	8·3	1·3	14·3	2·3
24	2 21·0	2 21·4	2 14·6	2·4	0·4	8·4	1·3	14·4	2·3
25	2 21·3	2 21·6	2 14·8	2·5	0·4	8·5	1·3	14·5	2·3
26	2 21·5	2 21·9	2 15·1	2·6	0·4	8·6	1·4	14·6	2·3
27	2 21·8	2 22·1	2 15·3	2·7	0·4	8·7	1·4	14·7	2·3
28	2 22·0	2 22·4	2 15·5	2·8	0·4	8·8	1·4	14·8	2·3
29	2 22·3	2 22·6	2 15·8	2·9	0·5	8·9	1·4	14·9	2·4
30	2 22·5	2 22·9	2 16·0	3·0	0·5	9·0	1·4	15·0	2·4
31	2 22·8	2 23·1	2 16·2	3·1	0·5	9·1	1·4	15·1	2·4
32	2 23·0	2 23·4	2 16·5	3·2	0·5	9·2	1·5	15·2	2·4
33	2 23·3	2 23·6	2 16·7	3·3	0·5	9·3	1·5	15·3	2·4
34	2 23·5	2 23·9	2 17·0	3·4	0·5	9·4	1·5	15·4	2·4
35	2 23·8	2 24·1	2 17·2	3·5	0·6	9·5	1·5	15·5	2·5
36	2 24·0	2 24·4	2 17·4	3·6	0·6	9·6	1·5	15·6	2·5
37	2 24·3	2 24·6	2 17·7	3·7	0·6	9·7	1·5	15·7	2·5
38	2 24·5	2 24·9	2 17·9	3·8	0·6	9·8	1·6	15·8	2·5
39	2 24·8	2 25·1	2 18·2	3·9	0·6	9·9	1·6	15·9	2·5
40	2 25·0	2 25·4	2 18·4	4·0	0·6	10·0	1·6	16·0	2·5
41	2 25·3	2 25·6	2 18·6	4·1	0·6	10·1	1·6	16·1	2·5
42	2 25·5	2 25·9	2 18·9	4·2	0·7	10·2	1·6	16·2	2·6
43	2 25·8	2 26·1	2 19·1	4·3	0·7	10·3	1·6	16·3	2·6
44	2 26·0	2 26·4	2 19·3	4·4	0·7	10·4	1·6	16·4	2·6
45	2 26·3	2 26·7	2 19·6	4·5	0·7	10·5	1·7	16·5	2·6
46	2 26·5	2 26·9	2 19·8	4·6	0·7	10·6	1·7	16·6	2·6
47	2 26·8	2 27·2	2 20·1	4·7	0·7	10·7	1·7	16·7	2·6
48	2 27·0	2 27·4	2 20·3	4·8	0·8	10·8	1·7	16·8	2·7
49	2 27·3	2 27·7	2 20·5	4·9	0·8	10·9	1·7	16·9	2·7
50	2 27·5	2 27·9	2 20·8	5·0	0·8	11·0	1·7	17·0	2·7
51	2 27·8	2 28·2	2 21·0	5·1	0·8	11·1	1·8	17·1	2·7
52	2 28·0	2 28·4	2 21·3	5·2	0·8	11·2	1·8	17·2	2·7
53	2 28·3	2 28·7	2 21·5	5·3	0·8	11·3	1·8	17·3	2·7
54	2 28·5	2 28·9	2 21·7	5·4	0·9	11·4	1·8	17·4	2·8
55	2 28·8	2 29·2	2 22·0	5·5	0·9	11·5	1·8	17·5	2·8
56	2 29·0	2 29·4	2 22·2	5·6	0·9	11·6	1·8	17·6	2·8
57	2 29·3	2 29·7	2 22·5	5·7	0·9	11·7	1·9	17·7	2·8
58	2 29·5	2 29·9	2 22·7	5·8	0·9	11·8	1·9	17·8	2·8
59	2 29·8	2 30·2	2 22·9	5·9	0·9	11·9	1·9	17·9	2·8
60	2 30·0	2 30·4	2 23·2	6·0	1·0	12·0	1·9	18·0	2·9

10m	SUN PLANETS	ARIES	MOON	v or d	Corrn	v or d	Corrn	v or d	Corrn
s	° ′	° ′	° ′	′	′	′	′	′	′
00	2 30·0	2 30·4	2 23·2	0·0	0·0	6·0	1·1	12·0	2·1
01	2 30·3	2 30·7	2 23·4	0·1	0·0	6·1	1·1	12·1	2·1
02	2 30·5	2 30·9	2 23·6	0·2	0·0	6·2	1·1	12·2	2·1
03	2 30·8	2 31·2	2 23·9	0·3	0·1	6·3	1·1	12·3	2·2
04	2 31·0	2 31·4	2 24·1	0·4	0·1	6·4	1·1	12·4	2·2
05	2 31·3	2 31·7	2 24·4	0·5	0·1	6·5	1·1	12·5	2·2
06	2 31·5	2 31·9	2 24·6	0·6	0·1	6·6	1·2	12·6	2·2
07	2 31·8	2 32·2	2 24·8	0·7	0·1	6·7	1·2	12·7	2·2
08	2 32·0	2 32·4	2 25·1	0·8	0·1	6·8	1·2	12·8	2·2
09	2 32·3	2 32·7	2 25·3	0·9	0·2	6·9	1·2	12·9	2·3
10	2 32·5	2 32·9	2 25·6	1·0	0·2	7·0	1·2	13·0	2·3
11	2 32·8	2 33·2	2 25·8	1·1	0·2	7·1	1·2	13·1	2·3
12	2 33·0	2 33·4	2 26·0	1·2	0·2	7·2	1·3	13·2	2·3
13	2 33·3	2 33·7	2 26·3	1·3	0·2	7·3	1·3	13·3	2·3
14	2 33·5	2 33·9	2 26·5	1·4	0·2	7·4	1·3	13·4	2·3
15	2 33·8	2 34·2	2 26·7	1·5	0·3	7·5	1·3	13·5	2·4
16	2 34·0	2 34·4	2 27·0	1·6	0·3	7·6	1·3	13·6	2·4
17	2 34·3	2 34·7	2 27·2	1·7	0·3	7·7	1·3	13·7	2·4
18	2 34·5	2 34·9	2 27·5	1·8	0·3	7·8	1·4	13·8	2·4
19	2 34·8	2 35·2	2 27·7	1·9	0·3	7·9	1·4	13·9	2·4
20	2 35·0	2 35·4	2 27·9	2·0	0·4	8·0	1·4	14·0	2·5
21	2 35·3	2 35·7	2 28·2	2·1	0·4	8·1	1·4	14·1	2·5
22	2 35·5	2 35·9	2 28·4	2·2	0·4	8·2	1·4	14·2	2·5
23	2 35·8	2 36·2	2 28·7	2·3	0·4	8·3	1·5	14·3	2·5
24	2 36·0	2 36·4	2 28·9	2·4	0·4	8·4	1·5	14·4	2·5
25	2 36·3	2 36·7	2 29·1	2·5	0·4	8·5	1·5	14·5	2·5
26	2 36·5	2 36·9	2 29·4	2·6	0·5	8·6	1·5	14·6	2·6
27	2 36·8	2 37·2	2 29·6	2·7	0·5	8·7	1·5	14·7	2·6
28	2 37·0	2 37·4	2 29·8	2·8	0·5	8·8	1·5	14·8	2·6
29	2 37·3	2 37·7	2 30·1	2·9	0·5	8·9	1·6	14·9	2·6
30	2 37·5	2 37·9	2 30·3	3·0	0·5	9·0	1·6	15·0	2·6
31	2 37·8	2 38·2	2 30·6	3·1	0·5	9·1	1·6	15·1	2·6
32	2 38·0	2 38·4	2 30·8	3·2	0·6	9·2	1·6	15·2	2·7
33	2 38·3	2 38·7	2 31·0	3·3	0·6	9·3	1·6	15·3	2·7
34	2 38·5	2 38·9	2 31·3	3·4	0·6	9·4	1·6	15·4	2·7
35	2 38·8	2 39·2	2 31·5	3·5	0·6	9·5	1·7	15·5	2·7
36	2 39·0	2 39·4	2 31·8	3·6	0·6	9·6	1·7	15·6	2·7
37	2 39·3	2 39·7	2 32·0	3·7	0·6	9·7	1·7	15·7	2·7
38	2 39·5	2 39·9	2 32·2	3·8	0·7	9·8	1·7	15·8	2·8
39	2 39·8	2 40·2	2 32·5	3·9	0·7	9·9	1·7	15·9	2·8
40	2 40·0	2 40·4	2 32·7	4·0	0·7	10·0	1·8	16·0	2·8
41	2 40·3	2 40·7	2 32·9	4·1	0·7	10·1	1·8	16·1	2·8
42	2 40·5	2 40·9	2 33·2	4·2	0·7	10·2	1·8	16·2	2·8
43	2 40·8	2 41·2	2 33·4	4·3	0·8	10·3	1·8	16·3	2·9
44	2 41·0	2 41·4	2 33·7	4·4	0·8	10·4	1·8	16·4	2·9
45	2 41·3	2 41·7	2 33·9	4·5	0·8	10·5	1·8	16·5	2·9
46	2 41·5	2 41·9	2 34·1	4·6	0·8	10·6	1·9	16·6	2·9
47	2 41·8	2 42·2	2 34·4	4·7	0·8	10·7	1·9	16·7	2·9
48	2 42·0	2 42·4	2 34·6	4·8	0·8	10·8	1·9	16·8	2·9
49	2 42·3	2 42·7	2 34·9	4·9	0·9	10·9	1·9	16·9	3·0
50	2 42·5	2 42·9	2 35·1	5·0	0·9	11·0	1·9	17·0	3·0
51	2 42·8	2 43·2	2 35·3	5·1	0·9	11·1	1·9	17·1	3·0
52	2 43·0	2 43·4	2 35·6	5·2	0·9	11·2	2·0	17·2	3·0
53	2 43·3	2 43·7	2 35·8	5·3	0·9	11·3	2·0	17·3	3·0
54	2 43·5	2 43·9	2 36·1	5·4	0·9	11·4	2·0	17·4	3·0
55	2 43·8	2 44·2	2 36·3	5·5	1·0	11·5	2·0	17·5	3·1
56	2 44·0	2 44·4	2 36·5	5·6	1·0	11·6	2·0	17·6	3·1
57	2 44·3	2 44·7	2 36·8	5·7	1·0	11·7	2·0	17·7	3·1
58	2 44·5	2 45·0	2 37·0	5·8	1·0	11·8	2·1	17·8	3·1
59	2 44·8	2 45·2	2 37·2	5·9	1·0	11·9	2·1	17·9	3·1
60	2 45·0	2 45·5	2 37·5	6·0	1·1	12·0	2·1	18·0	3·2

11m	SUN PLANETS	ARIES	MOON	v or d	Corrn	v or d	Corrn	v or d	Corrn
s	° ′	° ′	° ′	′	′	′	′	′	′
00	2 45·0	2 45·5	2 37·5	0·0	0·0	6·0	1·2	12·0	2·3
01	2 45·3	2 45·7	2 37·7	0·1	0·0	6·1	1·2	12·1	2·3
02	2 45·5	2 46·0	2 38·0	0·2	0·0	6·2	1·2	12·2	2·3
03	2 45·8	2 46·2	2 38·2	0·3	0·1	6·3	1·2	12·3	2·4
04	2 46·0	2 46·5	2 38·4	0·4	0·1	6·4	1·2	12·4	2·4
05	2 46·3	2 46·7	2 38·7	0·5	0·1	6·5	1·2	12·5	2·4
06	2 46·5	2 47·0	2 38·9	0·6	0·1	6·6	1·3	12·6	2·4
07	2 46·8	2 47·2	2 39·2	0·7	0·1	6·7	1·3	12·7	2·4
08	2 47·0	2 47·5	2 39·4	0·8	0·2	6·8	1·3	12·8	2·5
09	2 47·3	2 47·7	2 39·6	0·9	0·2	6·9	1·3	12·9	2·5
10	2 47·5	2 48·0	2 39·9	1·0	0·2	7·0	1·3	13·0	2·5
11	2 47·8	2 48·2	2 40·1	1·1	0·2	7·1	1·4	13·1	2·5
12	2 48·0	2 48·5	2 40·3	1·2	0·2	7·2	1·4	13·2	2·5
13	2 48·3	2 48·7	2 40·6	1·3	0·2	7·3	1·4	13·3	2·5
14	2 48·5	2 49·0	2 40·8	1·4	0·3	7·4	1·4	13·4	2·6
15	2 48·8	2 49·2	2 41·1	1·5	0·3	7·5	1·4	13·5	2·6
16	2 49·0	2 49·5	2 41·3	1·6	0·3	7·6	1·5	13·6	2·6
17	2 49·3	2 49·7	2 41·5	1·7	0·3	7·7	1·5	13·7	2·6
18	2 49·5	2 50·0	2 41·8	1·8	0·3	7·8	1·5	13·8	2·6
19	2 49·8	2 50·2	2 42·0	1·9	0·4	7·9	1·5	13·9	2·7
20	2 50·0	2 50·5	2 42·3	2·0	0·4	8·0	1·5	14·0	2·7
21	2 50·3	2 50·7	2 42·5	2·1	0·4	8·1	1·6	14·1	2·7
22	2 50·5	2 51·0	2 42·7	2·2	0·4	8·2	1·6	14·2	2·7
23	2 50·8	2 51·2	2 43·0	2·3	0·4	8·3	1·6	14·3	2·7
24	2 51·0	2 51·5	2 43·2	2·4	0·5	8·4	1·6	14·4	2·8
25	2 51·3	2 51·7	2 43·4	2·5	0·5	8·5	1·6	14·5	2·8
26	2 51·5	2 52·0	2 43·7	2·6	0·5	8·6	1·6	14·6	2·8
27	2 51·8	2 52·2	2 43·9	2·7	0·5	8·7	1·7	14·7	2·8
28	2 52·0	2 52·5	2 44·2	2·8	0·5	8·8	1·7	14·8	2·8
29	2 52·3	2 52·7	2 44·4	2·9	0·6	8·9	1·7	14·9	2·9
30	2 52·5	2 53·0	2 44·6	3·0	0·6	9·0	1·7	15·0	2·9
31	2 52·8	2 53·2	2 44·9	3·1	0·6	9·1	1·7	15·1	2·9
32	2 53·0	2 53·5	2 45·1	3·2	0·6	9·2	1·8	15·2	2·9
33	2 53·3	2 53·7	2 45·4	3·3	0·6	9·3	1·8	15·3	2·9
34	2 53·5	2 54·0	2 45·6	3·4	0·7	9·4	1·8	15·4	3·0
35	2 53·8	2 54·2	2 45·8	3·5	0·7	9·5	1·8	15·5	3·0
36	2 54·0	2 54·5	2 46·1	3·6	0·7	9·6	1·8	15·6	3·0
37	2 54·3	2 54·7	2 46·3	3·7	0·7	9·7	1·9	15·7	3·0
38	2 54·5	2 55·0	2 46·6	3·8	0·7	9·8	1·9	15·8	3·0
39	2 54·8	2 55·2	2 46·8	3·9	0·7	9·9	1·9	15·9	3·0
40	2 55·0	2 55·5	2 47·0	4·0	0·8	10·0	1·9	16·0	3·1
41	2 55·3	2 55·7	2 47·3	4·1	0·8	10·1	1·9	16·1	3·1
42	2 55·5	2 56·0	2 47·5	4·2	0·8	10·2	2·0	16·2	3·1
43	2 55·8	2 56·2	2 47·7	4·3	0·8	10·3	2·0	16·3	3·1
44	2 56·0	2 56·5	2 48·0	4·4	0·8	10·4	2·0	16·4	3·1
45	2 56·3	2 56·7	2 48·2	4·5	0·9	10·5	2·0	16·5	3·2
46	2 56·5	2 57·0	2 48·5	4·6	0·9	10·6	2·0	16·6	3·2
47	2 56·8	2 57·2	2 48·7	4·7	0·9	10·7	2·1	16·7	3·2
48	2 57·0	2 57·5	2 48·9	4·8	0·9	10·8	2·1	16·8	3·2
49	2 57·3	2 57·7	2 49·2	4·9	0·9	10·9	2·1	16·9	3·2
50	2 57·5	2 58·0	2 49·4	5·0	1·0	11·0	2·1	17·0	3·3
51	2 57·8	2 58·2	2 49·7	5·1	1·0	11·1	2·1	17·1	3·3
52	2 58·0	2 58·5	2 49·9	5·2	1·0	11·2	2·1	17·2	3·3
53	2 58·3	2 58·7	2 50·1	5·3	1·0	11·3	2·2	17·3	3·3
54	2 58·5	2 59·0	2 50·4	5·4	1·0	11·4	2·2	17·4	3·3
55	2 58·8	2 59·2	2 50·6	5·5	1·1	11·5	2·2	17·5	3·4
56	2 59·0	2 59·5	2 50·8	5·6	1·1	11·6	2·2	17·6	3·4
57	2 59·3	2 59·7	2 51·1	5·7	1·1	11·7	2·2	17·7	3·4
58	2 59·5	3 00·0	2 51·3	5·8	1·1	11·8	2·3	17·8	3·4
59	2 59·8	3 00·2	2 51·6	5·9	1·1	11·9	2·3	17·9	3·4
60	3 00·0	3 00·5	2 51·8	6·0	1·2	12·0	2·3	18·0	3·5

12m INCREMENTS AND CORRECTIONS 13m

12m	SUN PLANETS	ARIES	MOON	v or d	Corrn	v or d	Corrn	v or d	Corrn
s	° ′	° ′	° ′	′	′	′	′	′	′
00	3 00·0	3 00·5	2 51·8	0·0	0·0	6·0	1·3	12·0	2·5
01	3 00·3	3 00·7	2 52·0	0·1	0·0	6·1	1·3	12·1	2·5
02	3 00·5	3 01·0	2 52·3	0·2	0·0	6·2	1·3	12·2	2·5
03	3 00·8	3 01·2	2 52·5	0·3	0·1	6·3	1·3	12·3	2·6
04	3 01·0	3 01·5	2 52·8	0·4	0·1	6·4	1·3	12·4	2·6
05	3 01·3	3 01·7	2 53·0	0·5	0·1	6·5	1·4	12·5	2·6
06	3 01·5	3 02·0	2 53·2	0·6	0·1	6·6	1·4	12·6	2·6
07	3 01·8	3 02·2	2 53·5	0·7	0·1	6·7	1·4	12·7	2·6
08	3 02·0	3 02·5	2 53·7	0·8	0·2	6·8	1·4	12·8	2·7
09	3 02·3	3 02·7	2 53·9	0·9	0·2	6·9	1·4	12·9	2·7
10	3 02·5	3 03·0	2 54·2	1·0	0·2	7·0	1·5	13·0	2·7
11	3 02·8	3 03·3	2 54·4	1·1	0·2	7·1	1·5	13·1	2·7
12	3 03·0	3 03·5	2 54·7	1·2	0·3	7·2	1·5	13·2	2·8
13	3 03·3	3 03·8	2 54·9	1·3	0·3	7·3	1·5	13·3	2·8
14	3 03·5	3 04·0	2 55·1	1·4	0·3	7·4	1·5	13·4	2·8
15	3 03·8	3 04·3	2 55·4	1·5	0·3	7·5	1·6	13·5	2·8
16	3 04·0	3 04·5	2 55·6	1·6	0·3	7·6	1·6	13·6	2·8
17	3 04·3	3 04·8	2 55·9	1·7	0·4	7·7	1·6	13·7	2·9
18	3 04·5	3 05·0	2 56·1	1·8	0·4	7·8	1·6	13·8	2·9
19	3 04·8	3 05·3	2 56·3	1·9	0·4	7·9	1·6	13·9	2·9
20	3 05·0	3 05·5	2 56·6	2·0	0·4	8·0	1·7	14·0	2·9
21	3 05·3	3 05·8	2 56·8	2·1	0·4	8·1	1·7	14·1	2·9
22	3 05·5	3 06·0	2 57·0	2·2	0·5	8·2	1·7	14·2	3·0
23	3 05·8	3 06·3	2 57·3	2·3	0·5	8·3	1·7	14·3	3·0
24	3 06·0	3 06·5	2 57·5	2·4	0·5	8·4	1·8	14·4	3·0
25	3 06·3	3 06·8	2 57·8	2·5	0·5	8·5	1·8	14·5	3·0
26	3 06·5	3 07·0	2 58·0	2·6	0·5	8·6	1·8	14·6	3·0
27	3 06·8	3 07·3	2 58·2	2·7	0·6	8·7	1·8	14·7	3·1
28	3 07·0	3 07·5	2 58·5	2·8	0·6	8·8	1·8	14·8	3·1
29	3 07·3	3 07·8	2 58·7	2·9	0·6	8·9	1·9	14·9	3·1
30	3 07·5	3 08·0	2 59·0	3·0	0·6	9·0	1·9	15·0	3·1
31	3 07·8	3 08·3	2 59·2	3·1	0·6	9·1	1·9	15·1	3·1
32	3 08·0	3 08·5	2 59·4	3·2	0·7	9·2	1·9	15·2	3·2
33	3 08·3	3 08·8	2 59·7	3·3	0·7	9·3	1·9	15·3	3·2
34	3 08·5	3 09·0	2 59·9	3·4	0·7	9·4	2·0	15·4	3·2
35	3 08·8	3 09·3	3 00·2	3·5	0·7	9·5	2·0	15·5	3·2
36	3 09·0	3 09·5	3 00·4	3·6	0·8	9·6	2·0	15·6	3·3
37	3 09·3	3 09·8	3 00·6	3·7	0·8	9·7	2·0	15·7	3·3
38	3 09·5	3 10·0	3 00·9	3·8	0·8	9·8	2·0	15·8	3·3
39	3 09·8	3 10·3	3 01·1	3·9	0·8	9·9	2·1	15·9	3·3
40	3 10·0	3 10·5	3 01·3	4·0	0·8	10·0	2·1	16·0	3·3
41	3 10·3	3 10·8	3 01·6	4·1	0·9	10·1	2·1	16·1	3·4
42	3 10·5	3 11·0	3 01·8	4·2	0·9	10·2	2·1	16·2	3·4
43	3 10·8	3 11·3	3 02·1	4·3	0·9	10·3	2·1	16·3	3·4
44	3 11·0	3 11·5	3 02·3	4·4	0·9	10·4	2·2	16·4	3·4
45	3 11·3	3 11·8	3 02·5	4·5	0·9	10·5	2·2	16·5	3·4
46	3 11·5	3 12·0	3 02·8	4·6	1·0	10·6	2·2	16·6	3·5
47	3 11·8	3 12·3	3 03·0	4·7	1·0	10·7	2·2	16·7	3·5
48	3 12·0	3 12·5	3 03·3	4·8	1·0	10·8	2·3	16·8	3·5
49	3 12·3	3 12·8	3 03·5	4·9	1·0	10·9	2·3	16·9	3·5
50	3 12·5	3 13·0	3 03·7	5·0	1·0	11·0	2·3	17·0	3·5
51	3 12·8	3 13·3	3 04·0	5·1	1·1	11·1	2·3	17·1	3·6
52	3 13·0	3 13·5	3 04·2	5·2	1·1	11·2	2·3	17·2	3·6
53	3 13·3	3 13·8	3 04·4	5·3	1·1	11·3	2·4	17·3	3·6
54	3 13·5	3 14·0	3 04·7	5·4	1·1	11·4	2·4	17·4	3·6
55	3 13·8	3 14·3	3 04·9	5·5	1·1	11·5	2·4	17·5	3·6
56	3 14·0	3 14·5	3 05·2	5·6	1·2	11·6	2·4	17·6	3·7
57	3 14·3	3 14·8	3 05·4	5·7	1·2	11·7	2·4	17·7	3·7
58	3 14·5	3 15·0	3 05·6	5·8	1·2	11·8	2·5	17·8	3·7
59	3 14·8	3 15·3	3 05·9	5·9	1·2	11·9	2·5	17·9	3·7
60	3 15·0	3 15·5	3 06·1	6·0	1·3	12·0	2·5	18·0	3·8

13m	SUN PLANETS	ARIES	MOON	v or d	Corrn	v or d	Corrn	v or d	Corrn
s	° ′	° ′	° ′	′	′	′	′	′	′
00	3 15·0	3 15·5	3 06·1	0·0	0·0	6·0	1·4	12·0	2·7
01	3 15·3	3 15·8	3 06·4	0·1	0·0	6·1	1·4	12·1	2·7
02	3 15·5	3 16·0	3 06·6	0·2	0·0	6·2	1·4	12·2	2·7
03	3 15·8	3 16·3	3 06·8	0·3	0·1	6·3	1·4	12·3	2·8
04	3 16·0	3 16·5	3 07·1	0·4	0·1	6·4	1·4	12·4	2·8
05	3 16·3	3 16·8	3 07·3	0·5	0·1	6·5	1·5	12·5	2·8
06	3 16·5	3 17·0	3 07·5	0·6	0·1	6·6	1·5	12·6	2·8
07	3 16·8	3 17·3	3 07·8	0·7	0·2	6·7	1·5	12·7	2·9
08	3 17·0	3 17·5	3 08·0	0·8	0·2	6·8	1·5	12·8	2·9
09	3 17·3	3 17·8	3 08·3	0·9	0·2	6·9	1·6	12·9	2·9
10	3 17·5	3 18·0	3 08·5	1·0	0·2	7·0	1·6	13·0	2·9
11	3 17·8	3 18·3	3 08·7	1·1	0·2	7·1	1·6	13·1	2·9
12	3 18·0	3 18·5	3 09·0	1·2	0·3	7·2	1·6	13·2	3·0
13	3 18·3	3 18·8	3 09·2	1·3	0·3	7·3	1·6	13·3	3·0
14	3 18·5	3 19·0	3 09·5	1·4	0·3	7·4	1·7	13·4	3·0
15	3 18·8	3 19·3	3 09·7	1·5	0·3	7·5	1·7	13·5	3·0
16	3 19·0	3 19·5	3 09·9	1·6	0·4	7·6	1·7	13·6	3·1
17	3 19·3	3 19·8	3 10·2	1·7	0·4	7·7	1·7	13·7	3·1
18	3 19·5	3 20·0	3 10·4	1·8	0·4	7·8	1·8	13·8	3·1
19	3 19·8	3 20·3	3 10·7	1·9	0·4	7·9	1·8	13·9	3·1
20	3 20·0	3 20·5	3 10·9	2·0	0·5	8·0	1·8	14·0	3·2
21	3 20·3	3 20·8	3 11·1	2·1	0·5	8·1	1·8	14·1	3·2
22	3 20·5	3 21·0	3 11·4	2·2	0·5	8·2	1·8	14·2	3·2
23	3 20·8	3 21·3	3 11·6	2·3	0·5	8·3	1·9	14·3	3·2
24	3 21·0	3 21·6	3 11·8	2·4	0·5	8·4	1·9	14·4	3·2
25	3 21·3	3 21·8	3 12·1	2·5	0·6	8·5	1·9	14·5	3·3
26	3 21·5	3 22·1	3 12·3	2·6	0·6	8·6	1·9	14·6	3·3
27	3 21·8	3 22·3	3 12·6	2·7	0·6	8·7	2·0	14·7	3·3
28	3 22·0	3 22·6	3 12·8	2·8	0·6	8·8	2·0	14·8	3·3
29	3 22·3	3 22·8	3 13·0	2·9	0·7	8·9	2·0	14·9	3·4
30	3 22·5	3 23·1	3 13·3	3·0	0·7	9·0	2·0	15·0	3·4
31	3 22·8	3 23·3	3 13·5	3·1	0·7	9·1	2·0	15·1	3·4
32	3 23·0	3 23·6	3 13·8	3·2	0·7	9·2	2·1	15·2	3·4
33	3 23·3	3 23·8	3 14·0	3·3	0·7	9·3	2·1	15·3	3·4
34	3 23·5	3 24·1	3 14·2	3·4	0·8	9·4	2·1	15·4	3·5
35	3 23·8	3 24·3	3 14·5	3·5	0·8	9·5	2·1	15·5	3·5
36	3 24·0	3 24·6	3 14·7	3·6	0·8	9·6	2·2	15·6	3·5
37	3 24·3	3 24·8	3 14·9	3·7	0·8	9·7	2·2	15·7	3·5
38	3 24·5	3 25·1	3 15·2	3·8	0·9	9·8	2·2	15·8	3·6
39	3 24·8	3 25·3	3 15·4	3·9	0·9	9·9	2·2	15·9	3·6
40	3 25·0	3 25·6	3 15·7	4·0	0·9	10·0	2·3	16·0	3·6
41	3 25·3	3 25·8	3 15·9	4·1	0·9	10·1	2·3	16·1	3·6
42	3 25·5	3 26·1	3 16·1	4·2	0·9	10·2	2·3	16·2	3·6
43	3 25·8	3 26·3	3 16·4	4·3	1·0	10·3	2·3	16·3	3·7
44	3 26·0	3 26·6	3 16·6	4·4	1·0	10·4	2·3	16·4	3·7
45	3 26·3	3 26·8	3 16·9	4·5	1·0	10·5	2·4	16·5	3·7
46	3 26·5	3 27·1	3 17·1	4·6	1·0	10·6	2·4	16·6	3·7
47	3 26·8	3 27·3	3 17·3	4·7	1·1	10·7	2·4	16·7	3·8
48	3 27·0	3 27·6	3 17·6	4·8	1·1	10·8	2·4	16·8	3·8
49	3 27·3	3 27·8	3 17·8	4·9	1·1	10·9	2·5	16·9	3·8
50	3 27·5	3 28·1	3 18·0	5·0	1·1	11·0	2·5	17·0	3·8
51	3 27·8	3 28·3	3 18·3	5·1	1·1	11·1	2·5	17·1	3·8
52	3 28·0	3 28·6	3 18·5	5·2	1·2	11·2	2·5	17·2	3·9
53	3 28·3	3 28·8	3 18·8	5·3	1·2	11·3	2·5	17·3	3·9
54	3 28·5	3 29·1	3 19·0	5·4	1·2	11·4	2·6	17·4	3·9
55	3 28·8	3 29·3	3 19·2	5·5	1·2	11·5	2·6	17·5	3·9
56	3 29·0	3 29·6	3 19·5	5·6	1·3	11·6	2·6	17·6	4·0
57	3 29·3	3 29·8	3 19·7	5·7	1·3	11·7	2·6	17·7	4·0
58	3 29·5	3 30·1	3 20·0	5·8	1·3	11·8	2·7	17·8	4·0
59	3 29·8	3 30·3	3 20·2	5·9	1·3	11·9	2·7	17·9	4·0
60	3 30·0	3 30·6	3 20·4	6·0	1·4	12·0	2·7	18·0	4·1

14m s	SUN PLANETS ° ′	ARIES ° ′	MOON ° ′	v or d ′	Corrn ′	v or d ′	Corrn ′	v or d ′	Corrn ′
00	3 30·0	3 30·6	3 20·4	0·0	0·0	6·0	1·5	12·0	2·9
01	3 30·3	3 30·8	3 20·7	0·1	0·0	6·1	1·5	12·1	2·9
02	3 30·5	3 31·1	3 20·9	0·2	0·0	6·2	1·5	12·2	2·9
03	3 30·8	3 31·3	3 21·1	0·3	0·1	6·3	1·5	12·3	3·0
04	3 31·0	3 31·6	3 21·4	0·4	0·1	6·4	1·5	12·4	3·0
05	3 31·3	3 31·8	3 21·6	0·5	0·1	6·5	1·6	12·5	3·0
06	3 31·5	3 32·1	3 21·9	0·6	0·1	6·6	1·6	12·6	3·0
07	3 31·8	3 32·3	3 22·1	0·7	0·2	6·7	1·6	12·7	3·1
08	3 32·0	3 32·6	3 22·3	0·8	0·2	6·8	1·6	12·8	3·1
09	3 32·3	3 32·8	3 22·6	0·9	0·2	6·9	1·7	12·9	3·1
10	3 32·5	3 33·1	3 22·8	1·0	0·2	7·0	1·7	13·0	3·1
11	3 32·8	3 33·3	3 23·1	1·1	0·3	7·1	1·7	13·1	3·2
12	3 33·0	3 33·6	3 23·3	1·2	0·3	7·2	1·7	13·2	3·2
13	3 33·3	3 33·8	3 23·5	1·3	0·3	7·3	1·8	13·3	3·2
14	3 33·5	3 34·1	3 23·8	1·4	0·3	7·4	1·8	13·4	3·2
15	3 33·8	3 34·3	3 24·0	1·5	0·4	7·5	1·8	13·5	3·3
16	3 34·0	3 34·6	3 24·3	1·6	0·4	7·6	1·8	13·6	3·3
17	3 34·3	3 34·8	3 24·5	1·7	0·4	7·7	1·9	13·7	3·3
18	3 34·5	3 35·1	3 24·7	1·8	0·4	7·8	1·9	13·8	3·3
19	3 34·8	3 35·3	3 25·0	1·9	0·5	7·9	1·9	13·9	3·4
20	3 35·0	3 35·6	3 25·2	2·0	0·5	8·0	1·9	14·0	3·4
21	3 35·3	3 35·8	3 25·4	2·1	0·5	8·1	2·0	14·1	3·4
22	3 35·5	3 36·1	3 25·7	2·2	0·5	8·2	2·0	14·2	3·4
23	3 35·8	3 36·3	3 25·9	2·3	0·6	8·3	2·0	14·3	3·5
24	3 36·0	3 36·6	3 26·2	2·4	0·6	8·4	2·0	14·4	3·5
25	3 36·3	3 36·8	3 26·4	2·5	0·6	8·5	2·1	14·5	3·5
26	3 36·5	3 37·1	3 26·6	2·6	0·6	8·6	2·1	14·6	3·5
27	3 36·8	3 37·3	3 26·9	2·7	0·7	8·7	2·1	14·7	3·6
28	3 37·0	3 37·6	3 27·1	2·8	0·7	8·8	2·1	14·8	3·6
29	3 37·3	3 37·8	3 27·4	2·9	0·7	8·9	2·2	14·9	3·6
30	3 37·5	3 38·1	3 27·6	3·0	0·7	9·0	2·2	15·0	3·6
31	3 37·8	3 38·3	3 27·8	3·1	0·7	9·1	2·2	15·1	3·6
32	3 38·0	3 38·6	3 28·1	3·2	0·8	9·2	2·2	15·2	3·7
33	3 38·3	3 38·8	3 28·3	3·3	0·8	9·3	2·2	15·3	3·7
34	3 38·5	3 39·1	3 28·5	3·4	0·8	9·4	2·3	15·4	3·7
35	3 38·8	3 39·3	3 28·8	3·5	0·8	9·5	2·3	15·5	3·7
36	3 39·0	3 39·6	3 29·0	3·6	0·9	9·6	2·3	15·6	3·8
37	3 39·3	3 39·9	3 29·3	3·7	0·9	9·7	2·3	15·7	3·8
38	3 39·5	3 40·1	3 29·5	3·8	0·9	9·8	2·4	15·8	3·8
39	3 39·8	3 40·4	3 29·7	3·9	0·9	9·9	2·4	15·9	3·8
40	3 40·0	3 40·6	3 30·0	4·0	1·0	10·0	2·4	16·0	3·9
41	3 40·3	3 40·9	3 30·2	4·1	1·0	10·1	2·4	16·1	3·9
42	3 40·5	3 41·1	3 30·5	4·2	1·0	10·2	2·5	16·2	3·9
43	3 40·8	3 41·4	3 30·7	4·3	1·0	10·3	2·5	16·3	3·9
44	3 41·0	3 41·6	3 30·9	4·4	1·1	10·4	2·5	16·4	4·0
45	3 41·3	3 41·9	3 31·2	4·5	1·1	10·5	2·5	16·5	4·0
46	3 41·5	3 42·1	3 31·4	4·6	1·1	10·6	2·6	16·6	4·0
47	3 41·8	3 42·4	3 31·6	4·7	1·1	10·7	2·6	16·7	4·0
48	3 42·0	3 42·6	3 31·9	4·8	1·2	10·8	2·6	16·8	4·1
49	3 42·3	3 42·9	3 32·1	4·9	1·2	10·9	2·6	16·9	4·1
50	3 42·5	3 43·1	3 32·4	5·0	1·2	11·0	2·7	17·0	4·1
51	3 42·8	3 43·4	3 32·6	5·1	1·2	11·1	2·7	17·1	4·1
52	3 43·0	3 43·6	3 32·8	5·2	1·3	11·2	2·7	17·2	4·2
53	3 43·3	3 43·9	3 33·1	5·3	1·3	11·3	2·7	17·3	4·2
54	3 43·5	3 44·1	3 33·3	5·4	1·3	11·4	2·8	17·4	4·2
55	3 43·8	3 44·4	3 33·6	5·5	1·3	11·5	2·8	17·5	4·2
56	3 44·0	3 44·6	3 33·8	5·6	1·4	11·6	2·8	17·6	4·3
57	3 44·3	3 44·9	3 34·0	5·7	1·4	11·7	2·8	17·7	4·3
58	3 44·5	3 45·1	3 34·3	5·8	1·4	11·8	2·9	17·8	4·3
59	3 44·8	3 45·4	3 34·5	5·9	1·4	11·9	2·9	17·9	4·3
60	3 45·0	3 45·6	3 34·8	6·0	1·5	12·0	2·9	18·0	4·4

15m s	SUN PLANETS ° ′	ARIES ° ′	MOON ° ′	v or d ′	Corrn ′	v or d ′	Corrn ′	v or d ′	Corrn ′
00	3 45·0	3 45·6	3 34·8	0·0	0·0	6·0	1·6	12·0	3·1
01	3 45·3	3 45·9	3 35·0	0·1	0·0	6·1	1·6	12·1	3·1
02	3 45·5	3 46·1	3 35·2	0·2	0·1	6·2	1·6	12·2	3·2
03	3 45·8	3 46·4	3 35·5	0·3	0·1	6·3	1·6	12·3	3·2
04	3 46·0	3 46·6	3 35·7	0·4	0·1	6·4	1·7	12·4	3·2
05	3 46·3	3 46·9	3 35·9	0·5	0·1	6·5	1·7	12·5	3·2
06	3 46·5	3 47·1	3 36·2	0·6	0·2	6·6	1·7	12·6	3·3
07	3 46·8	3 47·4	3 36·4	0·7	0·2	6·7	1·7	12·7	3·3
08	3 47·0	3 47·6	3 36·7	0·8	0·2	6·8	1·8	12·8	3·3
09	3 47·3	3 47·9	3 36·9	0·9	0·2	6·9	1·8	12·9	3·3
10	3 47·5	3 48·1	3 37·1	1·0	0·3	7·0	1·8	13·0	3·4
11	3 47·8	3 48·4	3 37·4	1·1	0·3	7·1	1·8	13·1	3·4
12	3 48·0	3 48·6	3 37·6	1·2	0·3	7·2	1·9	13·2	3·4
13	3 48·3	3 48·9	3 37·9	1·3	0·3	7·3	1·9	13·3	3·4
14	3 48·5	3 49·1	3 38·1	1·4	0·4	7·4	1·9	13·4	3·5
15	3 48·8	3 49·4	3 38·3	1·5	0·4	7·5	1·9	13·5	3·5
16	3 49·0	3 49·6	3 38·6	1·6	0·4	7·6	2·0	13·6	3·5
17	3 49·3	3 49·9	3 38·8	1·7	0·4	7·7	2·0	13·7	3·5
18	3 49·5	3 50·1	3 39·0	1·8	0·5	7·8	2·0	13·8	3·6
19	3 49·8	3 50·4	3 39·3	1·9	0·5	7·9	2·0	13·9	3·6
20	3 50·0	3 50·6	3 39·5	2·0	0·5	8·0	2·1	14·0	3·6
21	3 50·3	3 50·9	3 39·8	2·1	0·5	8·1	2·1	14·1	3·6
22	3 50·5	3 51·1	3 40·0	2·2	0·6	8·2	2·1	14·2	3·7
23	3 50·8	3 51·4	3 40·2	2·3	0·6	8·3	2·1	14·3	3·7
24	3 51·0	3 51·6	3 40·5	2·4	0·6	8·4	2·2	14·4	3·7
25	3 51·3	3 51·9	3 40·7	2·5	0·6	8·5	2·2	14·5	3·7
26	3 51·5	3 52·1	3 41·0	2·6	0·7	8·6	2·2	14·6	3·8
27	3 51·8	3 52·4	3 41·2	2·7	0·7	8·7	2·2	14·7	3·8
28	3 52·0	3 52·6	3 41·4	2·8	0·7	8·8	2·3	14·8	3·8
29	3 52·3	3 52·9	3 41·7	2·9	0·7	8·9	2·3	14·9	3·8
30	3 52·5	3 53·1	3 41·9	3·0	0·8	9·0	2·3	15·0	3·9
31	3 52·8	3 53·4	3 42·1	3·1	0·8	9·1	2·4	15·1	3·9
32	3 53·0	3 53·6	3 42·4	3·2	0·8	9·2	2·4	15·2	3·9
33	3 53·3	3 53·9	3 42·6	3·3	0·9	9·3	2·4	15·3	4·0
34	3 53·5	3 54·1	3 42·9	3·4	0·9	9·4	2·4	15·4	4·0
35	3 53·8	3 54·4	3 43·1	3·5	0·9	9·5	2·5	15·5	4·0
36	3 54·0	3 54·6	3 43·3	3·6	0·9	9·6	2·5	15·6	4·0
37	3 54·3	3 54·9	3 43·6	3·7	1·0	9·7	2·5	15·7	4·1
38	3 54·5	3 55·1	3 43·8	3·8	1·0	9·8	2·5	15·8	4·1
39	3 54·8	3 55·4	3 44·1	3·9	1·0	9·9	2·6	15·9	4·1
40	3 55·0	3 55·6	3 44·3	4·0	1·0	10·0	2·6	16·0	4·1
41	3 55·3	3 55·9	3 44·5	4·1	1·1	10·1	2·6	16·1	4·2
42	3 55·5	3 56·1	3 44·8	4·2	1·1	10·2	2·6	16·2	4·2
43	3 55·8	3 56·4	3 45·0	4·3	1·1	10·3	2·7	16·3	4·2
44	3 56·0	3 56·6	3 45·2	4·4	1·1	10·4	2·7	16·4	4·2
45	3 56·3	3 56·9	3 45·5	4·5	1·2	10·5	2·7	16·5	4·3
46	3 56·5	3 57·1	3 45·7	4·6	1·2	10·6	2·7	16·6	4·3
47	3 56·8	3 57·4	3 46·0	4·7	1·2	10·7	2·8	16·7	4·3
48	3 57·0	3 57·6	3 46·2	4·8	1·2	10·8	2·8	16·8	4·3
49	3 57·3	3 57·9	3 46·4	4·9	1·3	10·9	2·8	16·9	4·4
50	3 57·5	3 58·2	3 46·7	5·0	1·3	11·0	2·8	17·0	4·4
51	3 57·8	3 58·4	3 46·9	5·1	1·3	11·1	2·9	17·1	4·4
52	3 58·0	3 58·7	3 47·2	5·2	1·3	11·2	2·9	17·2	4·4
53	3 58·3	3 58·9	3 47·4	5·3	1·4	11·3	2·9	17·3	4·5
54	3 58·5	3 59·2	3 47·6	5·4	1·4	11·4	2·9	17·4	4·5
55	3 58·8	3 59·4	3 47·9	5·5	1·4	11·5	3·0	17·5	4·5
56	3 59·0	3 59·7	3 48·1	5·6	1·4	11·6	3·0	17·6	4·5
57	3 59·3	3 59·9	3 48·4	5·7	1·5	11·7	3·0	17·7	4·6
58	3 59·5	4 00·2	3 48·6	5·8	1·5	11·8	3·0	17·8	4·6
59	3 59·8	4 00·4	3 48·8	5·9	1·5	11·9	3·1	17·9	4·6
60	4 00·0	4 00·7	3 49·1	6·0	1·6	12·0	3·1	18·0	4·7

16^m	SUN PLANFTS	ARIES	MOON	v or d	Corr^n	v or d	Corr^n	v or d	Corr^n
s	° ′	° ′	° ′	′	′	′	′	′	′
00	4 00·0	4 00·7	3 49·1	0·0	0·0	6·0	1·7	12·0	3·3
01	4 00·3	4 00·9	3 49·3	0·1	0·0	6·1	1·7	12·1	3·3
02	4 00·5	4 01·2	3 49·5	0·2	0·1	6·2	1·7	12·2	3·4
03	4 00·8	4 01·4	3 49·8	0·3	0·1	6·3	1·7	12·3	3·4
04	4 01·0	4 01·7	3 50·0	0·4	0·1	6·4	1·8	12·4	3·4
05	4 01·3	4 01·9	3 50·3	0·5	0·1	6·5	1·8	12·5	3·4
06	4 01·5	4 02·2	3 50·5	0·6	0·2	6·6	1·8	12·6	3·5
07	4 01·8	4 02·4	3 50·7	0·7	0·2	6·7	1·8	12·7	3·5
08	4 02·0	4 02·7	3 51·0	0·8	0·2	6·8	1·9	12·8	3·5
09	4 02·3	4 02·9	3 51·2	0·9	0·2	6·9	1·9	12·9	3·5
10	4 02·5	4 03·2	3 51·5	1·0	0·3	7·0	1·9	13·0	3·6
11	4 02·8	4 03·4	3 51·7	1·1	0·3	7·1	2·0	13·1	3·6
12	4 03·0	4 03·7	3 51·9	1·2	0·3	7·2	2·0	13·2	3·6
13	4 03·3	4 03·9	3 52·2	1·3	0·4	7·3	2·0	13·3	3·7
14	4 03·5	4 04·2	3 52·4	1·4	0·4	7·4	2·0	13·4	3·7
15	4 03·8	4 04·4	3 52·6	1·5	0·4	7·5	2·1	13·5	3·7
16	4 04·0	4 04·7	3 52·9	1·6	0·4	7·6	2·1	13·6	3·7
17	4 04·3	4 04·9	3 53·1	1·7	0·5	7·7	2·1	13·7	3·8
18	4 04·5	4 05·2	3 53·4	1·8	0·5	7·8	2·1	13·8	3·8
19	4 04·8	4 05·4	3 53·6	1·9	0·5	7·9	2·2	13·9	3·8
20	4 05·0	4 05·7	3 53·8	2·0	0·6	8·0	2·2	14·0	3·9
21	4 05·3	4 05·9	3 54·1	2·1	0·6	8·1	2·2	14·1	3·9
22	4 05·5	4 06·2	3 54·3	2·2	0·6	8·2	2·3	14·2	3·9
23	4 05·8	4 06·4	3 54·6	2·3	0·6	8·3	2·3	14·3	3·9
24	4 06·0	4 06·7	3 54·8	2·4	0·7	8·4	2·3	14·4	4·0
25	4 06·3	4 06·9	3 55·0	2·5	0·7	8·5	2·3	14·5	4·0
26	4 06·5	4 07·2	3 55·3	2·6	0·7	8·6	2·4	14·6	4·0
27	4 06·8	4 07·4	3 55·5	2·7	0·7	8·7	2·4	14·7	4·0
28	4 07·0	4 07·7	3 55·7	2·8	0·8	8·8	2·4	14·8	4·1
29	4 07·3	4 07·9	3 56·0	2·9	0·8	8·9	2·4	14·9	4·1
30	4 07·5	4 08·2	3 56·2	3·0	0·8	9·0	2·5	15·0	4·1
31	4 07·8	4 08·4	3 56·5	3·1	0·9	9·1	2·5	15·1	4·2
32	4 08·0	4 08·7	3 56·7	3·2	0·9	9·2	2·5	15·2	4·2
33	4 08·3	4 08·9	3 56·9	3·3	0·9	9·3	2·6	15·3	4·2
34	4 08·5	4 09·2	3 57·2	3·4	0·9	9·4	2·6	15·4	4·2
35	4 08·8	4 09·4	3 57·4	3·5	1·0	9·5	2·6	15·5	4·3
36	4 09·0	4 09·7	3 57·7	3·6	1·0	9·6	2·6	15·6	4·3
37	4 09·3	4 09·9	3 57·9	3·7	1·0	9·7	2·7	15·7	4·3
38	4 09·5	4 10·2	3 58·1	3·8	1·0	9·8	2·7	15·8	4·3
39	4 09·8	4 10·4	3 58·4	3·9	1·1	9·9	2·7	15·9	4·4
40	4 10·0	4 10·7	3 58·6	4·0	1·1	10·0	2·8	16·0	4·4
41	4 10·3	4 10·9	3 58·8	4·1	1·1	10·1	2·8	16·1	4·4
42	4 10·5	4 11·2	3 59·1	4·2	1·2	10·2	2·8	16·2	4·5
43	4 10·8	4 11·4	3 59·3	4·3	1·2	10·3	2·8	16·3	4·5
44	4 11·0	4 11·7	3 59·6	4·4	1·2	10·4	2·9	16·4	4·5
45	4 11·3	4 11·9	3 59·8	4·5	1·2	10·5	2·9	16·5	4·5
46	4 11·5	4 12·2	4 00·0	4·6	1·3	10·6	2·9	16·6	4·6
47	4 11·8	4 12·4	4 00·3	4·7	1·3	10·7	2·9	16·7	4·6
48	4 12·0	4 12·7	4 00·5	4·8	1·3	10·8	3·0	16·8	4·6
49	4 12·3	4 12·9	4 00·8	4·9	1·3	10·9	3·0	16·9	4·6
50	4 12·5	4 13·2	4 01·0	5·0	1·4	11·0	3·0	17·0	4·7
51	4 12·8	4 13·4	4 01·2	5·1	1·4	11·1	3·1	17·1	4·7
52	4 13·0	4 13·7	4 01·5	5·2	1·4	11·2	3·1	17·2	4·7
53	4 13·3	4 13·9	4 01·7	5·3	1·5	11·3	3·1	17·3	4·8
54	4 13·5	4 14·2	4 02·0	5·4	1·5	11·4	3·1	17·4	4·8
55	4 13·8	4 14·4	4 02·2	5·5	1·5	11·5	3·2	17·5	4·8
56	4 14·0	4 14·7	4 02·4	5·6	1·5	11·6	3·2	17·6	4·8
57	4 14·3	4 14·9	4 02·7	5·7	1·6	11·7	3·2	17·7	4·9
58	4 14·5	4 15·2	4 02·9	5·8	1·6	11·8	3·2	17·8	4·9
59	4 14·8	4 15·4	4 03·1	5·9	1·6	11·9	3·3	17·9	4·9
60	4 15·0	4 15·7	4 03·4	6·0	1·7	12·0	3·3	18·0	5·0

17^m	SUN PLANETS	ARIES	MOON	v or d	Corr^n	v or d	Corr^n	v or d	Corr^n
s	° ′	° ′	° ′	′	′	′	′	′	′
00	4 15·0	4 15·7	4 03·4	0·0	0·0	6·0	1·8	12·0	3·5
01	4 15·3	4 15·9	4 03·6	0·1	0·0	6·1	1·8	12·1	3·5
02	4 15·5	4 16·2	4 03·9	0·2	0·1	6·2	1·8	12·2	3·6
03	4 15·8	4 16·5	4 04·1	0·3	0·1	6·3	1·8	12·3	3·6
04	4 16·0	4 16·7	4 04·3	0·4	0·1	6·4	1·9	12·4	3·6
05	4 16·3	4 17·0	4 04·6	0·5	0·1	6·5	1·9	12·5	3·6
06	4 16·5	4 17·2	4 04·8	0·6	0·2	6·6	1·9	12·6	3·7
07	4 16·8	4 17·5	4 05·1	0·7	0·2	6·7	2·0	12·7	3·7
08	4 17·0	4 17·7	4 05·3	0·8	0·2	6·8	2·0	12·8	3·7
09	4 17·3	4 18·0	4 05·5	0·9	0·3	6·9	2·0	12·9	3·8
10	4 17·5	4 18·2	4 05·8	1·0	0·3	7·0	2·0	13·0	3·8
11	4 17·8	4 18·5	4 06·0	1·1	0·3	7·1	2·1	13·1	3·8
12	4 18·0	4 18·7	4 06·2	1·2	0·4	7·2	2·1	13·2	3·9
13	4 18·3	4 19·0	4 06·5	1·3	0·4	7·3	2·1	13·3	3·9
14	4 18·5	4 19·2	4 06·7	1·4	0·4	7·4	2·2	13·4	3·9
15	4 18·8	4 19·5	4 07·0	1·5	0·4	7·5	2·2	13·5	3·9
16	4 19·0	4 19·7	4 07·2	1·6	0·5	7·6	2·2	13·6	4·0
17	4 19·3	4 20·0	4 07·4	1·7	0·5	7·7	2·2	13·7	4·0
18	4 19·5	4 20·2	4 07·7	1·8	0·5	7·8	2·3	13·8	4·0
19	4 19·8	4 20·5	4 07·9	1·9	0·6	7·9	2·3	13·9	4·1
20	4 20·0	4 20·7	4 08·2	2·0	0·6	8·0	2·3	14·0	4·1
21	4 20·3	4 21·0	4 08·4	2·1	0·6	8·1	2·4	14·1	4·1
22	4 20·5	4 21·2	4 08·6	2·2	0·6	8·2	2·4	14·2	4·1
23	4 20·8	4 21·5	4 08·9	2·3	0·7	8·3	2·4	14·3	4·2
24	4 21·0	4 21·7	4 09·1	2·4	0·7	8·4	2·5	14·4	4·2
25	4 21·3	4 22·0	4 09·3	2·5	0·7	8·5	2·5	14·5	4·2
26	4 21·5	4 22·2	4 09·6	2·6	0·8	8·6	2·5	14·6	4·3
27	4 21·8	4 22·5	4 09·8	2·7	0·8	8·7	2·5	14·7	4·3
28	4 22·0	4 22·7	4 10·1	2·8	0·8	8·8	2·6	14·8	4·3
29	4 22·3	4 23·0	4 10·3	2·9	0·8	8·9	2·6	14·9	4·3
30	4 22·5	4 23·2	4 10·5	3·0	0·9	9·0	2·6	15·0	4·4
31	4 22·8	4 23·5	4 10·8	3·1	0·9	9·1	2·7	15·1	4·4
32	4 23·0	4 23·7	4 11·0	3·2	0·9	9·2	2·7	15·2	4·4
33	4 23·3	4 24·0	4 11·3	3·3	1·0	9·3	2·7	15·3	4·5
34	4 23·5	4 24·2	4 11·5	3·4	1·0	9·4	2·7	15·4	4·5
35	4 23·8	4 24·5	4 11·7	3·5	1·0	9·5	2·8	15·5	4·5
36	4 24·0	4 24·7	4 12·0	3·6	1·1	9·6	2·8	15·6	4·6
37	4 24·3	4 25·0	4 12·2	3·7	1·1	9·7	2·8	15·7	4·6
38	4 24·5	4 25·2	4 12·5	3·8	1·1	9·8	2·9	15·8	4·6
39	4 24·8	4 25·5	4 12·7	3·9	1·1	9·9	2·9	15·9	4·6
40	4 25·0	4 25·7	4 12·9	4·0	1·2	10·0	2·9	16·0	4·7
41	4 25·3	4 26·0	4 13·2	4·1	1·2	10·1	2·9	16·1	4·7
42	4 25·5	4 26·2	4 13·4	4·2	1·2	10·2	3·0	16·2	4·7
43	4 25·8	4 26·5	4 13·6	4·3	1·3	10·3	3·0	16·3	4·8
44	4 26·0	4 26·7	4 13·9	4·4	1·3	10·4	3·0	16·4	4·8
45	4 26·3	4 27·0	4 14·1	4·5	1·3	10·5	3·1	16·5	4·8
46	4 26·5	4 27·2	4 14·4	4·6	1·3	10·6	3·1	16·6	4·8
47	4 26·8	4 27·5	4 14·6	4·7	1·4	10·7	3·1	16·7	4·9
48	4 27·0	4 27·7	4 14·8	4·8	1·4	10·8	3·2	16·8	4·9
49	4 27·3	4 28·0	4 15·1	4·9	1·4	10·9	3·2	16·9	4·9
50	4 27·5	4 28·2	4 15·3	5·0	1·5	11·0	3·2	17·0	5·0
51	4 27·8	4 28·5	4 15·6	5·1	1·5	11·1	3·2	17·1	5·0
52	4 28·0	4 28·7	4 15·8	5·2	1·5	11·2	3·3	17·2	5·0
53	4 28·3	4 29·0	4 16·0	5·3	1·5	11·3	3·3	17·3	5·0
54	4 28·5	4 29·2	4 16·3	5·4	1·6	11·4	3·3	17·4	5·1
55	4 28·8	4 29·5	4 16·5	5·5	1·6	11·5	3·4	17·5	5·1
56	4 29·0	4 29·7	4 16·7	5·6	1·6	11·6	3·4	17·6	5·1
57	4 29·3	4 30·0	4 17·0	5·7	1·7	11·7	3·4	17·7	5·2
58	4 29·5	4 30·2	4 17·2	5·8	1·7	11·8	3·4	17·8	5·2
59	4 29·8	4 30·5	4 17·5	5·9	1·7	11·9	3·5	17·9	5·2
60	4 30·0	4 30·7	4 17·7	6·0	1·8	12·0	3·5	18·0	5·3

18^m	SUN PLANETS	ARIES	MOON	v or d	Corr^n	v or d	Corr^n	v or d	Corr^n
s	° ′	° ′	° ′	′	′	′	′	′	′
00	4 30·0	4 30·7	4 17·7	0·0	0·0	6·0	1·9	12·0	3·7
01	4 30·3	4 31·0	4 17·9	0·1	0·0	6·1	1·9	12·1	3·7
02	4 30·5	4 31·2	4 18·2	0·2	0·1	6·2	1·9	12·2	3·8
03	4 30·8	4 31·5	4 18·4	0·3	0·1	6·3	1·9	12·3	3·8
04	4 31·0	4 31·7	4 18·7	0·4	0·1	6·4	2·0	12·4	3·8
05	4 31·3	4 32·0	4 18·9	0·5	0·2	6·5	2·0	12·5	3·9
06	4 31·5	4 32·2	4 19·1	0·6	0·2	6·6	2·0	12·6	3·9
07	4 31·8	4 32·5	4 19·4	0·7	0·2	6·7	2·1	12·7	3·9
08	4 32·0	4 32·7	4 19·6	0·8	0·2	6·8	2·1	12·8	3·9
09	4 32·3	4 33·0	4 19·8	0·9	0·3	6·9	2·1	12·9	4·0
10	4 32·5	4 33·2	4 20·1	1·0	0·3	7·0	2·2	13·0	4·0
11	4 32·8	4 33·5	4 20·3	1·1	0·3	7·1	2·2	13·1	4·0
12	4 33·0	4 33·7	4 20·6	1·2	0·4	7·2	2·2	13·2	4·1
13	4 33·3	4 34·0	4 20·8	1·3	0·4	7·3	2·3	13·3	4·1
14	4 33·5	4 34·2	4 21·0	1·4	0·4	7·4	2·3	13·4	4·1
15	4 33·8	4 34·5	4 21·3	1·5	0·5	7·5	2·3	13·5	4·2
16	4 34·0	4 34·8	4 21·5	1·6	0·5	7·6	2·3	13·6	4·2
17	4 34·3	4 35·0	4 21·8	1·7	0·5	7·7	2·4	13·7	4·2
18	4 34·5	4 35·3	4 22·0	1·8	0·6	7·8	2·4	13·8	4·3
19	4 34·8	4 35·5	4 22·2	1·9	0·6	7·9	2·4	13·9	4·3
20	4 35·0	4 35·8	4 22·5	2·0	0·6	8·0	2·5	14·0	4·3
21	4 35·3	4 36·0	4 22·7	2·1	0·6	8·1	2·5	14·1	4·3
22	4 35·5	4 36·3	4 22·9	2·2	0·7	8·2	2·5	14·2	4·4
23	4 35·8	4 36·5	4 23·2	2·3	0·7	8·3	2·6	14·3	4·4
24	4 36·0	4 36·8	4 23·4	2·4	0·7	8·4	2·6	14·4	4·4
25	4 36·3	4 37·0	4 23·7	2·5	0·8	8·5	2·6	14·5	4·5
26	4 36·5	4 37·3	4 23·9	2·6	0·8	8·6	2·7	14·6	4·5
27	4 36·8	4 37·5	4 24·1	2·7	0·8	8·7	2·7	14·7	4·5
28	4 37·0	4 37·8	4 24·4	2·8	0·9	8·8	2·7	14·8	4·6
29	4 37·3	4 38·0	4 24·6	2·9	0·9	8·9	2·7	14·9	4·6
30	4 37·5	4 38·3	4 24·9	3·0	0·9	9·0	2·8	15·0	4·6
31	4 37·8	4 38·5	4 25·1	3·1	1·0	9·1	2·8	15·1	4·7
32	4 38·0	4 38·8	4 25·3	3·2	1·0	9·2	2·8	15·2	4·7
33	4 38·3	4 39·0	4 25·6	3·3	1·0	9·3	2·9	15·3	4·7
34	4 38·5	4 39·3	4 25·8	3·4	1·0	9·4	2·9	15·4	4·7
35	4 38·8	4 39·5	4 26·1	3·5	1·1	9·5	2·9	15·5	4·8
36	4 39·0	4 39·8	4 26·3	3·6	1·1	9·6	3·0	15·6	4·8
37	4 39·3	4 40·0	4 26·5	3·7	1·1	9·7	3·0	15·7	4·8
38	4 39·5	4 40·3	4 26·8	3·8	1·2	9·8	3·0	15·8	4·9
39	4 39·8	4 40·5	4 27·0	3·9	1·2	9·9	3·1	15·9	4·9
40	4 40·0	4 40·8	4 27·2	4·0	1·2	10·0	3·1	16·0	4·9
41	4 40·3	4 41·0	4 27·5	4·1	1·3	10·1	3·1	16·1	5·0
42	4 40·5	4 41·3	4 27·7	4·2	1·3	10·2	3·1	16·2	5·0
43	4 40·8	4 41·5	4 28·0	4·3	1·3	10·3	3·2	16·3	5·0
44	4 41·0	4 41·8	4 28·2	4·4	1·4	10·4	3·2	16·4	5·1
45	4 41·3	4 42·0	4 28·4	4·5	1·4	10·5	3·2	16·5	5·1
46	4 41·5	4 42·3	4 28·7	4·6	1·4	10·6	3·3	16·6	5·1
47	4 41·8	4 42·5	4 28·9	4·7	1·4	10·7	3·3	16·7	5·1
48	4 42·0	4 42·8	4 29·2	4·8	1·5	10·8	3·3	16·8	5·2
49	4 42·3	4 43·0	4 29·4	4·9	1·5	10·9	3·4	16·9	5·2
50	4 42·5	4 43·3	4 29·6	5·0	1·5	11·0	3·4	17·0	5·2
51	4 42·8	4 43·5	4 29·9	5·1	1·6	11·1	3·4	17·1	5·3
52	4 43·0	4 43·8	4 30·1	5·2	1·6	11·2	3·5	17·2	5·3
53	4 43·3	4 44·0	4 30·3	5·3	1·6	11·3	3·5	17·3	5·3
54	4 43·5	4 44·3	4 30·6	5·4	1·7	11·4	3·5	17·4	5·4
55	4 43·8	4 44·5	4 30·8	5·5	1·7	11·5	3·5	17·5	5·4
56	4 44·0	4 44·8	4 31·1	5·6	1·7	11·6	3·6	17·6	5·4
57	4 44·3	4 45·0	4 31·3	5·7	1·8	11·7	3·6	17·7	5·5
58	4 44·5	4 45·3	4 31·5	5·8	1·8	11·8	3·6	17·8	5·5
59	4 44·8	4 45·5	4 31·8	5·9	1·8	11·9	3·7	17·9	5·5
60	4 45·0	4 45·8	4 32·0	6·0	1·9	12·0	3·7	18·0	5·6

19^m	SUN PLANETS	ARIES	MOON	v or d	Corr^n	v or d	Corr^n	v or d	Corr^n
s	° ′	° ′	° ′	′	′	′	′	′	′
00	4 45·0	4 45·8	4 32·0	0·0	0·0	6·0	2·0	12·0	3·9
01	4 45·3	4 46·0	4 32·3	0·1	0·0	6·1	2·0	12·1	3·9
02	4 45·5	4 46·3	4 32·5	0·2	0·1	6·2	2·0	12·2	4·0
03	4 45·8	4 46·5	4 32·7	0·3	0·1	6·3	2·0	12·3	4·0
04	4 46·0	4 46·8	4 33·0	0·4	0·1	6·4	2·1	12·4	4·0
05	4 46·3	4 47·0	4 33·2	0·5	0·2	6·5	2·1	12·5	4·1
06	4 46·5	4 47·3	4 33·4	0·6	0·2	6·6	2·1	12·6	4·1
07	4 46·8	4 47·5	4 33·7	0·7	0·2	6·7	2·2	12·7	4·1
08	4 47·0	4 47·8	4 33·9	0·8	0·3	6·8	2·2	12·8	4·2
09	4 47·3	4 48·0	4 34·2	0·9	0·3	6·9	2·2	12·9	4·2
10	4 47·5	4 48·3	4 34·4	1·0	0·3	7·0	2·3	13·0	4·2
11	4 47·8	4 48·5	4 34·6	1·1	0·4	7·1	2·3	13·1	4·3
12	4 48·0	4 48·8	4 34·9	1·2	0·4	7·2	2·3	13·2	4·3
13	4 48·3	4 49·0	4 35·1	1·3	0·4	7·3	2·4	13·3	4·3
14	4 48·5	4 49·3	4 35·4	1·4	0·5	7·4	2·4	13·4	4·4
15	4 48·8	4 49·5	4 35·6	1·5	0·5	7·5	2·4	13·5	4·4
16	4 49·0	4 49·8	4 35·8	1·6	0·5	7·6	2·5	13·6	4·4
17	4 49·3	4 50·0	4 36·1	1·7	0·6	7·7	2·5	13·7	4·5
18	4 49·5	4 50·3	4 36·3	1·8	0·6	7·8	2·5	13·8	4·5
19	4 49·8	4 50·5	4 36·6	1·9	0·6	7·9	2·6	13·9	4·5
20	4 50·0	4 50·8	4 36·8	2·0	0·7	8·0	2·6	14·0	4·6
21	4 50·3	4 51·0	4 37·0	2·1	0·7	8·1	2·6	14·1	4·6
22	4 50·5	4 51·3	4 37·3	2·2	0·7	8·2	2·7	14·2	4·6
23	4 50·8	4 51·5	4 37·5	2·3	0·7	8·3	2·7	14·3	4·6
24	4 51·0	4 51·8	4 37·7	2·4	0·8	8·4	2·7	14·4	4·7
25	4 51·3	4 52·0	4 38·0	2·5	0·8	8·5	2·8	14·5	4·7
26	4 51·5	4 52·3	4 38·2	2·6	0·8	8·6	2·8	14·6	4·7
27	4 51·8	4 52·5	4 38·5	2·7	0·9	8·7	2·8	14·7	4·8
28	4 52·0	4 52·8	4 38·7	2·8	0·9	8·8	2·9	14·8	4·8
29	4 52·3	4 53·1	4 38·9	2·9	0·9	8·9	2·9	14·9	4·8
30	4 52·5	4 53·3	4 39·2	3·0	1·0	9·0	2·9	15·0	4·9
31	4 52·8	4 53·6	4 39·4	3·1	1·0	9·1	3·0	15·[illegible]	4·9
32	4 53·0	4 53·8	4 39·7	3·2	1·0	9·2	3·0	15·2	4·9
33	4 53·3	4 54·1	4 39·9	3·3	1·1	9·3	3·0	15·3	5·0
34	4 53·5	4 54·3	4 40·1	3·4	1·1	9·4	3·1	15·4	5·0
35	4 53·8	4 54·6	4 40·4	3·5	1·1	9·5	3·1	15·5	5·0
36	4 54·0	4 54·8	4 40·6	3·6	1·2	9·6	3·1	15·6	5·1
37	4 54·3	4 55·1	4 40·8	3·7	1·2	9·7	3·2	15·7	5·1
38	4 54·5	4 55·3	4 41·1	3·8	1·2	9·8	3·2	15·8	5·1
39	4 54·8	4 55·6	4 41·3	3·9	1·3	9·9	3·2	15·9	5·2
40	4 55·0	4 55·8	4 41·6	4·0	1·3	10·0	3·3	16·0	5·2
41	4 55·3	4 56·1	4 41·8	4·1	1·3	10·1	3·3	16·1	5·2
42	4 55·5	4 56·3	4 42·0	4·2	1·4	10·2	3·3	16·2	5·3
43	4 55·8	4 56·6	4 42·3	4·3	1·4	10·3	3·3	16·3	5·3
44	4 56·0	4 56·8	4 42·5	4·4	1·4	10·4	3·4	16·4	5·3
45	4 56·3	4 57·1	4 42·8	4·5	1·5	10·5	3·4	16·5	5·4
46	4 56·5	4 57·3	4 43·0	4·6	1·5	10·6	3·4	16·6	5·4
47	4 56·8	4 57·6	4 43·2	4·7	1·5	10·7	3·5	16·7	5·4
48	4 57·0	4 57·8	4 43·5	4·8	1·6	10·8	3·5	16·8	5·5
49	4 57·3	4 58·1	4 43·7	4·9	1·6	10·9	3·5	16·9	5·5
50	4 57·5	4 58·3	4 43·9	5·0	1·6	11·0	3·6	17·0	5·5
51	4 57·8	4 58·6	4 44·2	5·1	1·7	11·1	3·6	17·1	5·6
52	4 58·0	4 58·8	4 44·4	5·2	1·7	11·2	3·6	17·2	5·6
53	4 58·3	4 59·1	4 44·7	5·3	1·7	11·3	3·7	17·3	5·6
54	4 58·5	4 59·3	4 44·9	5·4	1·8	11·4	3·7	17·4	5·7
55	4 58·8	4 59·6	4 45·1	5·5	1·8	11·5	3·7	17·5	5·7
56	4 59·0	4 59·8	4 45·4	5·6	1·8	11·6	3·8	17·6	5·7
57	4 59·3	5 00·1	4 45·6	5·7	1·9	11·7	3·8	17·7	5·8
58	4 59·5	5 00·3	4 45·9	5·8	1·9	11·8	3·8	17·8	5·8
59	4 59·8	5 00·6	4 46·1	5·9	1·9	11·9	3·9	17·9	5·8
60	5 00·0	5 00·8	4 46·3	6·0	2·0	12·0	3·9	18·0	5·9

20ᵐ	SUN PLANETS	ARIES	MOON	v or d	Corrⁿ	v or d	Corrⁿ	v or d	Corrⁿ
s	° ′	° ′	° ′	′	′	′	′	′	′
00	5 00·0	5 00·8	4 46·3	0·0	0·0	6·0	2·1	12·0	4·1
01	5 00·3	5 01·1	4 46·6	0·1	0·0	6·1	2·1	12·1	4·1
02	5 00·5	5 01·3	4 46·8	0·2	0·1	6·2	2·1	12·2	4·2
03	5 00·8	5 01·6	4 47·0	0·3	0·1	6·3	2·2	12·3	4·2
04	5 01·0	5 01·8	4 47·3	0·4	0·1	6·4	2·2	12·4	4·2
05	5 01·3	5 02·1	4 47·5	0·5	0·2	6·5	2·2	12·5	4·3
06	5 01·5	5 02·3	4 47·8	0·6	0·2	6·6	2·3	12·6	4·3
07	5 01·8	5 02·6	4 48·0	0·7	0·2	6·7	2·3	12·7	4·3
08	5 02·0	5 02·8	4 48·2	0·8	0·3	6·8	2·3	12·8	4·4
09	5 02·3	5 03·1	4 48·5	0·9	0·3	6·9	2·4	12·9	4·4
10	5 02·5	5 03·3	4 48·7	1·0	0·3	7·0	2·4	13·0	4·4
11	5 02·8	5 03·6	4 49·0	1·1	0·4	7·1	2·4	13·1	4·5
12	5 03·0	5 03·8	4 49·2	1·2	0·4	7·2	2·5	13·2	4·5
13	5 03·3	5 04·1	4 49·4	1·3	0·4	7·3	2·5	13·3	4·5
14	5 03·5	5 04·3	4 49·7	1·4	0·5	7·4	2·5	13·4	4·6
15	5 03·8	5 04·6	4 49·9	1·5	0·5	7·5	2·6	13·5	4·6
16	5 04·0	5 04·8	4 50·2	1·6	0·5	7·6	2·6	13·6	4·6
17	5 04·3	5 05·1	4 50·4	1·7	0·6	7·7	2·6	13·7	4·7
18	5 04·5	5 05·3	4 50·6	1·8	0·6	7·8	2·7	13·8	4·7
19	5 04·8	5 05·6	4 50·9	1·9	0·6	7·9	2·7	13·9	4·7
20	5 05·0	5 05·8	4 51·1	2·0	0·7	8·0	2·7	14·0	4·8
21	5 05·3	5 06·1	4 51·3	2·1	0·7	8·1	2·8	14·1	4·8
22	5 05·5	5 06·3	4 51·6	2·2	0·8	8·2	2·8	14·2	4·9
23	5 05·8	5 06·6	4 51·8	2·3	0·8	8·3	2·8	14·3	4·9
24	5 06·0	5 06·8	4 52·1	2·4	0·8	8·4	2·9	14·4	4·9
25	5 06·3	5 07·1	4 52·3	2·5	0·9	8·5	2·9	14·5	5·0
26	5 06·5	5 07·3	4 52·5	2·6	0·9	8·6	2·9	14·6	5·0
27	5 06·8	5 07·6	4 52·8	2·7	0·9	8·7	3·0	14·7	5·0
28	5 07·0	5 07·8	4 53·0	2·8	1·0	8·8	3·0	14·8	5·1
29	5 07·3	5 08·1	4 53·3	2·9	1·0	8·9	3·0	14·9	5·1
30	5 07·5	5 08·3	4 53·5	3·0	1·0	9·0	3·1	15·0	5·1
31	5 07·8	5 08·6	4 53·7	3·1	1·1	9·1	3·1	15·1	5·2
32	5 08·0	5 08·8	4 54·0	3·2	1·1	9·2	3·1	15·2	5·2
33	5 08·3	5 09·1	4 54·2	3·3	1·1	9·3	3·2	15·3	5·2
34	5 08·5	5 09·3	4 54·4	3·4	1·2	9·4	3·2	15·4	5·3
35	5 08·8	5 09·6	4 54·7	3·5	1·2	9·5	3·2	15·5	5·3
36	5 09·0	5 09·8	4 54·9	3·6	1·2	9·6	3·3	15·6	5·3
37	5 09·3	5 10·1	4 55·2	3·7	1·3	9·7	3·3	15·7	5·4
38	5 09·5	5 10·3	4 55·4	3·8	1·3	9·8	3·3	15·8	5·4
39	5 09·8	5 10·6	4 55·6	3·9	1·3	9·9	3·4	15·9	5·4
40	5 10·0	5 10·8	4 55·9	4·0	1·4	10·0	3·4	16·0	5·5
41	5 10·3	5 11·1	4 56·1	4·1	1·4	10·1	3·5	16·1	5·5
42	5 10·5	5 11·4	4 56·4	4·2	1·4	10·2	3·5	16·2	5·5
43	5 10·8	5 11·6	4 56·6	4·3	1·5	10·3	3·5	16·3	5·6
44	5 11·0	5 11·9	4 56·8	4·4	1·5	10·4	3·6	16·4	5·6
45	5 11·3	5 12·1	4 57·1	4·5	1·5	10·5	3·6	16·5	5·6
46	5 11·5	5 12·4	4 57·3	4·6	1·6	10·6	3·6	16·6	5·7
47	5 11·8	5 12·6	4 57·5	4·7	1·6	10·7	3·7	16·7	5·7
48	5 12·0	5 12·9	4 57·8	4·8	1·6	10·8	3·7	16·8	5·7
49	5 12·3	5 13·1	4 58·0	4·9	1·7	10·9	3·7	16·9	5·8
50	5 12·5	5 13·4	4 58·3	5·0	1·7	11·0	3·8	17·0	5·8
51	5 12·8	5 13·6	4 58·5	5·1	1·7	11·1	3·8	17·1	5·8
52	5 13·0	5 13·9	4 58·7	5·2	1·8	11·2	3·8	17·2	5·9
53	5 13·3	5 14·1	4 59·0	5·3	1·8	11·3	3·9	17·3	5·9
54	5 13·5	5 14·4	4 59·2	5·4	1·8	11·4	3·9	17·4	5·9
55	5 13·8	5 14·6	4 59·5	5·5	1·9	11·5	3·9	17·5	6·0
56	5 14·0	5 14·9	4 59·7	5·6	1·9	11·6	4·0	17·6	6·0
57	5 14·3	5 15·1	4 59·9	5·7	1·9	11·7	4·0	17·7	6·0
58	5 14·5	5 15·4	5 00·2	5·8	2·0	11·8	4·0	17·8	6·1
59	5 14·8	5 15·6	5 00·4	5·9	2·0	11·9	4·1	17·9	6·1
60	5 15·0	5 15·9	5 00·7	6·0	2·1	12·0	4·1	18·0	6·2

21ᵐ	SUN PLANETS	ARIES	MOON	v or d	Corrⁿ	v or d	Corrⁿ	v or d	Corrⁿ
s	° ′	° ′	° ′	′	′	′	′	′	′
00	5 15·0	5 15·9	5 00·7	0·0	0·0	6·0	2·2	12·0	4·3
01	5 15·3	5 16·1	5 00·9	0·1	0·0	6·1	2·2	12·1	4·3
02	5 15·5	5 16·4	5 01·1	0·2	0·1	6·2	2·2	12·2	4·4
03	5 15·8	5 16·6	5 01·4	0·3	0·1	6·3	2·3	12·3	4·4
04	5 16·0	5 16·9	5 01·6	0·4	0·1	6·4	2·3	12·4	4·4
05	5 16·3	5 17·1	5 01·8	0·5	0·2	6·5	2·3	12·5	4·5
06	5 16·5	5 17·4	5 02·1	0·6	0·2	6·6	2·4	12·6	4·5
07	5 16·8	5 17·6	5 02·3	0·7	0·3	6·7	2·4	12·7	4·6
08	5 17·0	5 17·9	5 02·6	0·8	0·3	6·8	2·4	12·8	4·6
09	5 17·3	5 18·1	5 02·8	0·9	0·3	6·9	2·5	12·9	4·6
10	5 17·5	5 18·4	5 03·0	1·0	0·4	7·0	2·5	13·0	4·7
11	5 17·8	5 18·6	5 03·3	1·1	0·4	7·1	2·5	13·1	4·7
12	5 18·0	5 18·9	5 03·5	1·2	0·4	7·2	2·6	13·2	4·7
13	5 18·3	5 19·1	5 03·8	1·3	0·5	7·3	2·6	13·3	4·8
14	5 18·5	5 19·4	5 04·0	1·4	0·5	7·4	2·7	13·4	4·8
15	5 18·8	5 19·6	5 04·2	1·5	0·5	7·5	2·7	13·5	4·8
16	5 19·0	5 19·9	5 04·5	1·6	0·6	7·6	2·7	13·6	4·9
17	5 19·3	5 20·1	5 04·7	1·7	0·6	7·7	2·8	13·7	4·9
18	5 19·5	5 20·4	5 04·9	1·8	0·6	7·8	2·8	13·8	4·9
19	5 19·8	5 20·6	5 05·2	1·9	0·7	7·9	2·8	13·9	5·0
20	5 20·0	5 20·9	5 05·4	2·0	0·7	8·0	2·9	14·0	5·0
21	5 20·3	5 21·1	5 05·7	2·1	0·8	8·1	2·9	14·1	5·1
22	5 20·5	5 21·4	5 05·9	2·2	0·8	8·2	2·9	14·2	5·1
23	5 20·8	5 21·6	5 06·1	2·3	0·8	8·3	3·0	14·3	5·1
24	5 21·0	5 21·9	5 06·4	2·4	0·9	8·4	3·0	14·4	5·2
25	5 21·3	5 22·1	5 06·6	2·5	0·9	8·5	3·0	14·5	5·2
26	5 21·5	5 22·4	5 06·9	2·6	0·9	8·6	3·1	14·6	5·2
27	5 21·8	5 22·6	5 07·1	2·7	1·0	8·7	3·1	14·7	5·3
28	5 22·0	5 22·9	5 07·3	2·8	1·0	8·8	3·2	14·8	5·3
29	5 22·3	5 23·1	5 07·6	2·9	1·0	8·9	3·2	14·9	5·3
30	5 22·5	5 23·4	5 07·8	3·0	1·1	9·0	3·2	15·0	5·4
31	5 22·8	5 23·6	5 08·0	3·1	1·1	9·1	3·3	15·1	5·4
32	5 23·0	5 23·9	5 08·3	3·2	1·1	9·2	3·3	15·2	5·4
33	5 23·3	5 24·1	5 08·5	3·3	1·2	9·3	3·3	15·3	5·5
34	5 23·5	5 24·4	5 08·8	3·4	1·2	9·4	3·4	15·4	5·5
35	5 23·8	5 24·6	5 09·0	3·5	1·3	9·5	3·4	15·5	5·6
36	5 24·0	5 24·9	5 09·2	3·6	1·3	9·6	3·4	15·6	5·6
37	5 24·3	5 25·1	5 09·5	3·7	1·3	9·7	3·5	15·7	5·6
38	5 24·5	5 25·4	5 09·7	3·8	1·4	9·8	3·5	15·8	5·7
39	5 24·8	5 25·6	5 10·0	3·9	1·4	9·9	3·5	15·9	5·7
40	5 25·0	5 25·9	5 10·2	4·0	1·4	10·0	3·6	16·0	5·7
41	5 25·3	5 26·1	5 10·4	4·1	1·5	10·1	3·6	16·1	5·8
42	5 25·5	5 26·4	5 10·7	4·2	1·5	10·2	3·7	16·2	5·8
43	5 25·8	5 26·6	5 10·9	4·3	1·5	10·3	3·7	16·3	5·8
44	5 26·0	5 26·9	5 11·1	4·4	1·6	10·4	3·7	16·4	5·9
45	5 26·3	5 27·1	5 11·4	4·5	1·6	10·5	3·8	16·5	5·9
46	5 26·5	5 27·4	5 11·6	4·6	1·6	10·6	3·8	16·6	5·9
47	5 26·8	5 27·6	5 11·9	4·7	1·7	10·7	3·8	16·7	6·0
48	5 27·0	5 27·9	5 12·1	4·8	1·7	10·8	3·9	16·8	6·0
49	5 27·3	5 28·1	5 12·3	4·9	1·8	10·9	3·9	16·9	6·1
50	5 27·5	5 28·4	5 12·6	5·0	1·8	11·0	3·9	17·0	6·1
51	5 27·8	5 28·6	5 12·8	5·1	1·8	11·1	4·0	17·1	6·1
52	5 28·0	5 28·9	5 13·1	5·2	1·9	11·2	4·0	17·2	6·2
53	5 28·3	5 29·1	5 13·3	5·3	1·9	11·3	4·0	17·3	6·2
54	5 28·5	5 29·4	5 13·5	5·4	1·9	11·4	4·1	17·4	6·2
55	5 28·8	5 29·7	5 13·8	5·5	2·0	11·5	4·1	17·5	6·3
56	5 29·0	5 29·9	5 14·0	5·6	2·0	11·6	4·2	17·6	6·3
57	5 29·3	5 30·2	5 14·3	5·7	2·0	11·7	4·2	17·7	6·3
58	5 29·5	5 30·4	5 14·5	5·8	2·1	11·8	4·2	17·8	6·4
59	5 29·8	5 30·7	5 14·7	5·9	2·1	11·9	4·3	17·9	6·4
60	5 30·0	5 30·9	5 15·0	6·0	2·2	12·0	4·3	18·0	6·5

22ᵐ	SUN PLANETS	ARIES	MOON	v or d	Corrⁿ	v or d	Corrⁿ	v or d	Corrⁿ
s	° ′	° ′	° ′	′	′	′	′	′	′
00	5 30·0	5 30·9	5 15·0	0·0	0·0	6·0	2·3	12·0	4·5
01	5 30·3	5 31·2	5 15·2	0·1	0·0	6·1	2·3	12·1	4·5
02	5 30·5	5 31·4	5 15·4	0·2	0·1	6·2	2·3	12·2	4·6
03	5 30·8	5 31·7	5 15·7	0·3	0·1	6·3	2·4	12·3	4·6
04	5 31·0	5 31·9	5 15·9	0·4	0·2	6·4	2·4	12·4	4·7
05	5 31·3	5 32·2	5 16·2	0·5	0·2	6·5	2·4	12·5	4·7
06	5 31·5	5 32·4	5 16·4	0·6	0·2	6·6	2·5	12·6	4·7
07	5 31·8	5 32·7	5 16·6	0·7	0·3	6·7	2·5	12·7	4·8
08	5 32·0	5 32·9	5 16·9	0·8	0·3	6·8	2·6	12·8	4·8
09	5 32·3	5 33·2	5 17·1	0·9	0·3	6·9	2·6	12·9	4·8
10	5 32·5	5 33·4	5 17·4	1·0	0·4	7·0	2·6	13·0	4·9
11	5 32·8	5 33·7	5 17·6	1·1	0·4	7·1	2·7	13·1	4·9
12	5 33·0	5 33·9	5 17·8	1·2	0·5	7·2	2·7	13·2	5·0
13	5 33·3	5 34·2	5 18·1	1·3	0·5	7·3	2·7	13·3	5·0
14	5 33·5	5 34·4	5 18·3	1·4	0·5	7·4	2·8	13·4	5·0
15	5 33·8	5 34·7	5 18·5	1·5	0·6	7·5	2·8	13·5	5·1
16	5 34·0	5 34·9	5 18·8	1·6	0·6	7·6	2·9	13·6	5·1
17	5 34·3	5 35·2	5 19·0	1·7	0·6	7·7	2·9	13·7	5·1
18	5 34·5	5 35·4	5 19·3	1·8	0·7	7·8	2·9	13·8	5·2
19	5 34·8	5 35·7	5 19·5	1·9	0·7	7·9	3·0	13·9	5·2
20	5 35·0	5 35·9	5 19·7	2·0	0·8	8·0	3·0	14·0	5·3
21	5 35·3	5 36·2	5 20·0	2·1	0·8	8·1	3·0	14·1	5·3
22	5 35·5	5 36·4	5 20·2	2·2	0·8	8·2	3·1	14·2	5·3
23	5 35·8	5 36·7	5 20·5	2·3	0·9	8·3	3·1	14·3	5·4
24	5 36·0	5 36·9	5 20·7	2·4	0·9	8·4	3·2	14·4	5·4
25	5 36·3	5 37·2	5 20·9	2·5	0·9	8·5	3·2	14·5	5·4
26	5 36·5	5 37·4	5 21·2	2·6	1·0	8·6	3·2	14·6	5·5
27	5 36·8	5 37·7	5 21·4	2·7	1·0	8·7	3·3	14·7	5·5
28	5 37·0	5 37·9	5 21·6	2·8	1·1	8·8	3·3	14·8	5·6
29	5 37·3	5 38·2	5 21·9	2·9	1·1	8·9	3·3	14·9	5·6
30	5 37·5	5 38·4	5 22·1	3·0	1·1	9·0	3·4	15·0	5·6
31	5 37·8	5 38·7	5 22·4	3·1	1·2	9·1	3·4	15·1	5·7
32	5 38·0	5 38·9	5 22·6	3·2	1·2	9·2	3·5	15·2	5·7
33	5 38·3	5 39·2	5 22·8	3·3	1·2	9·3	3·5	15·3	5·7
34	5 38·5	5 39·4	5 23·1	3·4	1·3	9·4	3·5	15·4	5·8
35	5 38·8	5 39·7	5 23·3	3·5	1·3	9·5	3·6	15·5	5·8
36	5 39·0	5 39·9	5 23·6	3·6	1·4	9·6	3·6	15·6	5·9
37	5 39·3	5 40·2	5 23·8	3·7	1·4	9·7	3·6	15·7	5·9
38	5 39·5	5 40·4	5 24·0	3·8	1·4	9·8	3·7	15·8	5·9
39	5 39·8	5 40·7	5 24·3	3·9	1·5	9·9	3·7	15·9	6·0
40	5 40·0	5 40·9	5 24·5	4·0	1·5	10·0	3·8	16·0	6·0
41	5 40·3	5 41·2	5 24·7	4·1	1·5	10·1	3·8	16·1	6·0
42	5 40·5	5 41·4	5 25·0	4·2	1·6	10·2	3·8	16·2	6·1
43	5 40·8	5 41·7	5 25·2	4·3	1·6	10·3	3·9	16·3	6·1
44	5 41·0	5 41·9	5 25·5	4·4	1·7	10·4	3·9	16·4	6·2
45	5 41·3	5 42·2	5 25·7	4·5	1·7	10·5	3·9	16·5	6·2
46	5 41·5	5 42·4	5 25·9	4·6	1·7	10·6	4·0	16·6	6·2
47	5 41·8	5 42·7	5 26·2	4·7	1·8	10·7	4·0	16·7	6·3
48	5 42·0	5 42·9	5 26·4	4·8	1·8	10·8	4·1	16·8	6·3
49	5 42·3	5 43·2	5 26·7	4·9	1·8	10·9	4·1	16·9	6·3
50	5 42·5	5 43·4	5 26·9	5·0	1·9	11·0	4·1	17·0	6·4
51	5 42·8	5 43·7	5 27·1	5·1	1·9	11·1	4·2	17·1	6·4
52	5 43·0	5 43·9	5 27·4	5·2	2·0	11·2	4·2	17·2	6·5
53	5 43·3	5 44·2	5 27·6	5·3	2·0	11·3	4·2	17·3	6·5
54	5 43·5	5 44·4	5 27·9	5·4	2·0	11·4	4·3	17·4	6·5
55	5 43·8	5 44·7	5 28·1	5·5	2·1	11·5	4·3	17·5	6·6
56	5 44·0	5 44·9	5 28·3	5·6	2·1	11·6	4·4	17·6	6·6
57	5 44·3	5 45·2	5 28·6	5·7	2·1	11·7	4·4	17·7	6·6
58	5 44·5	5 45·4	5 28·8	5·8	2·2	11·8	4·4	17·8	6·7
59	5 44·8	5 45·7	5 29·0	5·9	2·2	11·9	4·5	17·9	6·7
60	5 45·0	5 45·9	5 29·3	6·0	2·3	12·0	4·5	18·0	6·8

23ᵐ	SUN PLANETS	ARIES	MOON	v or d	Corrⁿ	v or d	Corrⁿ	v or d	Corrⁿ
s	° ′	° ′	° ′	′	′	′	′	′	′
00	5 45·0	5 45·9	5 29·3	0·0	0·0	6·0	2·4	12·0	4·7
01	5 45·3	5 46·2	5 29·5	0·1	0·0	6·1	2·4	12·1	4·7
02	5 45·5	5 46·4	5 29·8	0·2	0·1	6·2	2·4	12·2	4·8
03	5 45·8	5 46·7	5 30·0	0·3	0·1	6·3	2·5	12·3	4·8
04	5 46·0	5 46·9	5 30·2	0·4	0·2	6·4	2·5	12·4	4·9
05	5 46·3	5 47·2	5 30·5	0·5	0·2	6·5	2·5	12·5	4·9
06	5 46·5	5 47·4	5 30·7	0·6	0·2	6·6	2·6	12·6	4·9
07	5 46·8	5 47·7	5 31·0	0·7	0·3	6·7	2·6	12·7	5·0
08	5 47·0	5 48·0	5 31·2	0·8	0·3	6·8	2·7	12·8	5·0
09	5 47·3	5 48·2	5 31·4	0·9	0·4	6·9	2·7	12·9	5·1
10	5 47·5	5 48·5	5 31·7	1·0	0·4	7·0	2·7	13·0	5·1
11	5 47·8	5 48·7	5 31·9	1·1	0·4	7·1	2·8	13·1	5·1
12	5 48·0	5 49·0	5 32·1	1·2	0·5	7·2	2·8	13·2	5·2
13	5 48·3	5 49·2	5 32·4	1·3	0·5	7·3	2·9	13·3	5·2
14	5 48·5	5 49·5	5 32·6	1·4	0·5	7·4	2·9	13·4	5·2
15	5 48·8	5 49·7	5 32·9	1·5	0·6	7·5	2·9	13·5	5·3
16	5 49·0	5 50·0	5 33·1	1·6	0·6	7·6	3·0	13·6	5·3
17	5 49·3	5 50·2	5 33·3	1·7	0·7	7·7	3·0	13·7	5·4
18	5 49·5	5 50·5	5 33·6	1·8	0·7	7·8	3·1	13·8	5·4
19	5 49·8	5 50·7	5 33·8	1·9	0·7	7·9	3·1	13·9	5·4
20	5 50·0	5 51·0	5 34·1	2·0	0·8	8·0	3·1	14·0	5·5
21	5 50·3	5 51·2	5 34·3	2·1	0·8	8·1	3·2	14·1	5·5
22	5 50·5	5 51·5	5 34·5	2·2	0·9	8·2	3·2	14·2	5·6
23	5 50·8	5 51·7	5 34·8	2·3	0·9	8·3	3·3	14·3	5·6
24	5 51·0	5 52·0	5 35·0	2·4	0·9	8·4	3·3	14·4	5·6
25	5 51·3	5 52·2	5 35·2	2·5	1·0	8·5	3·3	14·5	5·7
26	5 51·5	5 52·5	5 35·5	2·6	1·0	8·6	3·4	14·6	5·7
27	5 51·8	5 52·7	5 35·7	2·7	1·1	8·7	3·4	14·7	5·8
28	5 52·0	5 53·0	5 36·0	2·8	1·1	8·8	3·4	14·8	5·8
29	5 52·3	5 53·2	5 36·2	2·9	1·1	8·9	3·5	14·9	5·8
30	5 52·5	5 53·5	5 36·4	3·0	1·2	9·0	3·5	15·0	5·9
31	5 52·8	5 53·7	5 36·7	3·1	1·2	9·1	3·6	15·1	5·9
32	5 53·0	5 54·0	5 36·9	3·2	1·3	9·2	3·6	15·2	6·0
33	5 53·3	5 54·2	5 37·2	3·3	1·3	9·3	3·6	15·3	6·0
34	5 53·5	5 54·5	5 37·4	3·4	1·3	9·4	3·7	15·4	6·0
35	5 53·8	5 54·7	5 37·6	3·5	1·4	9·5	3·7	15·5	6·1
36	5 54·0	5 55·0	5 37·9	3·6	1·4	9·6	3·8	15·6	6·1
37	5 54·3	5 55·2	5 38·1	3·7	1·4	9·7	3·8	15·7	6·1
38	5 54·5	5 55·5	5 38·4	3·8	1·5	9·8	3·8	15·8	6·2
39	5 54·8	5 55·7	5 38·6	3·9	1·5	9·9	3·9	15·9	6·2
40	5 55·0	5 56·0	5 38·8	4·0	1·6	10·0	3·9	16·0	6·3
41	5 55·3	5 56·2	5 39·1	4·1	1·6	10·1	4·0	16·1	6·3
42	5 55·5	5 56·5	5 39·3	4·2	1·6	10·2	4·0	16·2	6·3
43	5 55·8	5 56·7	5 39·5	4·3	1·7	10·3	4·0	16·3	6·4
44	5 56·0	5 57·0	5 39·8	4·4	1·7	10·4	4·1	16·4	6·4
45	5 56·3	5 57·2	5 40·0	4·5	1·8	10·5	4·1	16·5	6·5
46	5 56·5	5 57·5	5 40·3	4·6	1·8	10·6	4·2	16·6	6·5
47	5 56·8	5 57·7	5 40·5	4·7	1·8	10·7	4·2	16·7	6·5
48	5 57·0	5 58·0	5 40·7	4·8	1·9	10·8	4·2	16·8	6·6
49	5 57·3	5 58·2	5 41·0	4·9	1·9	10·9	4·3	16·9	6·6
50	5 57·5	5 58·5	5 41·2	5·0	2·0	11·0	4·3	17·0	6·7
51	5 57·8	5 58·7	5 41·5	5·1	2·0	11·1	4·3	17·1	6·7
52	5 58·0	5 59·0	5 41·7	5·2	2·0	11·2	4·4	17·2	6·7
53	5 58·3	5 59·2	5 41·9	5·3	2·1	11·3	4·4	17·3	6·8
54	5 58·5	5 59·5	5 42·2	5·4	2·1	11·4	4·5	17·4	6·8
55	5 58·8	5 59·7	5 42·4	5·5	2·2	11·5	4·5	17·5	6·9
56	5 59·0	6 00·0	5 42·6	5·6	2·2	11·6	4·5	17·6	6·9
57	5 59·3	6 00·2	5 42·9	5·7	2·2	11·7	4·6	17·7	6·9
58	5 59·5	6 00·5	5 43·1	5·8	2·3	11·8	4·6	17·8	7·0
59	5 59·8	6 00·7	5 43·4	5·9	2·3	11·9	4·7	17·9	7·0
60	6 00·0	6 01·0	5 43·6	6·0	2·4	12·0	4·7	18·0	7·1

24m	SUN PLANETS	ARIES	MOON	v or d	Corrn	v or d	Corrn	v or d	Corrn
s	° ′	° ′	° ′	′	′	′	′	′	′
00	6 00·0	6 01·0	5 43·6	0·0	0·0	6·0	2·5	12·0	4·9
01	6 00·3	6 01·2	5 43·8	0·1	0·0	6·1	2·5	12·1	4·9
02	6 00·5	6 01·5	5 44·1	0·2	0·1	6·2	2·5	12·2	5·0
03	6 00·8	6 01·7	5 44·3	0·3	0·1	6·3	2·6	12·3	5·0
04	6 01·0	6 02·0	5 44·6	0·4	0·2	6·4	2·6	12·4	5·1
05	6 01·3	6 02·2	5 44·8	0·5	0·2	6·5	2·7	12·5	5·1
06	6 01·5	6 02·5	5 45·0	0·6	0·2	6·6	2·7	12·6	5·1
07	6 01·8	6 02·7	5 45·3	0·7	0·3	6·7	2·7	12·7	5·2
08	6 02·0	6 03·0	5 45·5	0·8	0·3	6·8	2·8	12·8	5·2
09	6 02·3	6 03·2	5 45·7	0·9	0·4	6·9	2·8	12·9	5·3
10	6 02·5	6 03·5	5 46·0	1·0	0·4	7·0	2·9	13·0	5·3
11	6 02·8	6 03·7	5 46·2	1·1	0·4	7·1	2·9	13·1	5·3
12	6 03·0	6 04·0	5 46·5	1·2	0·5	7·2	2·9	13·2	5·4
13	6 03·3	6 04·2	5 46·7	1·3	0·5	7·3	3·0	13·3	5·4
14	6 03·5	6 04·5	5 46·9	1·4	0·6	7·4	3·0	13·4	5·5
15	6 03·8	6 04·7	5 47·2	1·5	0·6	7·5	3·1	13·5	5·5
16	6 04·0	6 05·0	5 47·4	1·6	0·7	7·6	3·1	13·6	5·6
17	6 04·3	6 05·2	5 47·7	1·7	0·7	7·7	3·1	13·7	5·6
18	6 04·5	6 05·5	5 47·9	1·8	0·7	7·8	3·2	13·8	5·6
19	6 04·8	6 05·7	5 48·1	1·9	0·8	7·9	3·2	13·9	5·7
20	6 05·0	6 06·0	5 48·4	2·0	0·8	8·0	3·3	14·0	5·7
21	6 05·3	6 06·3	5 48·6	2·1	0·9	8·1	3·3	14·1	5·8
22	6 05·5	6 06·5	5 48·8	2·2	0·9	8·2	3·3	14·2	5·8
23	6 05·8	6 06·8	5 49·1	2·3	0·9	8·3	3·4	14·3	5·8
24	6 06·0	6 07·0	5 49·3	2·4	1·0	8·4	3·4	14·4	5·9
25	6 06·3	6 07·3	5 49·6	2·5	1·0	8·5	3·5	14·5	5·9
26	6 06·5	6 07·5	5 49·8	2·6	1·1	8·6	3·5	14·6	6·0
27	6 06·8	6 07·8	5 50·0	2·7	1·1	8·7	3·6	14·7	6·0
28	6 07·0	6 08·0	5 50·3	2·8	1·1	8·8	3·6	14·8	6·0
29	6 07·3	6 08·3	5 50·5	2·9	1·2	8·9	3·6	14·9	6·1
30	6 07·5	6 08·5	5 50·8	3·0	1·2	9·0	3·7	15·0	6·1
31	6 07·8	6 08·8	5 51·0	3·1	1·3	9·1	3·7	15·1	6·2
32	6 08·0	6 09·0	5 51·2	3·2	1·3	9·2	3·8	15·2	6·2
33	6 08·3	6 09·3	5 51·5	3·3	1·3	9·3	3·8	15·3	6·2
34	6 08·5	6 09·5	5 51·7	3·4	1·4	9·4	3·8	15·4	6·3
35	6 08·8	6 09·8	5 52·0	3·5	1·4	9·5	3·9	15·5	6·3
36	6 09·0	6 10·0	5 52·2	3·6	1·5	9·6	3·9	15·6	6·4
37	6 09·3	6 10·3	5 52·4	3·7	1·5	9·7	4·0	15·7	6·4
38	6 09·5	6 10·5	5 52·7	3·8	1·6	9·8	4·0	15·8	6·5
39	6 09·8	6 10·8	5 52·9	3·9	1·6	9·9	4·0	15·9	6·5
40	6 10·0	6 11·0	5 53·1	4·0	1·6	10·0	4·1	16·0	6·5
41	6 10·3	6 11·3	5 53·4	4·1	1·7	10·1	4·1	16·1	6·6
42	6 10·5	6 11·5	5 53·6	4·2	1·7	10·2	4·2	16·2	6·6
43	6 10·8	6 11·8	5 53·9	4·3	1·8	10·3	4·2	16·3	6·7
44	6 11·0	6 12·0	5 54·1	4·4	1·8	10·4	4·2	16·4	6·7
45	6 11·3	6 12·3	5 54·3	4·5	1·8	10·5	4·3	16·5	6·7
46	6 11·5	6 12·5	5 54·6	4·6	1·9	10·6	4·3	16·6	6·8
47	6 11·8	6 12·8	5 54·8	4·7	1·9	10·7	4·4	16·7	6·8
48	6 12·0	6 13·0	5 55·1	4·8	2·0	10·8	4·4	16·8	6·9
49	6 12·3	6 13·3	5 55·3	4·9	2·0	10·9	4·5	16·9	6·9
50	6 12·5	6 13·5	5 55·5	5·0	2·0	11·0	4·5	17·0	6·9
51	6 12·8	6 13·8	5 55·8	5·1	2·1	11·1	4·5	17·1	7·0
52	6 13·0	6 14·0	5 56·0	5·2	2·1	11·2	4·6	17·2	7·0
53	6 13·3	6 14·3	5 56·2	5·3	2·2	11·3	4·6	17·3	7·1
54	6 13·5	6 14·5	5 56·5	5·4	2·2	11·4	4·7	17·4	7·1
55	6 13·8	6 14·8	5 56·7	5·5	2·2	11·5	4·7	17·5	7·1
56	6 14·0	6 15·0	5 57·0	5·6	2·3	11·6	4·7	17·6	7·2
57	6 14·3	6 15·3	5 57·2	5·7	2·3	11·7	4·8	17·7	7·2
58	6 14·5	6 15·5	5 57·4	5·8	2·4	11·8	4·8	17·8	7·3
59	6 14·8	6 15·8	5 57·7	5·9	2·4	11·9	4·9	17·9	7·3
60	6 15·0	6 16·0	5 57·9	6·0	2·5	12·0	4·9	18·0	7·4

25m	SUN PLANETS	ARIES	MOON	v or d	Corrn	v or d	Corrn	v or d	Corrn
s	° ′	° ′	° ′	′	′	′	′	′	′
00	6 15·0	6 16·0	5 57·9	0·0	0·0	6·0	2·6	12·0	5·1
01	6 15·3	6 16·3	5 58·2	0·1	0·0	6·1	2·6	12·1	5·1
02	6 15·5	6 16·5	5 58·4	0·2	0·1	6·2	2·6	12·2	5·2
03	6 15·8	6 16·8	5 58·6	0·3	0·1	6·3	2·7	12·3	5·2
04	6 16·0	6 17·0	5 58·9	0·4	0·2	6·4	2·7	12·4	5·3
05	6 16·3	6 17·3	5 59·1	0·5	0·2	6·5	2·8	12·5	5·3
06	6 16·5	6 17·5	5 59·3	0·6	0·3	6·6	2·8	12·6	5·4
07	6 16·8	6 17·8	5 59·6	0·7	0·3	6·7	2·8	12·7	5·4
08	6 17·0	6 18·0	5 59·8	0·8	0·3	6·8	2·9	12·8	5·4
09	6 17·3	6 18·3	6 00·1	0·9	0·4	6·9	2·9	12·9	5·5
10	6 17·5	6 18·5	6 00·3	1·0	0·4	7·0	3·0	13·0	5·5
11	6 17·8	6 18·8	6 00·5	1·1	0·5	7·1	3·0	13·1	5·6
12	6 18·0	6 19·0	6 00·8	1·2	0·5	7·2	3·1	13·2	5·6
13	6 18·3	6 19·3	6 01·0	1·3	0·6	7·3	3·1	13·3	5·7
14	6 18·5	6 19·5	6 01·3	1·4	0·6	7·4	3·1	13·4	5·7
15	6 18·8	6 19·8	6 01·5	1·5	0·6	7·5	3·2	13·5	5·7
16	6 19·0	6 20·0	6 01·7	1·6	0·7	7·6	3·2	13·6	5·8
17	6 19·3	6 20·3	6 02·0	1·7	0·7	7·7	3·3	13·7	5·8
18	6 19·5	6 20·5	6 02·2	1·8	0·8	7·8	3·3	13·8	5·9
19	6 19·8	6 20·8	6 02·5	1·9	0·8	7·9	3·4	13·9	5·9
20	6 20·0	6 21·0	6 02·7	2·0	0·9	8·0	3·4	14·0	6·0
21	6 20·3	6 21·3	6 02·9	2·1	0·9	8·1	3·4	14·1	6·0
22	6 20·5	6 21·5	6 03·2	2·2	0·9	8·2	3·5	14·2	6·0
23	6 20·8	6 21·8	6 03·4	2·3	1·0	8·3	3·5	14·3	6·1
24	6 21·0	6 22·0	6 03·6	2·4	1·0	8·4	3·6	14·4	6·1
25	6 21·3	6 22·3	6 03·9	2·5	1·1	8·5	3·6	14·5	6·2
26	6 21·5	6 22·5	6 04·1	2·6	1·1	8·6	3·7	14·6	6·2
27	6 21·8	6 22·8	6 04·4	2·7	1·1	8·7	3·7	14·7	6·2
28	6 22·0	6 23·0	6 04·6	2·8	1·2	8·8	3·7	14·8	6·3
29	6 22·3	6 23·3	6 04·8	2·9	1·2	8·9	3·8	14·9	6·3
30	6 22·5	6 23·5	6 05·1	3·0	1·3	9·0	3·8	15·0	6·4
31	6 22·8	6 23·8	6 05·3	3·1	1·3	9·1	3·9	15·1	6·4
32	6 23·0	6 24·0	6 05·6	3·2	1·4	9·2	3·9	15·2	6·5
33	6 23·3	6 24·3	6 05·8	3·3	1·4	9·3	4·0	15·3	6·5
34	6 23·5	6 24·5	6 06·0	3·4	1·4	9·4	4·0	15·4	6·5
35	6 23·8	6 24·8	6 06·3	3·5	1·5	9·5	4·0	15·5	6·6
36	6 24·0	6 25·1	6 06·5	3·6	1·5	9·6	4·1	15·6	6·6
37	6 24·3	6 25·3	6 06·7	3·7	1·6	9·7	4·1	15·7	6·7
38	6 24·5	6 25·6	6 07·0	3·8	1·6	9·8	4·2	15·8	6·7
39	6 24·8	6 25·8	6 07·2	3·9	1·7	9·9	4·2	15·9	6·8
40	6 25·0	6 26·1	6 07·5	4·0	1·7	10·0	4·3	16·0	6·8
41	6 25·3	6 26·3	6 07·7	4·1	1·7	10·1	4·3	16·1	6·8
42	6 25·5	6 26·6	6 07·9	4·2	1·8	10·2	4·3	16·2	6·9
43	6 25·8	6 26·8	6 08·2	4·3	1·8	10·3	4·4	16·3	6·9
44	6 26·0	6 27·1	6 08·4	4·4	1·9	10·4	4·4	16·4	7·0
45	6 26·3	6 27·3	6 08·7	4·5	1·9	10·5	4·5	16·5	7·0
46	6 26·5	6 27·6	6 08·9	4·6	2·0	10·6	4·5	16·6	7·1
47	6 26·8	6 27·8	6 09·1	4·7	2·0	10·7	4·5	16·7	7·1
48	6 27·0	6 28·1	6 09·4	4·8	2·0	10·8	4·6	16·8	7·1
49	6 27·3	6 28·3	6 09·6	4·9	2·1	10·9	4·6	16·9	7·2
50	6 27·5	6 28·6	6 09·8	5·0	2·1	11·0	4·7	17·0	7·2
51	6 27·8	6 28·8	6 10·1	5·1	2·2	11·1	4·7	17·1	7·3
52	6 28·0	6 29·1	6 10·3	5·2	2·2	11·2	4·8	17·2	7·3
53	6 28·3	6 29·3	6 10·6	5·3	2·3	11·3	4·8	17·3	7·4
54	6 28·5	6 29·6	6 10·8	5·4	2·3	11·4	4·8	17·4	7·4
55	6 28·8	6 29·8	6 11·0	5·5	2·3	11·5	4·9	17·5	7·4
56	6 29·0	6 30·1	6 11·3	5·6	2·4	11·6	4·9	17·6	7·5
57	6 29·3	6 30·3	6 11·5	5·7	2·4	11·7	5·0	17·7	7·5
58	6 29·5	6 30·6	6 11·8	5·8	2·5	11·8	5·0	17·8	7·6
59	6 29·8	6 30·8	6 12·0	5·9	2·5	11·9	5·1	17·9	7·6
60	6 30·0	6 31·1	6 12·2	6·0	2·6	12·0	5·1	18·0	7·7

26ᵐ	SUN PLANETS	ARIES	MOON	v or d	Corrⁿ	v or d	Corrⁿ	v or d	Corrⁿ
s	° ′	° ′	° ′	′	′	′	′	′	′
00	6 30·0	6 31·1	6 12·2	0·0	0·0	6·0	2·7	12·0	5·3
01	6 30·3	6 31·3	6 12·5	0·1	0·0	6·1	2·7	12·1	5·3
02	6 30·5	6 31·6	6 12·7	0·2	0·1	6·2	2·7	12·2	5·4
03	6 30·8	6 31·8	6 12·9	0·3	0·1	6·3	2·8	12·3	5·4
04	6 31·0	6 32·1	6 13·2	0·4	0·2	6·4	2·8	12·4	5·5
05	6 31·3	6 32·3	6 13·4	0·5	0·2	6·5	2·9	12·5	5·5
06	6 31·5	6 32·6	6 13·7	0·6	0·3	6·6	2·9	12·6	5·6
07	6 31·8	6 32·8	6 13·9	0·7	0·3	6·7	3·0	12·7	5·6
08	6 32·0	6 33·1	6 14·1	0·8	0·4	6·8	3·0	12·8	5·7
09	6 32·3	6 33·3	6 14·4	0·9	0·4	6·9	3·0	12·9	5·7
10	6 32·5	6 33·6	6 14·6	1·0	0·4	7·0	3·1	13·0	5·7
11	6 32·8	6 33·8	6 14·9	1·1	0·5	7·1	3·1	13·1	5·8
12	6 33·0	6 34·1	6 15·1	1·2	0·5	7·2	3·2	13·2	5·8
13	6 33·3	6 34·3	6 15·3	1·3	0·6	7·3	3·2	13·3	5·9
14	6 33·5	6 34·6	6 15·6	1·4	0·6	7·4	3·3	13·4	5·9
15	6 33·8	6 34·8	6 15·8	1·5	0·7	7·5	3·3	13·5	6·0
16	6 34·0	6 35·1	6 16·1	1·6	0·7	7·6	3·4	13·6	6·0
17	6 34·3	6 35·3	6 16·3	1·7	0·8	7·7	3·4	13·7	6·1
18	6 34·5	6 35·6	6 16·5	1·8	0·8	7·8	3·4	13·8	6·1
19	6 34·8	6 35·8	6 16·8	1·9	0·8	7·9	3·5	13·9	6·1
20	6 35·0	6 36·1	6 17·0	2·0	0·9	8·0	3·5	14·0	6·2
21	6 35·3	6 36·3	6 17·2	2·1	0·9	8·1	3·6	14·1	6·2
22	6 35·5	6 36·6	6 17·5	2·2	1·0	8·2	3·6	14·2	6·3
23	6 35·8	6 36·8	6 17·7	2·3	1·0	8·3	3·7	14·3	6·3
24	6 36·0	6 37·1	6 18·0	2·4	1·1	8·4	3·7	14·4	6·4
25	6 36·3	6 37·3	6 18·2	2·5	1·1	8·5	3·8	14·5	6·4
26	6 36·5	6 37·6	6 18·4	2·6	1·1	8·6	3·8	14·6	6·4
27	6 36·8	6 37·8	6 18·7	2·7	1·2	8·7	3·8	14·7	6·5
28	6 37·0	6 38·1	6 18·9	2·8	1·2	8·8	3·9	14·8	6·5
29	6 37·3	6 38·3	6 19·2	2·9	1·3	8·9	3·9	14·9	6·6
30	6 37·5	6 38·6	6 19·4	3·0	1·3	9·0	4·0	15·0	6·6
31	6 37·8	6 38·8	6 19·6	3·1	1·4	9·1	4·0	15·1	6·7
32	6 38·0	6 39·1	6 19·9	3·2	1·4	9·2	4·1	15·2	6·7
33	6 38·3	6 39·3	6 20·1	3·3	1·5	9·3	4·1	15·3	6·8
34	6 38·5	6 39·6	6 20·3	3·4	1·5	9·4	4·2	15·4	6·8
35	6 38·8	6 39·8	6 20·6	3·5	1·5	9·5	4·2	15·5	6·8
36	6 39·0	6 40·1	6 20·8	3·6	1·6	9·6	4·2	15·6	6·9
37	6 39·3	6 40·3	6 21·1	3·7	1·6	9·7	4·3	15·7	6·9
38	6 39·5	6 40·6	6 21·3	3·8	1·7	9·8	4·3	15·8	7·0
39	6 39·8	6 40·8	6 21·5	3·9	1·7	9·9	4·4	15·9	7·0
40	6 40·0	6 41·1	6 21·8	4·0	1·8	10·0	4·4	16·0	7·1
41	6 40·3	6 41·3	6 22·0	4·1	1·8	10·1	4·5	16·1	7·1
42	6 40·5	6 41·6	6 22·3	4·2	1·9	10·2	4·5	16·2	7·2
43	6 40·8	6 41·8	6 22·5	4·3	1·9	10·3	4·5	16·3	7·2
44	6 41·0	6 42·1	6 22·7	4·4	1·9	10·4	4·6	16·4	7·2
45	6 41·3	6 42·3	6 23·0	4·5	2·0	10·5	4·6	16·5	7·3
46	6 41·5	6 42·6	6 23·2	4·6	2·0	10·6	4·7	16·6	7·3
47	6 41·8	6 42·8	6 23·4	4·7	2·1	10·7	4·7	16·7	7·4
48	6 42·0	6 43·1	6 23·7	4·8	2·1	10·8	4·8	16·8	7·4
49	6 42·3	6 43·4	6 23·9	4·9	2·2	10·9	4·8	16·9	7·5
50	6 42·5	6 43·6	6 24·2	5·0	2·2	11·0	4·9	17·0	7·5
51	6 42·8	6 43·9	6 24·4	5·1	2·3	11·1	4·9	17·1	7·6
52	6 43·0	6 44·1	6 24·6	5·2	2·3	11·2	4·9	17·2	7·6
53	6 43·3	6 44·4	6 24·9	5·3	2·3	11·3	5·0	17·3	7·6
54	6 43·5	6 44·6	6 25·1	5·4	2·4	11·4	5·0	17·4	7·7
55	6 43·8	6 44·9	6 25·4	5·5	2·4	11·5	5·1	17·5	7·7
56	6 44·0	6 45·1	6 25·6	5·6	2·5	11·6	5·1	17·6	7·8
57	6 44·3	6 45·4	6 25·8	5·7	2·5	11·7	5·2	17·7	7·8
58	6 44·5	6 45·6	6 26·1	5·8	2·6	11·8	5·2	17·8	7·9
59	6 44·8	6 45·9	6 26·3	5·9	2·6	11·9	5·3	17·9	7·9
60	6 45·0	6 46·1	6 26·6	6·0	2·7	12·0	5·3	18·0	8·0

27ᵐ	SUN PLANETS	ARIES	MOON	v or d	Corrⁿ	v or d	Corrⁿ	v or d	Corrⁿ
s	° ′	° ′	° ′	′	′	′	′	′	′
00	6 45·0	6 46·1	6 26·6	0·0	0·0	6·0	2·8	12·0	5·5
01	6 45·3	6 46·4	6 26·8	0·1	0·0	6·1	2·8	12·1	5·5
02	6 45·5	6 46·6	6 27·0	0·2	0·1	6·2	2·8	12·2	5·6
03	6 45·8	6 46·9	6 27·3	0·3	0·1	6·3	2·9	12·3	5·6
04	6 46·0	6 47·1	6 27·5	0·4	0·2	6·4	2·9	12·4	5·7
05	6 46·3	6 47·4	6 27·7	0·5	0·2	6·5	3·0	12·5	5·7
06	6 46·5	6 47·6	6 28·0	0·6	0·3	6·6	3·0	12·6	5·8
07	6 46·8	6 47·9	6 28·2	0·7	0·3	6·7	3·1	12·7	5·8
08	6 47·0	6 48·1	6 28·5	0·8	0·4	6·8	3·1	12·8	5·9
09	6 47·3	6 48·4	6 28·7	0·9	0·4	6·9	3·2	12·9	5·9
10	6 47·5	6 48·6	6 28·9	1·0	0·5	7·0	3·2	13·0	6·0
11	6 47·8	6 48·9	6 29·2	1·1	0·5	7·1	3·3	13·1	6·0
12	6 48·0	6 49·1	6 29·4	1·2	0·6	7·2	3·3	13·2	6·1
13	6 48·3	6 49·4	6 29·7	1·3	0·6	7·3	3·3	13·3	6·1
14	6 48·5	6 49·6	6 29·9	1·4	0·6	7·4	3·4	13·4	6·1
15	6 48·8	6 49·9	6 30·1	1·5	0·7	7·5	3·4	13·5	6·2
16	6 49·0	6 50·1	6 30·4	1·6	0·7	7·6	3·5	13·6	6·2
17	6 49·3	6 50·4	6 30·6	1·7	0·8	7·7	3·5	13·7	6·3
18	6 49·5	6 50·6	6 30·8	1·8	0·8	7·8	3·6	13·8	6·3
19	6 49·8	6 50·9	6 31·1	1·9	0·9	7·9	3·6	13·9	6·4
20	6 50·0	6 51·1	6 31·3	2·0	0·9	8·0	3·7	14·0	6·4
21	6 50·3	6 51·4	6 31·6	2·1	1·0	8·1	3·7	14·1	6·5
22	6 50·5	6 51·6	6 31·8	2·2	1·0	8·2	3·8	14·2	6·5
23	6 50·8	6 51·9	6 32·0	2·3	1·1	8·3	3·8	14·3	6·6
24	6 51·0	6 52·1	6 32·3	2·4	1·1	8·4	3·9	14·4	6·6
25	6 51·3	6 52·4	6 32·5	2·5	1·1	8·5	3·9	14·5	6·6
26	6 51·5	6 52·6	6 32·8	2·6	1·2	8·6	3·9	14·6	6·7
27	6 51·8	6 52·9	6 33·0	2·7	1·2	8·7	4·0	14·7	6·7
28	6 52·0	6 53·1	6 33·2	2·8	1·3	8·8	4·0	14·8	6·8
29	6 52·3	6 53·4	6 33·5	2·9	1·3	8·9	4·1	14·9	6·8
30	6 52·5	6 53·6	6 33·7	3·0	1·4	9·0	4·1	15·0	6·9
31	6 52·8	6 53·9	6 33·9	3·1	1·4	9·1	4·2	15·1	6·9
32	6 53·0	6 54·1	6 34·2	3·2	1·5	9·2	4·2	15·2	7·0
33	6 53·3	6 54·4	6 34·4	3·3	1·5	9·3	4·3	15·3	7·0
34	6 53·5	6 54·6	6 34·7	3·4	1·6	9·4	4·3	15·4	7·1
35	6 53·8	6 54·9	6 34·9	3·5	1·6	9·5	4·4	15·5	7·1
36	6 54·0	6 55·1	6 35·1	3·6	1·7	9·6	4·4	15·6	7·2
37	6 54·3	6 55·4	6 35·4	3·7	1·7	9·7	4·4	15·7	7·2
38	6 54·5	6 55·6	6 35·6	3·8	1·7	9·8	4·5	15·8	7·2
39	6 54·8	6 55·9	6 35·9	3·9	1·8	9·9	4·5	15·9	7·3
40	6 55·0	6 56·1	6 36·1	4·0	1·8	10·0	4·6	16·0	7·3
41	6 55·3	6 56·4	6 36·3	4·1	1·9	10·1	4·6	16·1	7·4
42	6 55·5	6 56·6	6 36·6	4·2	1·9	10·2	4·7	16·2	7·4
43	6 55·8	6 56·9	6 36·8	4·3	2·0	10·3	4·7	16·3	7·5
44	6 56·0	6 57·1	6 37·0	4·4	2·0	10·4	4·8	16·4	7·5
45	6 56·3	6 57·4	6 37·3	4·5	2·1	10·5	4·8	16·5	7·6
46	6 56·5	6 57·6	6 37·5	4·6	2·1	10·6	4·9	16·6	7·6
47	6 56·8	6 57·9	6 37·8	4·7	2·2	10·7	4·9	16·7	7·7
48	6 57·0	6 58·1	6 38·0	4·8	2·2	10·8	5·0	16·8	7·7
49	6 57·3	6 58·4	6 38·2	4·9	2·2	10·9	5·0	16·9	7·7
50	6 57·5	6 58·6	6 38·5	5·0	2·3	11·0	5·0	17·0	7·8
51	6 57·8	6 58·9	6 38·7	5·1	2·3	11·1	5·1	17·1	7·8
52	6 58·0	6 59·1	6 39·0	5·2	2·4	11·2	5·1	17·2	7·9
53	6 58·3	6 59·4	6 39·2	5·3	2·4	11·3	5·2	17·3	7·9
54	6 58·5	6 59·6	6 39·4	5·4	2·5	11·4	5·2	17·4	8·0
55	6 58·8	6 59·9	6 39·7	5·5	2·5	11·5	5·3	17·5	8·0
56	6 59·0	7 00·1	6 39·9	5·6	2·6	11·6	5·3	17·6	8·1
57	6 59·3	7 00·4	6 40·2	5·7	2·6	11·7	5·4	17·7	8·1
58	6 59·5	7 00·6	6 40·4	5·8	2·7	11·8	5·4	17·8	8·2
59	6 59·8	7 00·9	6 40·6	5·9	2·7	11·9	5·5	17·9	8·2
60	7 00·0	7 01·1	6 40·9	6·0	2·8	12·0	5·5	18·0	8·3

28^m	SUN PLANETS	ARIES	MOON	v or d	Corrⁿ	v or d	Corrⁿ	v or d	Corrⁿ
s	° ′	° ′	° ′	′	′	′	′	′	′
00	7 00·0	7 01·1	6 40·9	0·0	0·0	6·0	2·9	12·0	5·7
01	7 00·3	7 01·4	6 41·1	0·1	0·0	6·1	2·9	12·1	5·7
02	7 00·5	7 01·7	6 41·3	0·2	0·1	6·2	2·9	12·2	5·8
03	7 00·8	7 01·9	6 41·6	0·3	0·1	6·3	3·0	12·3	5·8
04	7 01·0	7 02·2	6 41·8	0·4	0·2	6·4	3·0	12·4	5·9
05	7 01·3	7 02·4	6 42·1	0·5	0·2	6·5	3·1	12·5	5·9
06	7 01·5	7 02·7	6 42·3	0·6	0·3	6·6	3·1	12·6	6·0
07	7 01·8	7 02·9	6 42·5	0·7	0·3	6·7	3·2	12·7	6·0
08	7 02·0	7 03·2	6 42·8	0·8	0·4	6·8	3·2	12·8	6·1
09	7 02·3	7 03·4	6 43·0	0·9	0·4	6·9	3·3	12·9	6·1
10	7 02·5	7 03·7	6 43·3	1·0	0·5	7·0	3·3	13·0	6·2
11	7 02·8	7 03·9	6 43·5	1·1	0·5	7·1	3·4	13·1	6·2
12	7 03·0	7 04·2	6 43·7	1·2	0·6	7·2	3·4	13·2	6·3
13	7 03·3	7 04·4	6 44·0	1·3	0·6	7·3	3·5	13·3	6·3
14	7 03·5	7 04·7	6 44·2	1·4	0·7	7·4	3·5	13·4	6·4
15	7 03·8	7 04·9	6 44·4	1·5	0·7	7·5	3·6	13·5	6·4
16	7 04·0	7 05·2	6 44·7	1·6	0·8	7·6	3·6	13·6	6·5
17	7 04·3	7 05·4	6 44·9	1·7	0·8	7·7	3·7	13·7	6·5
18	7 04·5	7 05·7	6 45·2	1·8	0·9	7·8	3·7	13·8	6·6
19	7 04·8	7 05·9	6 45·4	1·9	0·9	7·9	3·8	13·9	6·6
20	7 05·0	7 06·2	6 45·6	2·0	1·0	8·0	3·8	14·0	6·7
21	7 05·3	7 06·4	6 45·9	2·1	1·0	8·1	3·8	14·1	6·7
22	7 05·5	7 06·7	6 46·1	2·2	1·0	8·2	3·9	14·2	6·7
23	7 05·8	7 06·9	6 46·4	2·3	1·1	8·3	3·9	14·3	6·8
24	7 06·0	7 07·2	6 46·6	2·4	1·1	8·4	4·0	14·4	6·8
25	7 06·3	7 07·4	6 46·8	2·5	1·2	8·5	4·0	14·5	6·9
26	7 06·5	7 07·7	6 47·1	2·6	1·2	8·6	4·1	14·6	6·9
27	7 06·8	7 07·9	6 47·3	2·7	1·3	8·7	4·1	14·7	7·0
28	7 07·0	7 08·2	6 47·5	2·8	1·3	8·8	4·2	14·8	7·0
29	7 07·3	7 08·4	6 47·8	2·9	1·4	8·9	4·2	14·9	7·1
30	7 07·5	7 08·7	6 48·0	3·0	1·4	9·0	4·3	15·0	7·1
31	7 07·8	7 08·9	6 48·3	3·1	1·5	9·1	4·3	15·1	7·2
32	7 08·0	7 09·2	6 48·5	3·2	1·5	9·2	4·4	15·2	7·2
33	7 08·3	7 09·4	6 48·7	3·3	1·6	9·3	4·4	15·3	7·3
34	7 08·5	7 09·7	6 49·0	3·4	1·6	9·4	4·5	15·4	7·3
35	7 08·8	7 09·9	6 49·2	3·5	1·7	9·5	4·5	15·5	7·4
36	7 09·0	7 10·2	6 49·5	3·6	1·7	9·6	4·6	15·6	7·4
37	7 09·3	7 10·4	6 49·7	3·7	1·8	9·7	4·6	15·7	7·5
38	7 09·5	7 10·7	6 49·9	3·8	1·8	9·8	4·7	15·8	7·5
39	7 09·8	7 10·9	6 50·2	3·9	1·9	9·9	4·7	15·9	7·6
40	7 10·0	7 11·2	6 50·4	4·0	1·9	10·0	4·8	16·0	7·6
41	7 10·3	7 11·4	6 50·6	4·1	1·9	10·1	4·8	16·1	7·6
42	7 10·5	7 11·7	6 50·9	4·2	2·0	10·2	4·8	16·2	7·7
43	7 10·8	7 11·9	6 51·1	4·3	2·0	10·3	4·9	16·3	7·7
44	7 11·0	7 12·2	6 51·4	4·4	2·1	10·4	4·9	16·4	7·8
45	7 11·3	7 12·4	6 51·6	4·5	2·1	10·5	5·0	16·5	7·8
46	7 11·5	7 12·7	6 51·8	4·6	2·2	10·6	5·0	16·6	7·9
47	7 11·8	7 12·9	6 52·1	4·7	2·2	10·7	5·1	16·7	7·9
48	7 12·0	7 13·2	6 52·3	4·8	2·3	10·8	5·1	16·8	8·0
49	7 12·3	7 13·4	6 52·6	4·9	2·3	10·9	5·2	16·9	8·0
50	7 12·5	7 13·7	6 52·8	5·0	2·4	11·0	5·2	17·0	8·1
51	7 12·8	7 13·9	6 53·0	5·1	2·4	11·1	5·3	17·1	8·1
52	7 13·0	7 14·2	6 53·3	5·2	2·5	11·2	5·3	17·2	8·2
53	7 13·3	7 14·4	6 53·5	5·3	2·5	11·3	5·4	17·3	8·2
54	7 13·5	7 14·7	6 53·8	5·4	2·6	11·4	5·4	17·4	8·3
55	7 13·8	7 14·9	6 54·0	5·5	2·6	11·5	5·5	17·5	8·3
56	7 14·0	7 15·2	6 54·2	5·6	2·7	11·6	5·5	17·6	8·4
57	7 14·3	7 15·4	6 54·5	5·7	2·7	11·7	5·6	17·7	8·4
58	7 14·5	7 15·7	6 54·7	5·8	2·8	11·8	5·6	17·8	8·5
59	7 14·8	7 15·9	6 54·9	5·9	2·8	11·9	5·7	17·9	8·5
60	7 15·0	7 16·2	6 55·2	6·0	2·9	12·0	5·7	18·0	8·6

29^m	SUN PLANETS	ARIES	MOON	v or d	Corrⁿ	v or d	Corrⁿ	v or d	Corrⁿ
s	° ′	° ′	° ′	′	′	′	′	′	′
00	7 15·0	7 16·2	6 55·2	0·0	0·0	6·0	3·0	12·0	5·9
01	7 15·3	7 16·4	6 55·4	0·1	0·0	6·1	3·0	12·1	5·9
02	7 15·5	7 16·7	6 55·7	0·2	0·1	6·2	3·0	12·2	6·0
03	7 15·8	7 16·9	6 55·9	0·3	0·1	6·3	3·1	12·3	6·0
04	7 16·0	7 17·2	6 56·1	0·4	0·2	6·4	3·1	12·4	6·1
05	7 16·3	7 17·4	6 56·4	0·5	0·2	6·5	3·2	12·5	6·1
06	7 16·5	7 17·7	6 56·6	0·6	0·3	6·6	3·2	12·6	6·2
07	7 16·8	7 17·9	6 56·9	0·7	0·3	6·7	3·3	12·7	6·2
08	7 17·0	7 18·2	6 57·1	0·8	0·4	6·8	3·3	12·8	6·3
09	7 17·3	7 18·4	6 57·3	0·9	0·4	6·9	3·4	12·9	6·3
10	7 17·5	7 18·7	6 57·6	1·0	0·5	7·0	3·4	13·0	6·4
11	7 17·8	7 18·9	6 57·8	1·1	0·5	7·1	3·5	13·1	6·4
12	7 18·0	7 19·2	6 58·0	1·2	0·6	7·2	3·5	13·2	6·5
13	7 18·3	7 19·4	6 58·3	1·3	0·6	7·3	3·6	13·3	6·5
14	7 18·5	7 19·7	6 58·5	1·4	0·7	7·4	3·6	13·4	6·6
15	7 18·8	7 20·0	6 58·8	1·5	0·7	7·5	3·7	13·5	6·6
16	7 19·0	7 20·2	6 59·0	1·6	0·8	7·6	3·7	13·6	6·7
17	7 19·3	7 20·5	6 59·2	1·7	0·8	7·7	3·8	13·7	6·7
18	7 19·5	7 20·7	6 59·5	1·8	0·9	7·8	3·8	13·8	6·8
19	7 19·8	7 21·0	6 59·7	1·9	0·9	7·9	3·9	13·9	6·8
20	7 20·0	7 21·2	7 00·0	2·0	1·0	8·0	3·9	14·0	6·9
21	7 20·3	7 21·5	7 00·2	2·1	1·0	8·1	4·0	14·1	6·9
22	7 20·5	7 21·7	7 00·4	2·2	1·1	8·2	4·0	14·2	7·0
23	7 20·8	7 22·0	7 00·7	2·3	1·1	8·3	4·1	14·3	7·0
24	7 21·0	7 22·2	7 00·9	2·4	1·2	8·4	4·1	14·4	7·1
25	7 21·3	7 22·5	7 01·1	2·5	1·2	8·5	4·2	14·5	7·1
26	7 21·5	7 22·7	7 01·4	2·6	1·3	8·6	4·2	14·6	7·2
27	7 21·8	7 23·0	7 01·6	2·7	1·3	8·7	4·3	14·7	7·2
28	7 22·0	7 23·2	7 01·9	2·8	1·4	8·8	4·3	14·8	7·3
29	7 22·3	7 23·5	7 02·1	2·9	1·4	8·9	4·4	14·9	7·3
30	7 22·5	7 23·7	7 02·3	3·0	1·5	9·0	4·4	15·0	7·4
31	7 22·8	7 24·0	7 02·6	3·1	1·5	9·1	4·5	15·1	7·4
32	7 23·0	7 24·2	7 02·8	3·2	1·6	9·2	4·5	15·2	7·5
33	7 23·3	7 24·5	7 03·1	3·3	1·6	9·3	4·6	15·3	7·5
34	7 23·5	7 24·7	7 03·3	3·4	1·7	9·4	4·6	15·4	7·6
35	7 23·8	7 25·0	7 03·5	3·5	1·7	9·5	4·7	15·5	7·6
36	7 24·0	7 25·2	7 03·8	3·6	1·8	9·6	4·7	15·6	7·7
37	7 24·3	7 25·5	7 04·0	3·7	1·8	9·7	4·8	15·7	7·7
38	7 24·5	7 25·7	7 04·3	3·8	1·9	9·8	4·8	15·8	7·8
39	7 24·8	7 26·0	7 04·5	3·9	1·9	9·9	4·9	15·9	7·8
40	7 25·0	7 26·2	7 04·7	4·0	2·0	10·0	4·9	16·0	7·9
41	7 25·3	7 26·5	7 05·0	4·1	2·0	10·1	5·0	16·1	7·9
42	7 25·5	7 26·7	7 05·2	4·2	2·1	10·2	5·0	16·2	8·0
43	7 25·8	7 27·0	7 05·4	4·3	2·1	10·3	5·1	16·3	8·0
44	7 26·0	7 27·2	7 05·7	4·4	2·2	10·4	5·1	16·4	8·1
45	7 26·3	7 27·5	7 05·9	4·5	2·2	10·5	5·2	16·5	8·1
46	7 26·5	7 27·7	7 06·2	4·6	2·3	10·6	5·2	16·6	8·2
47	7 26·8	7 28·0	7 06·4	4·7	2·3	10·7	5·3	16·7	8·2
48	7 27·0	7 28·2	7 06·6	4·8	2·4	10·8	5·3	16·8	8·3
49	7 27·3	7 28·5	7 06·9	4·9	2·4	10·9	5·4	16·9	8·3
50	7 27·5	7 28·7	7 07·1	5·0	2·5	11·0	5·4	17·0	8·4
51	7 27·8	7 29·0	7 07·4	5·1	2·5	11·1	5·5	17·1	8·4
52	7 28·0	7 29·2	7 07·6	5·2	2·6	11·2	5·5	17·2	8·5
53	7 28·3	7 29·5	7 07·8	5·3	2·6	11·3	5·6	17·3	8·5
54	7 28·5	7 29·7	7 08·1	5·4	2·7	11·4	5·6	17·4	8·6
55	7 28·8	7 30·0	7 08·3	5·5	2·7	11·5	5·7	17·5	8·6
56	7 29·0	7 30·2	7 08·5	5·6	2·8	11·6	5·7	17·6	8·7
57	7 29·3	7 30·5	7 08·8	5·7	2·8	11·7	5·8	17·7	8·7
58	7 29·5	7 30·7	7 09·0	5·8	2·9	11·8	5·8	17·8	8·8
59	7 29·8	7 31·0	7 09·3	5·9	2·9	11·9	5·9	17·9	8·8
60	7 30·0	7 31·2	7 09·5	6·0	3·0	12·0	5·9	18·0	8·9

30m s	SUN PLANETS ° ′	ARIES ° ′	MOON ° ′	v or d ′	Corrn ′	v or d ′	Corrn ′	v or d ′	Corrn ′
00	7 30·0	7 31·2	7 09·5	0·0	0·0	6·0	3·1	12·0	6·1
01	7 30·3	7 31·5	7 09·7	0·1	0·1	6·1	3·1	12·1	6·2
02	7 30·5	7 31·7	7 10·0	0·2	0·1	6·2	3·2	12·2	6·2
03	7 30·8	7 32·0	7 10·2	0·3	0·2	6·3	3·2	12·3	6·3
04	7 31·0	7 32·2	7 10·5	0·4	0·2	6·4	3·3	12·4	6·3
05	7 31·3	7 32·5	7 10·7	0·5	0·3	6·5	3·3	12·5	6·4
06	7 31·5	7 32·7	7 10·9	0·6	0·3	6·6	3·4	12·6	6·4
07	7 31·8	7 33·0	7 11·2	0·7	0·4	6·7	3·4	12·7	6·5
08	7 32·0	7 33·2	7 11·4	0·8	0·4	6·8	3·5	12·8	6·5
09	7 32·3	7 33·5	7 11·6	0·9	0·5	6·9	3·5	12·9	6·6
10	7 32·5	7 33·7	7 11·9	1·0	0·5	7·0	3·6	13·0	6·6
11	7 32·8	7 34·0	7 12·1	1·1	0·6	7·1	3·6	13·1	6·7
12	7 33·0	7 34·2	7 12·4	1·2	0·6	7·2	3·7	13·2	6·7
13	7 33·3	7 34·5	7 12·6	1·3	0·7	7·3	3·7	13·3	6·8
14	7 33·5	7 34·7	7 12·8	1·4	0·7	7·4	3·8	13·4	6·8
15	7 33·8	7 35·0	7 13·1	1·5	0·8	7·5	3·8	13·5	6·9
16	7 34·0	7 35·2	7 13·3	1·6	0·8	7·6	3·9	13·6	6·9
17	7 34·3	7 35·5	7 13·6	1·7	0·9	7·7	3·9	13·7	7·0
18	7 34·5	7 35·7	7 13·8	1·8	0·9	7·8	4·0	13·8	7·0
19	7 34·8	7 36·0	7 14·0	1·9	1·0	7·9	4·0	13·9	7·1
20	7 35·0	7 36·2	7 14·3	2·0	1·0	8·0	4·1	14·0	7·1
21	7 35·3	7 36·5	7 14·5	2·1	1·1	8·1	4·1	14·1	7·2
22	7 35·5	7 36·7	7 14·7	2·2	1·1	8·2	4·2	14·2	7·2
23	7 35·8	7 37·0	7 15·0	2·3	1·2	8·3	4·2	14·3	7·3
24	7 36·0	7 37·2	7 15·2	2·4	1·2	8·4	4·3	14·4	7·3
25	7 36·3	7 37·5	7 15·5	2·5	1·3	8·5	4·3	14·5	7·4
26	7 36·5	7 37·7	7 15·7	2·6	1·3	8·6	4·4	14·6	7·4
27	7 36·8	7 38·0	7 15·9	2·7	1·4	8·7	4·4	14·7	7·5
28	7 37·0	7 38·3	7 16·2	2·8	1·4	8·8	4·5	14·8	7·5
29	7 37·3	7 38·5	7 16·4	2·9	1·5	8·9	4·5	14·9	7·6
30	7 37·5	7 38·8	7 16·7	3·0	1·5	9·0	4·6	15·0	7·6
31	7 37·8	7 39·0	7 16·9	3·1	1·6	9·1	4·6	15·1	7·7
32	7 38·0	7 39·3	7 17·1	3·2	1·6	9·2	4·7	15·2	7·7
33	7 38·3	7 39·5	7 17·4	3·3	1·7	9·3	4·7	15·3	7·8
34	7 38·5	7 39·8	7 17·6	3·4	1·7	9·4	4·8	15·4	7·8
35	7 38·8	7 40·0	7 17·9	3·5	1·8	9·5	4·8	15·5	7·9
36	7 39·0	7 40·3	7 18·1	3·6	1·8	9·6	4·9	15·6	7·9
37	7 39·3	7 40·5	7 18·3	3·7	1·9	9·7	4·9	15·7	8·0
38	7 39·5	7 40·8	7 18·6	3·8	1·9	9·8	5·0	15·8	8·0
39	7 39·8	7 41·0	7 18·8	3·9	2·0	9·9	5·0	15·9	8·1
40	7 40·0	7 41·3	7 19·0	4·0	2·0	10·0	5·1	16·0	8·1
41	7 40·3	7 41·5	7 19·3	4·1	2·1	10·1	5·1	16·1	8·2
42	7 40·5	7 41·8	7 19·5	4·2	2·1	10·2	5·2	16·2	8·2
43	7 40·8	7 42·0	7 19·8	4·3	2·2	10·3	5·2	16·3	8·3
44	7 41·0	7 42·3	7 20·0	4·4	2·2	10·4	5·3	16·4	8·3
45	7 41·3	7 42·5	7 20·2	4·5	2·3	10·5	5·3	16·5	8·4
46	7 41·5	7 42·8	7 20·5	4·6	2·3	10·6	5·4	16·6	8·4
47	7 41·8	7 43·0	7 20·7	4·7	2·4	10·7	5·4	16·7	8·5
48	7 42·0	7 43·3	7 21·0	4·8	2·4	10·8	5·5	16·8	8·5
49	7 42·3	7 43·5	7 21·2	4·9	2·5	10·9	5·5	16·9	8·6
50	7 42·5	7 43·8	7 21·4	5·0	2·5	11·0	5·6	17·0	8·6
51	7 42·8	7 44·0	7 21·7	5·1	2·6	11·1	5·6	17·1	8·7
52	7 43·0	7 44·3	7 21·9	5·2	2·6	11·2	5·7	17·2	8·7
53	7 43·3	7 44·5	7 22·1	5·3	2·7	11·3	5·7	17·3	8·8
54	7 43·5	7 44·8	7 22·4	5·4	2·7	11·4	5·8	17·4	8·8
55	7 43·8	7 45·0	7 22·6	5·5	2·8	11·5	5·8	17·5	8·9
56	7 44·0	7 45·3	7 22·9	5·6	2·8	11·6	5·9	17·6	8·9
57	7 44·3	7 45·5	7 23·1	5·7	2·9	11·7	5·9	17·7	9·0
58	7 44·5	7 45·8	7 23·3	5·8	2·9	11·8	6·0	17·8	9·0
59	7 44·8	7 46·0	7 23·6	5·9	3·0	11·9	6·0	17·9	9·1
60	7 45·0	7 46·3	7 23·8	6·0	3·1	12·0	6·1	18·0	9·2

31m s	SUN PLANETS ° ′	ARIES ° ′	MOON ° ′	v or d ′	Corrn ′	v or d ′	Corrn ′	v or d ′	Corrn ′
00	7 45·0	7 46·3	7 23·8	0·0	0·0	6·0	3·2	12·0	6·3
01	7 45·3	7 46·5	7 24·1	0·1	0·1	6·1	3·2	12·1	6·4
02	7 45·5	7 46·8	7 24·3	0·2	0·1	6·2	3·3	12·2	6·4
03	7 45·8	7 47·0	7 24·5	0·3	0·2	6·3	3·3	12·3	6·5
04	7 46·0	7 47·3	7 24·8	0·4	0·2	6·4	3·4	12·4	6·5
05	7 46·3	7 47·5	7 25·0	0·5	0·3	6·5	3·4	12·5	6·6
06	7 46·5	7 47·8	7 25·2	0·6	0·3	6·6	3·5	12·6	6·6
07	7 46·8	7 48·0	7 25·5	0·7	0·4	6·7	3·5	12·7	6·7
08	7 47·0	7 48·3	7 25·7	0·8	0·4	6·8	3·6	12·8	6·7
09	7 47·3	7 48·5	7 26·0	0·9	0·5	6·9	3·6	12·9	6·8
10	7 47·5	7 48·8	7 26·2	1·0	0·5	7·0	3·7	13·0	6·8
11	7 47·8	7 49·0	7 26·4	1·1	0·6	7·1	3·7	13·1	6·9
12	7 48·0	7 49·3	7 26·7	1·2	0·6	7·2	3·8	13·2	6·9
13	7 48·3	7 49·5	7 26·9	1·3	0·7	7·3	3·8	13·3	7·0
14	7 48·5	7 49·8	7 27·2	1·4	0·7	7·4	3·9	13·4	7·0
15	7 48·8	7 50·0	7 27·4	1·5	0·8	7·5	3·9	13·5	7·1
16	7 49·0	7 50·3	7 27·6	1·6	0·8	7·6	4·0	13·6	7·1
17	7 49·3	7 50·5	7 27·9	1·7	0·9	7·7	4·0	13·7	7·2
18	7 49·5	7 50·8	7 28·1	1·8	0·9	7·8	4·1	13·8	7·2
19	7 49·8	7 51·0	7 28·4	1·9	1·0	7·9	4·1	13·9	7·3
20	7 50·0	7 51·3	7 28·6	2·0	1·1	8·0	4·2	14·0	7·4
21	7 50·3	7 51·5	7 28·8	2·1	1·1	8·1	4·3	14·1	7·4
22	7 50·5	7 51·8	7 29·1	2·2	1·2	8·2	4·3	14·2	7·5
23	7 50·8	7 52·0	7 29·3	2·3	1·2	8·3	4·4	14·3	7·5
24	7 51·0	7 52·3	7 29·5	2·4	1·3	8·4	4·4	14·4	7·6
25	7 51·3	7 52·5	7 29·8	2·5	1·3	8·5	4·5	14·5	7·6
26	7 51·5	7 52·8	7 30·0	2·6	1·4	8·6	4·5	14·6	7·7
27	7 51·8	7 53·0	7 30·3	2·7	1·4	8·7	4·6	14·7	7·7
28	7 52·0	7 53·3	7 30·5	2·8	1·5	8·8	4·6	14·8	7·8
29	7 52·3	7 53·5	7 30·7	2·9	1·5	8·9	4·7	14·9	7·8
30	7 52·5	7 53·8	7 31·0	3·0	1·6	9·0	4·7	15·0	7·9
31	7 52·8	7 54·0	7 31·2	3·1	1·6	9·1	4·8	15·1	7·9
32	7 53·0	7 54·3	7 31·5	3·2	1·7	9·2	4·8	15·2	8·0
33	7 53·3	7 54·5	7 31·7	3·3	1·7	9·3	4·9	15·3	8·0
34	7 53·5	7 54·8	7 31·9	3·4	1·8	9·4	4·9	15·4	8·1
35	7 53·8	7 55·0	7 32·2	3·5	1·8	9·5	5·0	15·5	8·1
36	7 54·0	7 55·3	7 32·4	3·6	1·9	9·6	5·0	15·6	8·2
37	7 54·3	7 55·5	7 32·6	3·7	1·9	9·7	5·1	15·7	8·2
38	7 54·5	7 55·8	7 32·9	3·8	2·0	9·8	5·1	15·8	8·3
39	7 54·8	7 56·0	7 33·1	3·9	2·0	9·9	5·2	15·9	8·3
40	7 55·0	7 56·3	7 33·4	4·0	2·1	10·0	5·3	16·0	8·4
41	7 55·3	7 56·6	7 33·6	4·1	2·2	10·1	5·3	16·1	8·5
42	7 55·5	7 56·8	7 33·8	4·2	2·2	10·2	5·4	16·2	8·5
43	7 55·8	7 57·1	7 34·1	4·3	2·3	10·3	5·4	16·3	8·6
44	7 56·0	7 57·3	7 34·3	4·4	2·3	10·4	5·5	16·4	8·6
45	7 56·3	7 57·6	7 34·6	4·5	2·4	10·5	5·5	16·5	8·7
46	7 56·5	7 57·8	7 34·8	4·6	2·4	10·6	5·6	16·6	8·7
47	7 56·8	7 58·1	7 35·0	4·7	2·5	10·7	5·6	16·7	8·8
48	7 57·0	7 58·3	7 35·3	4·8	2·5	10·8	5·7	16·8	8·8
49	7 57·3	7 58·6	7 35·5	4·9	2·6	10·9	5·7	16·9	8·9
50	7 57·5	7 58·8	7 35·7	5·0	2·6	11·0	5·8	17·0	8·9
51	7 57·8	7 59·1	7 36·0	5·1	2·7	11·1	5·8	17·1	9·0
52	7 58·0	7 59·3	7 36·2	5·2	2·7	11·2	5·9	17·2	9·0
53	7 58·3	7 59·6	7 36·5	5·3	2·8	11·3	5·9	17·3	9·1
54	7 58·5	7 59·8	7 36·7	5·4	2·8	11·4	6·0	17·4	9·1
55	7 58·8	8 00·1	7 36·9	5·5	2·9	11·5	6·0	17·5	9·2
56	7 59·0	8 00·3	7 37·2	5·6	2·9	11·6	6·1	17·6	9·2
57	7 59·3	8 00·6	7 37·4	5·7	3·0	11·7	6·1	17·7	9·3
58	7 59·5	8 00·8	7 37·7	5·8	3·0	11·8	6·2	17·8	9·3
59	7 59·8	8 01·1	7 37·9	5·9	3·1	11·9	6·2	17·9	9·4
60	8 00·0	8 01·3	7 38·1	6·0	3·2	12·0	6·3	18·0	9·5

32m	SUN PLANETS	ARIES	MOON	v or d	Corrn	v or d	Corrn	v or d	Corrn
s	° ′	° ′	° ′	′	′	′	′	′	′
00	8 00·0	8 01·3	7 38·1	0·0	0·0	6·0	3·3	12·0	6·5
01	8 00·3	8 01·6	7 38·4	0·1	0·1	6·1	3·3	12·1	6·6
02	8 00·5	8 01·8	7 38·6	0·2	0·1	6·2	3·4	12·2	6·6
03	8 00·8	8 02·1	7 38·8	0·3	0·2	6·3	3·4	12·3	6·7
04	8 01·0	8 02·3	7 39·1	0·4	0·2	6·4	3·5	12·4	6·7
05	8 01·3	8 02·6	7 39·3	0·5	0·3	6·5	3·5	12·5	6·8
06	8 01·5	8 02·8	7 39·6	0·6	0·3	6·6	3·6	12·6	6·8
07	8 01·8	8 03·1	7 39·8	0·7	0·4	6·7	3·6	12·7	6·9
08	8 02·0	8 03·3	7 40·0	0·8	0·4	6·8	3·7	12·8	6·9
09	8 02·3	8 03·6	7 40·3	0·9	0·5	6·9	3·7	12·9	7·0
10	8 02·5	8 03·8	7 40·5	1·0	0·5	7·0	3·8	13·0	7·0
11	8 02·8	8 04·1	7 40·8	1·1	0·6	7·1	3·8	13·1	7·1
12	8 03·0	8 04·3	7 41·0	1·2	0·7	7·2	3·9	13·2	7·2
13	8 03·3	8 04·6	7 41·2	1·3	0·7	7·3	4·0	13·3	7·2
14	8 03·5	8 04·8	7 41·5	1·4	0·8	7·4	4·0	13·4	7·3
15	8 03·8	8 05·1	7 41·7	1·5	0·8	7·5	4·1	13·5	7·3
16	8 04·0	8 05·3	7 42·0	1·6	0·9	7·6	4·1	13·6	7·4
17	8 04·3	8 05·6	7 42·2	1·7	0·9	7·7	4·2	13·7	7·4
18	8 04·5	8 05·8	7 42·4	1·8	1·0	7·8	4·2	13·8	7·5
19	8 04·8	8 06·1	7 42·7	1·9	1·0	7·9	4·3	13·9	7·5
20	8 05·0	8 06·3	7 42·9	2·0	1·1	8·0	4·3	14·0	7·6
21	8 05·3	8 06·6	7 43·1	2·1	1·1	8·1	4·4	14·1	7·6
22	8 05·5	8 06·8	7 43·4	2·2	1·2	8·2	4·4	14·2	7·7
23	8 05·8	8 07·1	7 43·6	2·3	1·2	8·3	4·5	14·3	7·7
24	8 06·0	8 07·3	7 43·9	2·4	1·3	8·4	4·6	14·4	7·8
25	8 06·3	8 07·6	7 44·1	2·5	1·4	8·5	4·6	14·5	7·9
26	8 06·5	8 07·8	7 44·3	2·6	1·4	8·6	4·7	14·6	7·9
27	8 06·8	8 08·1	7 44·6	2·7	1·5	8·7	4·7	14·7	8·0
28	8 07·0	8 08·3	7 44·8	2·8	1·5	8·8	4·8	14·8	8·0
29	8 07·3	8 08·6	7 45·1	2·9	1·6	8·9	4·8	14·9	8·1
30	8 07·5	8 08·8	7 45·3	3·0	1·6	9·0	4·9	15·0	8·1
31	8 07·8	8 09·1	7 45·5	3·1	1·7	9·1	4·9	15·1	8·2
32	8 08·0	8 09·3	7 45·8	3·2	1·7	9·2	5·0	15·2	8·2
33	8 08·3	8 09·6	7 46·0	3·3	1·8	9·3	5·0	15·3	8·3
34	8 08·5	8 09·8	7 46·2	3·4	1·8	9·4	5·1	15·4	8·3
35	8 08·8	8 10·1	7 46·5	3·5	1·9	9·5	5·1	15·5	8·4
36	8 09·0	8 10·3	7 46·7	3·6	2·0	9·6	5·2	15·6	8·5
37	8 09·3	8 10·6	7 47·0	3·7	2·0	9·7	5·3	15·7	8·5
38	8 09·5	8 10·8	7 47·2	3·8	2·1	9·8	5·3	15·8	8·6
39	8 09·8	8 11·1	7 47·4	3·9	2·1	9·9	5·4	15·9	8·6
40	8 10·0	8 11·3	7 47·7	4·0	2·2	10·0	5·4	16·0	8·7
41	8 10·3	8 11·6	7 47·9	4·1	2·2	10·1	5·5	16·1	8·7
42	8 10·5	8 11·8	7 48·2	4·2	2·3	10·2	5·5	16·2	8·8
43	8 10·8	8 12·1	7 48·4	4·3	2·3	10·3	5·6	16·3	8·8
44	8 11·0	8 12·3	7 48·6	4·4	2·4	10·4	5·6	16·4	8·9
45	8 11·3	8 12·6	7 48·9	4·5	2·4	10·5	5·7	16·5	8·9
46	8 11·5	8 12·8	7 49·1	4·6	2·5	10·6	5·7	16·6	9·0
47	8 11·8	8 13·1	7 49·3	4·7	2·5	10·7	5·8	16·7	9·0
48	8 12·0	8 13·3	7 49·6	4·8	2·6	10·8	5·9	16·8	9·1
49	8 12·3	8 13·6	7 49·8	4·9	2·7	10·9	5·9	16·9	9·2
50	8 12·5	8 13·8	7 50·1	5·0	2·7	11·0	6·0	17·0	9·2
51	8 12·8	8 14·1	7 50·3	5·1	2·8	11·1	6·0	17·1	9·3
52	8 13·0	8 14·3	7 50·5	5·2	2·8	11·2	6·1	17·2	9·3
53	8 13·3	8 14·6	7 50·8	5·3	2·9	11·3	6·1	17·3	9·4
54	8 13·5	8 14·9	7 51·0	5·4	2·9	11·4	6·2	17·4	9·4
55	8 13·8	8 15·1	7 51·3	5·5	3·0	11·5	6·2	17·5	9·5
56	8 14·0	8 15·4	7 51·5	5·6	3·0	11·6	6·3	17·6	9·5
57	8 14·3	8 15·6	7 51·7	5·7	3·1	11·7	6·3	17·7	9·6
58	8 14·5	8 15·9	7 52·0	5·8	3·1	11·8	6·4	17·8	9·6
59	8 14·8	8 16·1	7 52·2	5·9	3·2	11·9	6·4	17·9	9·7
60	8 15·0	8 16·4	7 52·5	6·0	3·3	12·0	6·5	18·0	9·8

33m	SUN PLANETS	ARIES	MOON	v or d	Corrn	v or d	Corrn	v or d	Corrn
s	° ′	° ′	° ′	′	′	′	′	′	′
00	8 15·0	8 16·4	7 52·5	0·0	0·0	6·0	3·4	12·0	6·7
01	8 15·3	8 16·6	7 52·7	0·1	0·1	6·1	3·4	12·1	6·8
02	8 15·5	8 16·9	7 52·9	0·2	0·1	6·2	3·5	12·2	6·8
03	8 15·8	8 17·1	7 53·2	0·3	0·2	6·3	3·5	12·3	6·9
04	8 16·0	8 17·4	7 53·4	0·4	0·2	6·4	3·6	12·4	6·9
05	8 16·3	8 17·6	7 53·6	0·5	0·3	6·5	3·6	12·5	7·0
06	8 16·5	8 17·9	7 53·9	0·6	0·3	6·6	3·7	12·6	7·0
07	8 16·8	8 18·1	7 54·1	0·7	0·4	6·7	3·7	12·7	7·1
08	8 17·0	8 18·4	7 54·4	0·8	0·4	6·8	3·8	12·8	7·1
09	8 17·3	8 18·6	7 54·6	0·9	0·5	6·9	3·9	12·9	7·2
10	8 17·5	8 18·9	7 54·8	1·0	0·6	7·0	3·9	13·0	7·3
11	8 17·8	8 19·1	7 55·1	1·1	0·6	7·1	4·0	13·1	7·3
12	8 18·0	8 19·4	7 55·3	1·2	0·7	7·2	4·0	13·2	7·4
13	8 18·3	8 19·6	7 55·6	1·3	0·7	7·3	4·1	13·3	7·4
14	8 18·5	8 19·9	7 55·8	1·4	0·8	7·4	4·1	13·4	7·5
15	8 18·8	8 20·1	7 56·0	1·5	0·8	7·5	4·2	13·5	7·5
16	8 19·0	8 20·4	7 56·3	1·6	0·9	7·6	4·2	13·6	7·6
17	8 19·3	8 20·6	7 56·5	1·7	0·9	7·7	4·3	13·7	7·6
18	8 19·5	8 20·9	7 56·7	1·8	1·0	7·8	4·4	13·8	7·7
19	8 19·8	8 21·1	7 57·0	1·9	1·1	7·9	4·4	13·9	7·8
20	8 20·0	8 21·4	7 57·2	2·0	1·1	8·0	4·5	14·0	7·8
21	8 20·3	8 21·6	7 57·5	2·1	1·2	8·1	4·5	14·1	7·9
22	8 20·5	8 21·9	7 57·7	2·2	1·2	8·2	4·6	14·2	7·9
23	8 20·8	8 22·1	7 57·9	2·3	1·3	8·3	4·6	14·3	8·0
24	8 21·0	8 22·4	7 58·2	2·4	1·3	8·4	4·7	14·4	8·0
25	8 21·3	8 22·6	7 58·4	2·5	1·4	8·5	4·7	14·5	8·1
26	8 21·5	8 22·9	7 58·7	2·6	1·5	8·6	4·8	14·6	8·2
27	8 21·8	8 23·1	7 58·9	2·7	1·5	8·7	4·9	14·7	8·2
28	8 22·0	8 23·4	7 59·1	2·8	1·6	8·8	4·9	14·8	8·3
29	8 22·3	8 23·6	7 59·4	2·9	1·6	8·9	5·0	14·9	8·3
30	8 22·5	8 23·9	7 59·6	3·0	1·7	9·0	5·0	15·0	8·4
31	8 22·8	8 24·1	7 59·8	3·1	1·7	9·1	5·1	15·1	8·4
32	8 23·0	8 24·4	8 00·1	3·2	1·8	9·2	5·1	15·2	8·5
33	8 23·3	8 24·6	8 00·3	3·3	1·8	9·3	5·2	15·3	8·5
34	8 23·5	8 24·9	8 00·6	3·4	1·9	9·4	5·2	15·4	8·6
35	8 23·8	8 25·1	8 00·8	3·5	2·0	9·5	5·3	15·5	8·7
36	8 24·0	8 25·4	8 01·0	3·6	2·0	9·6	5·4	15·6	8·7
37	8 24·3	8 25·6	8 01·3	3·7	2·1	9·7	5·4	15·7	8·8
38	8 24·5	8 25·9	8 01·5	3·8	2·1	9·8	5·5	15·8	8·8
39	8 24·8	8 26·1	8 01·8	3·9	2·2	9·9	5·5	15·9	8·9
40	8 25·0	8 26·4	8 02·0	4·0	2·2	10·0	5·6	16·0	8·9
41	8 25·3	8 26·6	8 02·2	4·1	2·3	10·1	5·6	16·1	9·0
42	8 25·5	8 26·9	8 02·5	4·2	2·3	10·2	5·7	16·2	9·0
43	8 25·8	8 27·1	8 02·7	4·3	2·4	10·3	5·8	16·3	9·1
44	8 26·0	8 27·4	8 02·9	4·4	2·5	10·4	5·8	16·4	9·2
45	8 26·3	8 27·6	8 03·2	4·5	2·5	10·5	5·9	16·5	9·2
46	8 26·5	8 27·9	8 03·4	4·6	2·6	10·6	5·9	16·6	9·3
47	8 26·8	8 28·1	8 03·7	4·7	2·6	10·7	6·0	16·7	9·3
48	8 27·0	8 28·4	8 03·9	4·8	2·7	10·8	6·0	16·8	9·4
49	8 27·3	8 28·6	8 04·1	4·9	2·7	10·9	6·1	16·9	9·4
50	8 27·5	8 28·9	8 04·4	5·0	2·8	11·0	6·1	17·0	9·5
51	8 27·8	8 29·1	8 04·6	5·1	2·8	11·1	6·2	17·1	9·5
52	8 28·0	8 29·4	8 04·9	5·2	2·9	11·2	6·3	17·2	9·6
53	8 28·3	8 29·6	8 05·1	5·3	3·0	11·3	6·3	17·3	9·7
54	8 28·5	8 29·9	8 05·3	5·4	3·0	11·4	6·4	17·4	9·7
55	8 28·8	8 30·1	8 05·6	5·5	3·1	11·5	6·4	17·5	9·8
56	8 29·0	8 30·4	8 05·8	5·6	3·1	11·6	6·5	17·6	9·8
57	8 29·3	8 30·6	8 06·1	5·7	3·2	11·7	6·5	17·7	9·9
58	8 29·5	8 30·9	8 06·3	5·8	3·2	11·8	6·6	17·8	9·9
59	8 29·8	8 31·1	8 06·5	5·9	3·3	11·9	6·6	17·9	10·0
60	8 30·0	8 31·4	8 06·8	6·0	3·4	12·0	6·7	18·0	10·1

34ᵐ	SUN PLANETS	ARIES	MOON	v or d	Corrⁿ	v or d	Corrⁿ	v or d	Corrⁿ
s	° ′	° ′	° ′	′	′	′	′	′	′
00	8 30·0	8 31·4	8 06·8	0·0	0·0	6·0	3·5	12·0	6·9
01	8 30·3	8 31·6	8 07·0	0·1	0·1	6·1	3·5	12·1	7·0
02	8 30·5	8 31·9	8 07·2	0·2	0·1	6·2	3·6	12·2	7·0
03	8 30·8	8 32·1	8 07·5	0·3	0·2	6·3	3·6	12·3	7·1
04	8 31·0	8 32·4	8 07·7	0·4	0·2	6·4	3·7	12·4	7·1
05	8 31·3	8 32·6	8 08·0	0·5	0·3	6·5	3·7	12·5	7·2
06	8 31·5	8 32·9	8 08·2	0·6	0·3	6·6	3·8	12·6	7·2
07	8 31·8	8 33·2	8 08·4	0·7	0·4	6·7	3·9	12·7	7·3
08	8 32·0	8 33·4	8 08·7	0·8	0·5	6·8	3·9	12·8	7·4
09	8 32·3	8 33·7	8 08·9	0·9	0·5	6·9	4·0	12·9	7·4
10	8 32·5	8 33·9	8 09·2	1·0	0·6	7·0	4·0	13·0	7·5
11	8 32·8	8 34·2	8 09·4	1·1	0·6	7·1	4·1	13·1	7·5
12	8 33·0	8 34·4	8 09·6	1·2	0·7	7·2	4·1	13·2	7·6
13	8 33·3	8 34·7	8 09·9	1·3	0·7	7·3	4·2	13·3	7·6
14	8 33·5	8 34·9	8 10·1	1·4	0·8	7·4	4·3	13·4	7·7
15	8 33·8	8 35·2	8 10·3	1·5	0·9	7·5	4·3	13·5	7·8
16	8 34·0	8 35·4	8 10·6	1·6	0·9	7·6	4·4	13·6	7·8
17	8 34·3	8 35·7	8 10·8	1·7	1·0	7·7	4·4	13·7	7·9
18	8 34·5	8 35·9	8 11·1	1·8	1·0	7·8	4·5	13·8	7·9
19	8 34·8	8 36·2	8 11·3	1·9	1·1	7·9	4·5	13·9	8·0
20	8 35·0	8 36·4	8 11·5	2·0	1·2	8·0	4·6	14·0	8·1
21	8 35·3	8 36·7	8 11·8	2·1	1·2	8·1	4·7	14·1	8·1
22	8 35·5	8 36·9	8 12·0	2·2	1·3	8·2	4·7	14·2	8·2
23	8 35·8	8 37·2	8 12·3	2·3	1·3	8·3	4·8	14·3	8·2
24	8 36·0	8 37·4	8 12·5	2·4	1·4	8·4	4·8	14·4	8·3
25	8 36·3	8 37·7	8 12·7	2·5	1·4	8·5	4·9	14·5	8·3
26	8 36·5	8 37·9	8 13·0	2·6	1·5	8·6	4·9	14·6	8·4
27	8 36·8	8 38·2	8 13·2	2·7	1·6	8·7	5·0	14·7	8·5
28	8 37·0	8 38·4	8 13·4	2·8	1·6	8·8	5·1	14·8	8·5
29	8 37·3	8 38·7	8 13·7	2·9	1·7	8·9	5·1	14·9	8·6
30	8 37·5	8 38·9	8 13·9	3·0	1·7	9·0	5·2	15·0	8·6
31	8 37·8	8 39·2	8 14·2	3·1	1·8	9·1	5·2	15·1	8·7
32	8 38·0	8 39·4	8 14·4	3·2	1·8	9·2	5·3	15·2	8·7
33	8 38·3	8 39·7	8 14·6	3·3	1·9	9·3	5·3	15·3	8·8
34	8 38·5	8 39·9	8 14·9	3·4	2·0	9·4	5·4	15·4	8·9
35	8 38·8	8 40·2	8 15·1	3·5	2·0	9·5	5·5	15·5	8·9
36	8 39·0	8 40·4	8 15·4	3·6	2·1	9·6	5·5	15·6	9·0
37	8 39·3	8 40·7	8 15·6	3·7	2·1	9·7	5·6	15·7	9·0
38	8 39·5	8 40·9	8 15·8	3·8	2·2	9·8	5·6	15·8	9·1
39	8 39·8	8 41·2	8 16·1	3·9	2·2	9·9	5·7	15·9	9·1
40	8 40·0	8 41·4	8 16·3	4·0	2·3	10·0	5·8	16·0	9·2
41	8 40·3	8 41·7	8 16·5	4·1	2·4	10·1	5·8	16·1	9·3
42	8 40·5	8 41·9	8 16·8	4·2	2·4	10·2	5·9	16·2	9·3
43	8 40·8	8 42·2	8 17·0	4·3	2·5	10·3	5·9	16·3	9·4
44	8 41·0	8 42·4	8 17·3	4·4	2·5	10·4	6·0	16·4	9·4
45	8 41·3	8 42·7	8 17·5	4·5	2·6	10·5	6·0	16·5	9·5
46	8 41·5	8 42·9	8 17·7	4·6	2·6	10·6	6·1	16·6	9·5
47	8 41·8	8 43·2	8 18·0	4·7	2·7	10·7	6·2	16·7	9·6
48	8 42·0	8 43·4	8 18·2	4·8	2·8	10·8	6·2	16·8	9·7
49	8 42·3	8 43·7	8 18·5	4·9	2·8	10·9	6·3	16·9	9·7
50	8 42·5	8 43·9	8 18·7	5·0	2·9	11·0	6·3	17·0	9·8
51	8 42·8	8 44·2	8 18·9	5·1	2·9	11·1	6·4	17·1	9·8
52	8 43·0	8 44·4	8 19·2	5·2	3·0	11·2	6·4	17·2	9·9
53	8 43·3	8 44·7	8 19·4	5·3	3·0	11·3	6·5	17·3	9·9
54	8 43·5	8 44·9	8 19·7	5·4	3·1	11·4	6·6	17·4	10·0
55	8 43·8	8 45·2	8 19·9	5·5	3·2	11·5	6·6	17·5	10·1
56	8 44·0	8 45·4	8 20·1	5·6	3·2	11·6	6·7	17·6	10·1
57	8 44·3	8 45·7	8 20·4	5·7	3·3	11·7	6·7	17·7	10·2
58	8 44·5	8 45·9	8 20·6	5·8	3·3	11·8	6·8	17·8	10·2
59	8 44·8	8 46·2	8 20·8	5·9	3·4	11·9	6·8	17·9	10·3
60	8 45·0	8 46·4	8 21·1	6·0	3·5	12·0	6·9	18·0	10·4

35ᵐ	SUN PLANETS	ARIES	MOON	v or d	Corrⁿ	v or d	Corrⁿ	v or d	Corrⁿ
s	° ′	° ′	° ′	′	′	′	′	′	′
00	8 45·0	8 46·4	8 21·1	0·0	0·0	6·0	3·6	12·0	7·1
01	8 45·3	8 46·7	8 21·3	0·1	0·1	6·1	3·6	12·1	7·2
02	8 45·5	8 46·9	8 21·6	0·2	0·1	6·2	3·7	12·2	7·2
03	8 45·8	8 47·2	8 21·8	0·3	0·2	6·3	3·7	12·3	7·3
04	8 46·0	8 47·4	8 22·0	0·4	0·2	6·4	3·8	12·4	7·3
05	8 46·3	8 47·7	8 22·3	0·5	0·3	6·5	3·8	12·5	7·4
06	8 46·5	8 47·9	8 22·5	0·6	0·4	6·6	3·9	12·6	7·5
07	8 46·8	8 48·2	8 22·8	0·7	0·4	6·7	4·0	12·7	7·5
08	8 47·0	8 48·4	8 23·0	0·8	0·5	6·8	4·0	12·8	7·6
09	8 47·3	8 48·7	8 23·2	0·9	0·5	6·9	4·1	12·9	7·6
10	8 47·5	8 48·9	8 23·5	1·0	0·6	7·0	4·1	13·0	7·7
11	8 47·8	8 49·2	8 23·7	1·1	0·7	7·1	4·2	13·1	7·8
12	8 48·0	8 49·4	8 23·9	1·2	0·7	7·2	4·3	13·2	7·8
13	8 48·3	8 49·7	8 24·2	1·3	0·8	7·3	4·3	13·3	7·9
14	8 48·5	8 49·9	8 24·4	1·4	0·8	7·4	4·4	13·4	7·9
15	8 48·8	8 50·2	8 24·7	1·5	0·9	7·5	4·4	13·5	8·0
16	8 49·0	8 50·4	8 24·9	1·6	0·9	7·6	4·5	13·6	8·0
17	8 49·3	8 50·7	8 25·1	1·7	1·0	7·7	4·6	13·7	8·1
18	8 49·5	8 50·9	8 25·4	1·8	1·1	7·8	4·6	13·8	8·2
19	8 49·8	8 51·2	8 25·6	1·9	1·1	7·9	4·7	13·9	8·2
20	8 50·0	8 51·5	8 25·9	2·0	1·2	8·0	4·7	14·0	8·3
21	8 50·3	8 51·7	8 26·1	2·1	1·2	8·1	4·8	14·1	8·3
22	8 50·5	8 52·0	8 26·3	2·2	1·3	8·2	4·9	14·2	8·4
23	8 50·8	8 52·2	8 26·6	2·3	1·4	8·3	4·9	14·3	8·5
24	8 51·0	8 52·5	8 26·8	2·4	1·4	8·4	5·0	14·4	8·5
25	8 51·3	8 52·7	8 27·0	2·5	1·5	8·5	5·0	14·5	8·6
26	8 51·5	8 53·0	8 27·3	2·6	1·5	8·6	5·1	14·6	8·6
27	8 51·8	8 53·2	8 27·5	2·7	1·6	8·7	5·1	14·7	8·7
28	8 52·0	8 53·5	8 27·8	2·8	1·7	8·8	5·2	14·8	8·8
29	8 52·3	8 53·7	8 28·0	2·9	1·7	8·9	5·3	14·9	8·8
30	8 52·5	8 54·0	8 28·2	3·0	1·8	9·0	5·3	15·0	8·9
31	8 52·8	8 54·2	8 28·5	3·1	1·8	9·1	5·4	15·1	8·9
32	8 53·0	8 54·5	8 28·7	3·2	1·9	9·2	5·4	15·2	9·0
33	8 53·3	8 54·7	8 29·0	3·3	2·0	9·3	5·5	15·3	9·1
34	8 53·5	8 55·0	8 29·2	3·4	2·0	9·4	5·6	15·4	9·1
35	8 53·8	8 55·2	8 29·4	3·5	2·1	9·5	5·6	15·5	9·2
36	8 54·0	8 55·5	8 29·7	3·6	2·1	9·6	5·7	15·6	9·2
37	8 54·3	8 55·7	8 29·9	3·7	2·2	9·7	5·7	15·7	9·3
38	8 54·5	8 56·0	8 30·2	3·8	2·2	9·8	5·8	15·8	9·3
39	8 54·8	8 56·2	8 30·4	3·9	2·3	9·9	5·9	15·9	9·4
40	8 55·0	8 56·5	8 30·6	4·0	2·4	10·0	5·9	16·0	9·5
41	8 55·3	8 56·7	8 30·9	4·1	2·4	10·1	6·0	16·1	9·5
42	8 55·5	8 57·0	8 31·1	4·2	2·5	10·2	6·0	16·2	9·6
43	8 55·8	8 57·2	8 31·3	4·3	2·5	10·3	6·1	16·3	9·6
44	8 56·0	8 57·5	8 31·6	4·4	2·6	10·4	6·2	16·4	9·7
45	8 56·3	8 57·7	8 31·8	4·5	2·7	10·5	6·2	16·5	9·8
46	8 56·5	8 58·0	8 32·1	4·6	2·7	10·6	6·3	16·6	9·8
47	8 56·8	8 58·2	8 32·3	4·7	2·8	10·7	6·3	16·7	9·9
48	8 57·0	8 58·5	8 32·5	4·8	2·8	10·8	6·4	16·8	9·9
49	8 57·3	8 58·7	8 32·8	4·9	2·9	10·9	6·4	16·9	10·0
50	8 57·5	8 59·0	8 33·0	5·0	3·0	11·0	6·5	17·0	10·1
51	8 57·8	8 59·2	8 33·3	5·1	3·0	11·1	6·6	17·1	10·1
52	8 58·0	8 59·5	8 33·5	5·2	3·1	11·2	6·6	17·2	10·2
53	8 58·3	8 59·7	8 33·7	5·3	3·1	11·3	6·7	17·3	10·2
54	8 58·5	9 00·0	8 34·0	5·4	3·2	11·4	6·7	17·4	10·3
55	8 58·8	9 00·2	8 34·2	5·5	3·3	11·5	6·8	17·5	10·4
56	8 59·0	9 00·5	8 34·4	5·6	3·3	11·6	6·9	17·6	10·4
57	8 59·3	9 00·7	8 34·7	5·7	3·4	11·7	6·9	17·7	10·5
58	8 59·5	9 01·0	8 34·9	5·8	3·4	11·8	7·0	17·8	10·5
59	8 59·8	9 01·2	8 35·2	5·9	3·5	11·9	7·0	17·9	10·6
60	9 00·0	9 01·5	8 35·4	6·0	3·6	12·0	7·1	18·0	10·7

36^m	SUN PLANETS	ARIES	MOON	v or d	Corr^n	v or d	Corr^n	v or d	Corr^n
s	° ′	° ′	° ′	′	′	′	′	′	′
00	9 00·0	9 01·5	8 35·4	0·0	0·0	6·0	3·7	12·0	7·3
01	9 00·3	9 01·7	8 35·6	0·1	0·1	6·1	3·7	12·1	7·4
02	9 00·5	9 02·0	8 35·9	0·2	0·1	6·2	3·8	12·2	7·4
03	9 00·8	9 02·2	8 36·1	0·3	0·2	6·3	3·8	12·3	7·5
04	9 01·0	9 02·5	8 36·4	0·4	0·2	6·4	3·9	12·4	7·5
05	9 01·3	9 02·7	8 36·6	0·5	0·3	6·5	4·0	12·5	7·6
06	9 01·5	9 03·0	8 36·8	0·6	0·4	6·6	4·0	12·6	7·7
07	9 01·8	9 03·2	8 37·1	0·7	0·4	6·7	4·1	12·7	7·7
08	9 02·0	9 03·5	8 37·3	0·8	0·5	6·8	4·1	12·8	7·8
09	9 02·3	9 03·7	8 37·5	0·9	0·5	6·9	4·2	12·9	7·8
10	9 02·5	9 04·0	8 37·8	1·0	0·6	7·0	4·3	13·0	7·9
11	9 02·8	9 04·2	8 38·0	1·1	0·7	7·1	4·3	13·1	8·0
12	9 03·0	9 04·5	8 38·3	1·2	0·7	7·2	4·4	13·2	8·0
13	9 03·3	9 04·7	8 38·5	1·3	0·8	7·3	4·4	13·3	8·1
14	9 03·5	9 05·0	8 38·7	1·4	0·9	7·4	4·5	13·4	8·2
15	9 03·8	9 05·2	8 39·0	1·5	0·9	7·5	4·6	13·5	8·2
16	9 04·0	9 05·5	8 39·2	1·6	1·0	7·6	4·6	13·6	8·3
17	9 04·3	9 05·7	8 39·5	1·7	1·0	7·7	4·7	13·7	8·3
18	9 04·5	9 06·0	8 39·7	1·8	1·1	7·8	4·7	13·8	8·4
19	9 04·8	9 06·2	8 39·9	1·9	1·2	7·9	4·8	13·9	8·5
20	9 05·0	9 06·5	8 40·2	2·0	1·2	8·0	4·9	14·0	8·5
21	9 05·3	9 06·7	8 40·4	2·1	1·3	8·1	4·9	14·1	8·6
22	9 05·5	9 07·0	8 40·6	2·2	1·3	8·2	5·0	14·2	8·6
23	9 05·8	9 07·2	8 40·9	2·3	1·4	8·3	5·0	14·3	8·7
24	9 06·0	9 07·5	8 41·1	2·4	1·5	8·4	5·1	14·4	8·8
25	9 06·3	9 07·7	8 41·4	2·5	1·5	8·5	5·2	14·5	8·8
26	9 06·5	9 08·0	8 41·6	2·6	1·6	8·6	5·2	14·6	8·9
27	9 06·8	9 08·2	8 41·8	2·7	1·6	8·7	5·3	14·7	8·9
28	9 07·0	9 08·5	8 42·1	2·8	1·7	8·8	5·4	14·8	9·0
29	9 07·3	9 08·7	8 42·3	2·9	1·8	8·9	5·4	14·9	9·1
30	9 07·5	9 09·0	8 42·6	3·0	1·8	9·0	5·5	15·0	9·1
31	9 07·8	9 09·2	8 42·8	3·1	1·9	9·1	5·5	15·1	9·2
32	9 08·0	9 09·5	8 43·0	3·2	1·9	9·2	5·6	15·2	9·2
33	9 08·3	9 09·8	8 43·3	3·3	2·0	9·3	5·7	15·3	9·3
34	9 08·5	9 10·0	8 43·5	3·4	2·1	9·4	5·7	15·4	9·4
35	9 08·8	9 10·3	8 43·8	3·5	2·1	9·5	5·8	15·5	9·4
36	9 09·0	9 10·5	8 44·0	3·6	2·2	9·6	5·8	15·6	9·5
37	9 09·3	9 10·8	8 44·2	3·7	2·3	9·7	5·9	15·7	9·6
38	9 09·5	9 11·0	8 44·5	3·8	2·3	9·8	6·0	15·8	9·6
39	9 09·8	9 11·3	8 44·7	3·9	2·4	9·9	6·0	15·9	9·7
40	9 10·0	9 11·5	8 44·9	4·0	2·4	10·0	6·1	16·0	9·7
41	9 10·3	9 11·8	8 45·2	4·1	2·5	10·1	6·1	16·1	9·8
42	9 10·5	9 12·0	8 45·4	4·2	2·6	10·2	6·2	16·2	9·9
43	9 10·8	9 12·3	8 45·7	4·3	2·6	10·3	6·3	16·3	9·9
44	9 11·0	9 12·5	8 45·9	4·4	2·7	10·4	6·3	16·4	10·0
45	9 11·3	9 12·8	8 46·1	4·5	2·7	10·5	6·4	16·5	10·0
46	9 11·5	9 13·0	8 46·4	4·6	2·8	10·6	6·4	16·6	10·1
47	9 11·8	9 13·3	8 46·6	4·7	2·9	10·7	6·5	16·7	10·2
48	9 12·0	9 13·5	8 46·9	4·8	2·9	10·8	6·6	16·8	10·2
49	9 12·3	9 13·8	8 47·1	4·9	3·0	10·9	6·6	16·9	10·3
50	9 12·5	9 14·0	8 47·3	5·0	3·0	11·0	6·7	17·0	10·3
51	9 12·8	9 14·3	8 47·6	5·1	3·1	11·1	6·8	17·1	10·4
52	9 13·0	9 14·5	8 47·8	5·2	3·2	11·2	6·8	17·2	10·5
53	9 13·3	9 14·8	8 48·0	5·3	3·2	11·3	6·9	17·3	10·5
54	9 13·5	9 15·0	8 48·3	5·4	3·3	11·4	6·9	17·4	10·6
55	9 13·8	9 15·3	8 48·5	5·5	3·3	11·5	7·0	17·5	10·6
56	9 14·0	9 15·5	8 48·8	5·6	3·4	11·6	7·1	17·6	10·7
57	9 14·3	9 15·8	8 49·0	5·7	3·5	11·7	7·1	17·7	10·8
58	9 14·5	9 16·0	8 49·2	5·8	3·5	11·8	7·2	17·8	10·8
59	9 14·8	9 16·3	8 49·5	5·9	3·6	11·9	7·2	17·9	10·9
60	9 15·0	9 16·5	8 49·7	6·0	3·7	12·0	7·3	18·0	11·0

37^m	SUN PLANETS	ARIES	MOON	v or d	Corr^n	v or d	Corr^n	v or d	Corr^n
s	° ′	° ′	° ′	′	′	′	′	′	′
00	9 15·0	9 16·5	8 49·7	0·0	0·0	6·0	3·8	12·0	7·5
01	9 15·3	9 16·8	8 50·0	0·1	0·1	6·1	3·8	12·1	7·6
02	9 15·5	9 17·0	8 50·2	0·2	0·1	6·2	3·9	12·2	7·6
03	9 15·8	9 17·3	8 50·4	0·3	0·2	6·3	3·9	12·3	7·7
04	9 16·0	9 17·5	8 50·7	0·4	0·3	6·4	4·0	12·4	7·8
05	9 16·3	9 17·8	8 50·9	0·5	0·3	6·5	4·1	12·5	7·8
06	9 16·5	9 18·0	8 51·1	0·6	0·4	6·6	4·1	12·6	7·9
07	9 16·8	9 18·3	8 51·4	0·7	0·4	6·7	4·2	12·7	7·9
08	9 17·0	9 18·5	8 51·6	0·8	0·5	6·8	4·3	12·8	8·0
09	9 17·3	9 18·8	8 51·9	0·9	0·6	6·9	4·3	12·9	8·1
10	9 17·5	9 19·0	8 52·1	1·0	0·6	7·0	4·4	13·0	8·1
11	9 17·8	9 19·3	8 52·3	1·1	0·7	7·1	4·4	13·1	8·2
12	9 18·0	9 19·5	8 52·6	1·2	0·8	7·2	4·5	13·2	8·3
13	9 18·3	9 19·8	8 52·8	1·3	0·8	7·3	4·6	13·3	8·3
14	9 18·5	9 20·0	8 53·1	1·4	0·9	7·4	4·6	13·4	8·4
15	9 18·8	9 20·3	8 53·3	1·5	0·9	7·5	4·7	13·5	8·4
16	9 19·0	9 20·5	8 53·5	1·6	1·0	7·6	4·8	13·6	8·5
17	9 19·3	9 20·8	8 53·8	1·7	1·1	7·7	4·8	13·7	8·6
18	9 19·5	9 21·0	8 54·0	1·8	1·1	7·8	4·9	13·8	8·6
19	9 19·8	9 21·3	8 54·3	1·9	1·2	7·9	4·9	13·9	8·7
20	9 20·0	9 21·5	8 54·5	2·0	1·3	8·0	5·0	14·0	8·8
21	9 20·3	9 21·8	8 54·7	2·1	1·3	8·1	5·1	14·1	8·8
22	9 20·5	9 22·0	8 55·0	2·2	1·4	8·2	5·1	14·2	8·9
23	9 20·8	9 22·3	8 55·2	2·3	1·4	8·3	5·2	14·3	8·9
24	9 21·0	9 22·5	8 55·4	2·4	1·5	8·4	5·3	14·4	9·0
25	9 21·3	9 22·8	8 55·7	2·5	1·6	8·5	5·3	14·5	9·1
26	9 21·5	9 23·0	8 55·9	2·6	1·6	8·6	5·4	14·6	9·1
27	9 21·8	9 23·3	8 56·2	2·7	1·7	8·7	5·4	14·7	9·2
28	9 22·0	9 23·5	8 56·4	2·8	1·8	8·8	5·5	14·8	9·3
29	9 22·3	9 23·8	8 56·6	2·9	1·8	8·9	5·6	14·9	9·3
30	9 22·5	9 24·0	8 56·9	3·0	1·9	9·0	5·6	15·0	9·4
31	9 22·8	9 24·3	8 57·1	3·1	1·9	9·1	5·7	15·1	9·4
32	9 23·0	9 24·5	8 57·4	3·2	2·0	9·2	5·8	15·2	9·5
33	9 23·3	9 24·8	8 57·6	3·3	2·1	9·3	5·8	15·3	9·6
34	9 23·5	9 25·0	8 57·8	3·4	2·1	9·4	5·9	15·4	9·6
35	9 23·8	9 25·3	8 58·1	3·5	2·2	9·5	5·9	15·5	9·7
36	9 24·0	9 25·5	8 58·3	3·6	2·3	9·6	6·0	15·6	9·8
37	9 24·3	9 25·8	8 58·5	3·7	2·3	9·7	6·1	15·7	9·8
38	9 24·5	9 26·0	8 58·8	3·8	2·4	9·8	6·1	15·8	9·9
39	9 24·8	9 26·3	8 59·0	3·9	2·4	9·9	6·2	15·9	9·9
40	9 25·0	9 26·5	8 59·3	4·0	2·5	10·0	6·3	16·0	10·0
41	9 25·3	9 26·8	8 59·5	4·1	2·6	10·1	6·3	16·1	10·1
42	9 25·5	9 27·0	8 59·7	4·2	2·6	10·2	6·4	16·2	10·1
43	9 25·8	9 27·3	9 00·0	4·3	2·7	10·3	6·4	16·3	10·2
44	9 26·0	9 27·5	9 00·2	4·4	2·8	10·4	6·5	16·4	10·3
45	9 26·3	9 27·8	9 00·5	4·5	2·8	10·5	6·6	16·5	10·3
46	9 26·5	9 28·1	9 00·7	4·6	2·9	10·6	6·6	16·6	10·4
47	9 26·8	9 28·3	9 00·9	4·7	2·9	10·7	6·7	16·7	10·4
48	9 27·0	9 28·6	9 01·2	4·8	3·0	10·8	6·8	16·8	10·5
49	9 27·3	9 28·8	9 01·4	4·9	3·1	10·9	6·8	16·9	10·6
50	9 27·5	9 29·1	9 01·6	5·0	3·1	11·0	6·9	17·0	10·6
51	9 27·8	9 29·3	9 01·9	5·1	3·2	11·1	6·9	17·1	10·7
52	9 28·0	9 29·6	9 02·1	5·2	3·3	11·2	7·0	17·2	10·8
53	9 28·3	9 29·8	9 02·4	5·3	3·3	11·3	7·1	17·3	10·8
54	9 28·5	9 30·1	9 02·6	5·4	3·4	11·4	7·1	17·4	10·9
55	9 28·8	9 30·3	9 02·8	5·5	3·4	11·5	7·2	17·5	10·9
56	9 29·0	9 30·6	9 03·1	5·6	3·5	11·6	7·3	17·6	11·0
57	9 29·3	9 30·8	9 03·3	5·7	3·6	11·7	7·3	17·7	11·1
58	9 29·5	9 31·1	9 03·6	5·8	3·6	11·8	7·4	17·8	11·1
59	9 29·8	9 31·3	9 03·8	5·9	3·7	11·9	7·4	17·9	11·2
60	9 30·0	9 31·6	9 04·0	6·0	3·8	12·0	7·5	18·0	11·3

38m s	SUN PLANETS ° ′	ARIES ° ′	MOON ° ′	v or d ′	Corrⁿ ′	v or d ′	Corrⁿ ′	v or d ′	Corrⁿ ′
00	9 30·0	9 31·6	9 04·0	0·0	0·0	6·0	3·9	12·0	7·7
01	9 30·3	9 31·8	9 04·3	0·1	0·1	6·1	3·9	12·1	7·8
02	9 30·5	9 32·1	9 04·5	0·2	0·1	6·2	4·0	12·2	7·8
03	9 30·8	9 32·3	9 04·7	0·3	0·2	6·3	4·0	12·3	7·9
04	9 31·0	9 32·6	9 05·0	0·4	0·3	6·4	4·1	12·4	8·0
05	9 31·3	9 32·8	9 05·2	0·5	0·3	6·5	4·2	12·5	8·0
06	9 31·5	9 33·1	9 05·5	0·6	0·4	6·6	4·2	12·6	8·1
07	9 31·8	9 33·3	9 05·7	0·7	0·4	6·7	4·3	12·7	8·1
08	9 32·0	9 33·6	9 05·9	0·8	0·5	6·8	4·4	12·8	8·2
09	9 32·3	9 33·8	9 06·2	0·9	0·6	6·9	4·4	12·9	8·3
10	9 32·5	9 34·1	9 06·4	1·0	0·6	7·0	4·5	13·0	8·3
11	9 32·8	9 34·3	9 06·7	1·1	0·7	7·1	4·6	13·1	8·4
12	9 33·0	9 34·6	9 06·9	1·2	0·8	7·2	4·6	13·2	8·5
13	9 33·3	9 34·8	9 07·1	1·3	0·8	7·3	4·7	13·3	8·5
14	9 33·5	9 35·1	9 07·4	1·4	0·9	7·4	4·7	13·4	8·6
15	9 33·8	9 35·3	9 07·6	1·5	1·0	7·5	4·8	13·5	8·7
16	9 34·0	9 35·6	9 07·9	1·6	1·0	7·6	4·9	13·6	8·7
17	9 34·3	9 35·8	9 08·1	1·7	1·1	7·7	4·9	13·7	8·8
18	9 34·5	9 36·1	9 08·3	1·8	1·2	7·8	5·0	13·8	8·9
19	9 34·8	9 36·3	9 08·6	1·9	1·2	7·9	5·1	13·9	8·9
20	9 35·0	9 36·6	9 08·8	2·0	1·3	8·0	5·1	14·0	9·0
21	9 35·3	9 36·8	9 09·0	2·1	1·3	8·1	5·2	14·1	9·0
22	9 35·5	9 37·1	9 09·3	2·2	1·4	8·2	5·3	14·2	9·1
23	9 35·8	9 37·3	9 09·5	2·3	1·5	8·3	5·3	14·3	9·2
24	9 36·0	9 37·6	9 09·8	2·4	1·5	8·4	5·4	14·4	9·2
25	9 36·3	9 37·8	9 10·0	2·5	1·6	8·5	5·5	14·5	9·3
26	9 36·5	9 38·1	9 10·2	2·6	1·7	8·6	5·5	14·6	9·4
27	9 36·8	9 38·3	9 10·5	2·7	1·7	8·7	5·6	14·7	9·4
28	9 37·0	9 38·6	9 10·7	2·8	1·8	8·8	5·6	14·8	9·5
29	9 37·3	9 38·8	9 11·0	2·9	1·9	8·9	5·7	14·9	9·6
30	9 37·5	9 39·1	9 11·2	3·0	1·9	9·0	5·8	15·0	9·6
31	9 37·8	9 39·3	9 11·4	3·1	2·0	9·1	5·8	15·1	9·7
32	9 38·0	9 39·6	9 11·7	3·2	2·1	9·2	5·9	15·2	9·8
33	9 38·3	9 39·8	9 11·9	3·3	2·1	9·3	6·0	15·3	9·8
34	9 38·5	9 40·1	9 12·1	3·4	2·2	9·4	6·0	15·4	9·9
35	9 38·8	9 40·3	9 12·4	3·5	2·2	9·5	6·1	15·5	9·9
36	9 39·0	9 40·6	9 12·6	3·6	2·3	9·6	6·2	15·6	10·0
37	9 39·3	9 40·8	9 12·9	3·7	2·4	9·7	6·2	15·7	10·1
38	9 39·5	9 41·1	9 13·1	3·8	2·4	9·8	6·3	15·8	10·1
39	9 39·8	9 41·3	9 13·3	3·9	2·5	9·9	6·4	15·9	10·2
40	9 40·0	9 41·6	9 13·6	4·0	2·6	10·0	6·4	16·0	10·3
41	9 40·3	9 41·8	9 13·8	4·1	2·6	10·1	6·5	16·1	10·3
42	9 40·5	9 42·1	9 14·1	4·2	2·7	10·2	6·5	16·2	10·4
43	9 40·8	9 42·3	9 14·3	4·3	2·8	10·3	6·6	16·3	10·5
44	9 41·0	9 42·6	9 14·5	4·4	2·8	10·4	6·7	16·4	10·5
45	9 41·3	9 42·8	9 14·8	4·5	2·9	10·5	6·7	16·5	10·6
46	9 41·5	9 43·1	9 15·0	4·6	3·0	10·6	6·8	16·6	10·7
47	9 41·8	9 43·3	9 15·2	4·7	3·0	10·7	6·9	16·7	10·7
48	9 42·0	9 43·6	9 15·5	4·8	3·1	10·8	6·9	16·8	10·8
49	9 42·3	9 43·8	9 15·7	4·9	3·1	10·9	7·0	16·9	10·8
50	9 42·5	9 44·1	9 16·0	5·0	3·2	11·0	7·1	17·0	10·9
51	9 42·8	9 44·3	9 16·2	5·1	3·3	11·1	7·1	17·1	11·0
52	9 43·0	9 44·6	9 16·4	5·2	3·3	11·2	7·2	17·2	11·0
53	9 43·3	9 44·8	9 16·7	5·3	3·4	11·3	7·3	17·3	11·1
54	9 43·5	9 45·1	9 16·9	5·4	3·5	11·4	7·3	17·4	11·2
55	9 43·8	9 45·3	9 17·2	5·5	3·5	11·5	7·4	17·5	11·2
56	9 44·0	9 45·6	9 17·4	5·6	3·6	11·6	7·4	17·6	11·3
57	9 44·3	9 45·8	9 17·6	5·7	3·7	11·7	7·5	17·7	11·4
58	9 44·5	9 46·1	9 17·9	5·8	3·7	11·8	7·6	17·8	11·4
59	9 44·8	9 46·4	9 18·1	5·9	3·8	11·9	7·6	17·9	11·5
60	9 45·0	9 46·6	9 18·4	6·0	3·9	12·0	7·7	18·0	11·6

39m s	SUN PLANETS ° ′	ARIES ° ′	MOON ° ′	v or d ′	Corrⁿ ′	v or d ′	Corrⁿ ′	v or d ′	Corrⁿ ′
00	9 45·0	9 46·6	9 18·4	0·0	0·0	6·0	4·0	12·0	7·9
01	9 45·3	9 46·9	9 18·6	0·1	0·1	6·1	4·0	12·1	8·0
02	9 45·5	9 47·1	9 18·8	0·2	0·1	6·2	4·1	12·2	8·0
03	9 45·8	9 47·4	9 19·1	0·3	0·2	6·3	4·1	12·3	8·1
04	9 46·0	9 47·6	9 19·3	0·4	0·3	6·4	4·2	12·4	8·2
05	9 46·3	9 47·9	9 19·5	0·5	0·3	6·5	4·3	12·5	8·2
06	9 46·5	9 48·1	9 19·8	0·6	0·4	6·6	4·3	12·6	8·3
07	9 46·8	9 48·4	9 20·0	0·7	0·5	6·7	4·4	12·7	8·4
08	9 47·0	9 48·6	9 20·3	0·8	0·5	6·8	4·5	12·8	8·4
09	9 47·3	9 48·9	9 20·5	0·9	0·6	6·9	4·5	12·9	8·5
10	9 47·5	9 49·1	9 20·7	1·0	0·7	7·0	4·6	13·0	8·6
11	9 47·8	9 49·4	9 21·0	1·1	0·7	7·1	4·7	13·1	8·6
12	9 48·0	9 49·6	9 21·2	1·2	0·8	7·2	4·7	13·2	8·7
13	9 48·3	9 49·9	9 21·5	1·3	0·9	7·3	4·8	13·3	8·8
14	9 48·5	9 50·1	9 21·7	1·4	0·9	7·4	4·9	13·4	8·8
15	9 48·8	9 50·4	9 21·9	1·5	1·0	7·5	4·9	13·5	8·9
16	9 49·0	9 50·6	9 22·2	1·6	1·1	7·6	5·0	13·6	9·0
17	9 49·3	9 50·9	9 22·4	1·7	1·1	7·7	5·1	13·7	9·0
18	9 49·5	9 51·1	9 22·6	1·8	1·2	7·8	5·1	13·8	9·1
19	9 49·8	9 51·4	9 22·9	1·9	1·3	7·9	5·2	13·9	9·2
20	9 50·0	9 51·6	9 23·1	2·0	1·3	8·0	5·3	14·0	9·2
21	9 50·3	9 51·9	9 23·4	2·1	1·4	8·1	5·3	14·1	9·3
22	9 50·5	9 52·1	9 23·6	2·2	1·4	8·2	5·4	14·2	9·3
23	9 50·8	9 52·4	9 23·8	2·3	1·5	8·3	5·5	14·3	9·4
24	9 51·0	9 52·6	9 24·1	2·4	1·6	8·4	5·5	14·4	9·5
25	9 51·3	9 52·9	9 24·3	2·5	1·6	8·5	5·6	14·5	9·5
26	9 51·5	9 53·1	9 24·6	2·6	1·7	8·6	5·7	14·6	9·6
27	9 51·8	9 53·4	9 24·8	2·7	1·8	8·7	5·7	14·7	9·7
28	9 52·0	9 53·6	9 25·0	2·8	1·8	8·8	5·8	14·8	9·7
29	9 52·3	9 53·9	9 25·3	2·9	1·9	8·9	5·9	14·9	9·8
30	9 52·5	9 54·1	9 25·5	3·0	2·0	9·0	5·9	15·0	9·9
31	9 52·8	9 54·4	9 25·7	3·1	2·0	9·1	6·0	15·1	9·9
32	9 53·0	9 54·6	9 26·0	3·2	2·1	9·2	6·1	15·2	10·0
33	9 53·3	9 54·9	9 26·2	3·3	2·2	9·3	6·1	15·3	10·1
34	9 53·5	9 55·1	9 26·5	3·4	2·2	9·4	6·2	15·4	10·1
35	9 53·8	9 55·4	9 26·7	3·5	2·3	9·5	6·3	15·5	10·2
36	9 54·0	9 55·6	9 26·9	3·6	2·4	9·6	6·3	15·6	10·3
37	9 54·3	9 55·9	9 27·2	3·7	2·4	9·7	6·4	15·7	10·3
38	9 54·5	9 56·1	9 27·4	3·8	2·5	9·8	6·5	15·8	10·4
39	9 54·8	9 56·4	9 27·7	3·9	2·6	9·9	6·5	15·9	10·5
40	9 55·0	9 56·6	9 27·9	4·0	2·6	10·0	6·6	16·0	10·5
41	9 55·3	9 56·9	9 28·1	4·1	2·7	10·1	6·6	16·1	10·6
42	9 55·5	9 57·1	9 28·4	4·2	2·8	10·2	6·7	16·2	10·7
43	9 55·8	9 57·4	9 28·6	4·3	2·8	10·3	6·8	16·3	10·7
44	9 56·0	9 57·6	9 28·8	4·4	2·9	10·4	6·8	16·4	10·8
45	9 56·3	9 57·9	9 29·1	4·5	3·0	10·5	6·9	16·5	10·9
46	9 56·5	9 58·1	9 29·3	4·6	3·0	10·6	7·0	16·6	10·9
47	9 56·8	9 58·4	9 29·6	4·7	3·1	10·7	7·0	16·7	11·0
48	9 57·0	9 58·6	9 29·8	4·8	3·2	10·8	7·1	16·8	11·1
49	9 57·3	9 58·9	9 30·0	4·9	3·2	10·9	7·2	16·9	11·1
50	9 57·5	9 59·1	9 30·3	5·0	3·3	11·0	7·2	17·0	11·2
51	9 57·8	9 59·4	9 30·5	5·1	3·4	11·1	7·3	17·1	11·3
52	9 58·0	9 59·6	9 30·8	5·2	3·4	11·2	7·4	17·2	11·3
53	9 58·3	9 59·9	9 31·0	5·3	3·5	11·3	7·4	17·3	11·4
54	9 58·5	10 00·1	9 31·2	5·4	3·6	11·4	7·5	17·4	11·5
55	9 58·8	10 00·4	9 31·5	5·5	3·6	11·5	7·6	17·5	11·5
56	9 59·0	10 00·6	9 31·7	5·6	3·7	11·6	7·6	17·6	11·6
57	9 59·3	10 00·9	9 32·0	5·7	3·8	11·7	7·7	17·7	11·7
58	9 59·5	10 01·1	9 32·2	5·8	3·8	11·8	7·8	17·8	11·7
59	9 59·8	10 01·4	9 32·4	5·9	3·9	11·9	7·8	17·9	11·8
60	10 00·0	10 01·6	9 32·7	6·0	4·0	12·0	7·9	18·0	11·9

40^m	SUN PLANETS	ARIES	MOON	v or d	Corr^n	v or d	Corr^n	v or d	Corr^n
s	° ′	° ′	° ′	′	′	′	′	′	′
00	10 00·0	10 01·6	9 32·7	0·0	0·0	6·0	4·1	12·0	8·1
01	10 00·3	10 01·9	9 32·9	0·1	0·1	6·1	4·1	12·1	8·2
02	10 00·5	10 02·1	9 33·1	0·2	0·1	6·2	4·2	12·2	8·2
03	10 00·8	10 02·4	9 33·4	0·3	0·2	6·3	4·3	12·3	8·3
04	10 01·0	10 02·6	9 33·6	0·4	0·3	6·4	4·3	12·4	8·4
05	10 01·3	10 02·9	9 33·9	0·5	0·3	6·5	4·4	12·5	8·4
06	10 01·5	10 03·1	9 34·1	0·6	0·4	6·6	4·5	12·6	8·5
07	10 01·8	10 03·4	9 34·3	0·7	0·5	6·7	4·5	12·7	8·6
08	10 02·0	10 03·6	9 34·6	0·8	0·5	6·8	4·6	12·8	8·6
09	10 02·3	10 03·9	9 34·8	0·9	0·6	6·9	4·7	12·9	8·7
10	10 02·5	10 04·1	9 35·1	1·0	0·7	7·0	4·7	13·0	8·8
11	10 02·8	10 04·4	9 35·3	1·1	0·7	7·1	4·8	13·1	8·8
12	10 03·0	10 04·7	9 35·5	1·2	0·8	7·2	4·9	13·2	8·9
13	10 03·3	10 04·9	9 35·8	1·3	0·9	7·3	4·9	13·3	9·0
14	10 03·5	10 05·2	9 36·0	1·4	0·9	7·4	5·0	13·4	9·0
15	10 03·8	10 05·4	9 36·2	1·5	1·0	7·5	5·1	13·5	9·1
16	10 04·0	10 05·7	9 36·5	1·6	1·1	7·6	5·1	13·6	9·2
17	10 04·3	10 05·9	9 36·7	1·7	1·1	7·7	5·2	13·7	9·2
18	10 04·5	10 06·2	9 37·0	1·8	1·2	7·8	5·3	13·8	9·3
19	10 04·8	10 06·4	9 37·2	1·9	1·3	7·9	5·3	13·9	9·4
20	10 05·0	10 06·7	9 37·4	2·0	1·4	8·0	5·4	14·0	9·5
21	10 05·3	10 06·9	9 37·7	2·1	1·4	8·1	5·5	14·1	9·5
22	10 05·5	10 07·2	9 37·9	2·2	1·5	8·2	5·5	14·2	9·6
23	10 05·8	10 07·4	9 38·2	2·3	1·6	8·3	5·6	14·3	9·7
24	10 06·0	10 07·7	9 38·4	2·4	1·6	8·4	5·7	14·4	9·7
25	10 06·3	10 07·9	9 38·6	2·5	1·7	8·5	5·7	14·5	9·8
26	10 06·5	10 08·2	9 38·9	2·6	1·8	8·6	5·8	14·6	9·9
27	10 06·8	10 08·4	9 39·1	2·7	1·8	8·7	5·9	14·7	9·9
28	10 07·0	10 08·7	9 39·3	2·8	1·9	8·8	5·9	14·8	10·0
29	10 07·3	10 08·9	9 39·6	2·9	2·0	8·9	6·0	14·9	10·1
30	10 07·5	10 09·2	9 39·8	3·0	2·0	9·0	6·1	15·0	10·1
31	10 07·8	10 09·4	9 40·1	3·1	2·1	9·1	6·1	15·1	10·2
32	10 08·0	10 09·7	9 40·3	3·2	2·2	9·2	6·2	15·2	10·3
33	10 08·3	10 09·9	9 40·5	3·3	2·2	9·3	6·3	15·3	10·3
34	10 08·5	10 10·2	9 40·8	3·4	2·3	9·4	6·3	15·4	10·4
35	10 08·8	10 10·4	9 41·0	3·5	2·4	9·5	6·4	15·5	10·5
36	10 09·0	10 10·7	9 41·3	3·6	2·4	9·6	6·5	15·6	10·5
37	10 09·3	10 10·9	9 41·5	3·7	2·5	9·7	6·5	15·7	10·6
38	10 09·5	10 11·2	9 41·7	3·8	2·6	9·8	6·6	15·8	10·7
39	10 09·8	10 11·4	9 42·0	3·9	2·6	9·9	6·7	15·9	10·7
40	10 10·0	10 11·7	9 42·2	4·0	2·7	10·0	6·8	16·0	10·8
41	10 10·3	10 11·9	9 42·4	4·1	2·8	10·1	6·8	16·1	10·9
42	10 10·5	10 12·2	9 42·7	4·2	2·8	10·2	6·9	16·2	10·9
43	10 10·8	10 12·4	9 42·9	4·3	2·9	10·3	7·0	16·3	11·0
44	10 11·0	10 12·7	9 43·2	4·4	3·0	10·4	7·0	16·4	11·1
45	10 11·3	10 12·9	9 43·4	4·5	3·0	10·5	7·1	16·5	11·1
46	10 11·5	10 13·2	9 43·6	4·6	3·1	10·6	7·2	16·6	11·2
47	10 11·8	10 13·4	9 43·9	4·7	3·2	10·7	7·2	16·7	11·3
48	10 12·0	10 13·7	9 44·1	4·8	3·2	10·8	7·3	16·8	11·3
49	10 12·3	10 13·9	9 44·4	4·9	3·3	10·9	7·4	16·9	11·4
50	10 12·5	10 14·2	9 44·6	5·0	3·4	11·0	7·4	17·0	11·5
51	10 12·8	10 14·4	9 44·8	5·1	3·4	11·1	7·5	17·1	11·5
52	10 13·0	10 14·7	9 45·1	5·2	3·5	11·2	7·6	17·2	11·6
53	10 13·3	10 14·9	9 45·3	5·3	3·6	11·3	7·6	17·3	11·7
54	10 13·5	10 15·2	9 45·6	5·4	3·6	11·4	7·7	17·4	11·7
55	10 13·8	10 15·4	9 45·8	5·5	3·7	11·5	7·8	17·5	11·8
56	10 14·0	10 15·7	9 46·0	5·6	3·8	11·6	7·8	17·6	11·9
57	10 14·3	10 15·9	9 46·3	5·7	3·8	11·7	7·9	17·7	11·9
58	10 14·5	10 16·2	9 46·5	5·8	3·9	11·8	8·0	17·8	12·0
59	10 14·8	10 16·4	9 46·7	5·9	4·0	11·9	8·0	17·9	12·1
60	10 15·0	10 16·7	9 47·0	6·0	4·1	12·0	8·1	18·0	12·2

41^m	SUN PLANETS	ARIES	MOON	v or d	Corr^n	v or d	Corr^n	v or d	Corr^n
s	° ′	° ′	° ′	′	′	′	′	′	′
00	10 15·0	10 16·7	9 47·0	0·0	0·0	6·0	4·2	12·0	8·3
01	10 15·3	10 16·9	9 47·2	0·1	0·1	6·1	4·2	12·1	8·4
02	10 15·5	10 17·2	9 47·5	0·2	0·1	6·2	4·3	12·2	8·4
03	10 15·8	10 17·4	9 47·7	0·3	0·2	6·3	4·4	12·3	8·5
04	10 16·0	10 17·7	9 47·9	0·4	0·3	6·4	4·4	12·4	8·6
05	10 16·3	10 17·9	9 48·2	0·5	0·3	6·5	4·5	12·5	8·6
06	10 16·5	10 18·2	9 48·4	0·6	0·4	6·6	4·6	12·6	8·7
07	10 16·8	10 18·4	9 48·7	0·7	0·5	6·7	4·6	12·7	8·8
08	10 17·0	10 18·7	9 48·9	0·8	0·6	6·8	4·7	12·8	8·9
09	10 17·3	10 18·9	9 49·1	0·9	0·6	6·9	4·8	12·9	8·9
10	10 17·5	10 19·2	9 49·4	1·0	0·7	7·0	4·8	13·0	9·0
11	10 17·8	10 19·4	9 49·6	1·1	0·8	7·1	4·9	13·1	9·1
12	10 18·0	10 19·7	9 49·8	1·2	0·8	7·2	5·0	13·2	9·1
13	10 18·3	10 19·9	9 50·1	1·3	0·9	7·3	5·0	13·3	9·2
14	10 18·5	10 20·2	9 50·3	1·4	1·0	7·4	5·1	13·4	9·3
15	10 18·8	10 20·4	9 50·6	1·5	1·0	7·5	5·2	13·5	9·3
16	10 19·0	10 20·7	9 50·8	1·6	1·1	7·6	5·3	13·6	9·4
17	10 19·3	10 20·9	9 51·0	1·7	1·2	7·7	5·3	13·7	9·5
18	10 19·5	10 21·2	9 51·3	1·8	1·2	7·8	5·4	13·8	9·5
19	10 19·8	10 21·4	9 51·5	1·9	1·3	7·9	5·5	13·9	9·6
20	10 20·0	10 21·7	9 51·8	2·0	1·4	8·0	5·5	14·0	9·7
21	10 20·3	10 21·9	9 52·0	2·1	1·5	8·1	5·6	14·1	9·8
22	10 20·5	10 22·2	9 52·2	2·2	1·5	8·2	5·7	14·2	9·8
23	10 20·8	10 22·4	9 52·5	2·3	1·6	8·3	5·7	14·3	9·9
24	10 21·0	10 22·7	9 52·7	2·4	1·7	8·4	5·8	14·4	10·0
25	10 21·3	10 23·0	9 52·9	2·5	1·7	8·5	5·9	14·5	10·0
26	10 21·5	10 23·2	9 53·2	2·6	1·8	8·6	5·9	14·6	10·1
27	10 21·8	10 23·5	9 53·4	2·7	1·9	8·7	6·0	14·7	10·2
28	10 22·0	10 23·7	9 53·7	2·8	1·9	8·8	6·1	14·8	10·2
29	10 22·3	10 24·0	9 53·9	2·9	2·0	8·9	6·2	14·9	10·3
30	10 22·5	10 24·2	9 54·1	3·0	2·1	9·0	6·2	15·0	10·4
31	10 22·8	10 24·5	9 54·4	3·1	2·1	9·1	6·3	15·1	10·4
32	10 23·0	10 24·7	9 54·6	3·2	2·2	9·2	6·4	15·2	10·5
33	10 23·3	10 25·0	9 54·9	3·3	2·3	9·3	6·4	15·3	10·6
34	10 23·5	10 25·2	9 55·1	3·4	2·4	9·4	6·5	15·4	10·7
35	10 23·8	10 25·5	9 55·3	3·5	2·4	9·5	6·6	15·5	10·7
36	10 24·0	10 25·7	9 55·6	3·6	2·5	9·6	6·6	15·6	10·8
37	10 24·3	10 26·0	9 55·8	3·7	2·6	9·7	6·7	15·7	10·9
38	10 24·5	10 26·2	9 56·1	3·8	2·6	9·8	6·8	15·8	10·9
39	10 24·8	10 26·5	9 56·3	3·9	2·7	9·9	6·8	15·9	11·0
40	10 25·0	10 26·7	9 56·5	4·0	2·8	10·0	6·9	16·0	11·1
41	10 25·3	10 27·0	9 56·8	4·1	2·8	10·1	7·0	16·1	11·1
42	10 25·5	10 27·2	9 57·0	4·2	2·9	10·2	7·1	16·2	11·2
43	10 25·8	10 27·5	9 57·2	4·3	3·0	10·3	7·1	16·3	11·3
44	10 26·0	10 27·7	9 57·5	4·4	3·0	10·4	7·2	16·4	11·3
45	10 26·3	10 28·0	9 57·7	4·5	3·1	10·5	7·3	16·5	11·4
46	10 26·5	10 28·2	9 58·0	4·6	3·2	10·6	7·3	16·6	11·5
47	10 26·8	10 28·5	9 58·2	4·7	3·3	10·7	7·4	16·7	11·6
48	10 27·0	10 28·7	9 58·4	4·8	3·3	10·8	7·5	16·8	11·6
49	10 27·3	10 29·0	9 58·7	4·9	3·4	10·9	7·5	16·9	11·7
50	10 27·5	10 29·2	9 58·9	5·0	3·5	11·0	7·6	17·0	11·8
51	10 27·8	10 29·5	9 59·2	5·1	3·5	11·1	7·7	17·1	11·8
52	10 28·0	10 29·7	9 59·4	5·2	3·6	11·2	7·7	17·2	11·9
53	10 28·3	10 30·0	9 59·6	5·3	3·7	11·3	7·8	17·3	12·0
54	10 28·5	10 30·2	9 59·9	5·4	3·7	11·4	7·9	17·4	12·0
55	10 28·8	10 30·5	10 00·1	5·5	3·8	11·5	8·0	17·5	12·1
56	10 29·0	10 30·7	10 00·3	5·6	3·9	11·6	8·0	17·6	12·2
57	10 29·3	10 31·0	10 00·6	5·7	3·9	11·7	8·1	17·7	12·2
58	10 29·5	10 31·2	10 00·8	5·8	4·0	11·8	8·2	17·8	12·3
59	10 29·8	10 31·5	10 01·1	5·9	4·1	11·9	8·2	17·9	12·4
60	10 30·0	10 31·7	10 01·3	6·0	4·2	12·0	8·3	18·0	12·5

42m s	SUN PLANETS ° ′	ARIES ° ′	MOON ° ′	v or d ′	Corrⁿ ′	v or d ′	Corrⁿ ′	v or d ′	Corrⁿ ′
00	10 30·0	10 31·7	10 01·3	0·0	0·0	6·0	4·3	12·0	8·5
01	10 30·3	10 32·0	10 01·5	0·1	0·1	6·1	4·3	12·1	8·6
02	10 30·5	10 32·2	10 01·8	0·2	0·1	6·2	4·4	12·2	8·6
03	10 30·8	10 32·5	10 02·0	0·3	0·2	6·3	4·5	12·3	8·7
04	10 31·0	10 32·7	10 02·3	0·4	0·3	6·4	4·5	12·4	8·8
05	10 31·3	10 33·0	10 02·5	0·5	0·4	6·5	4·6	12·5	8·9
06	10 31·5	10 33·2	10 02·7	0·6	0·4	6·6	4·7	12·6	8·9
07	10 31·8	10 33·5	10 03·0	0·7	0·5	6·7	4·7	12·7	9·0
08	10 32·0	10 33·7	10 03·2	0·8	0·6	6·8	4·8	12·8	9·1
09	10 32·3	10 34·0	10 03·4	0·9	0·6	6·9	4·9	12·9	9·1
10	10 32·5	10 34·2	10 03·7	1·0	0·7	7·0	5·0	13·0	9·2
11	10 32·8	10 34·5	10 03·9	1·1	0·8	7·1	5·0	13·1	9·3
12	10 33·0	10 34·7	10 04·2	1·2	0·9	7·2	5·1	13·2	9·4
13	10 33·3	10 35·0	10 04·4	1·3	0·9	7·3	5·2	13·3	9·4
14	10 33·5	10 35·2	10 04·6	1·4	1·0	7·4	5·2	13·4	9·5
15	10 33·8	10 35·5	10 04·9	1·5	1·1	7·5	5·3	13·5	9·6
16	10 34·0	10 35·7	10 05·1	1·6	1·1	7·6	5·4	13·6	9·6
17	10 34·3	10 36·0	10 05·4	1·7	1·2	7·7	5·5	13·7	9·7
18	10 34·5	10 36·2	10 05·6	1·8	1·3	7·8	5·5	13·8	9·8
19	10 34·8	10 36·5	10 05·8	1·9	1·3	7·9	5·6	13·9	9·8
20	10 35·0	10 36·7	10 06·1	2·0	1·4	8·0	5·7	14·0	9·9
21	10 35·3	10 37·0	10 06·3	2·1	1·5	8·1	5·7	14·1	10·0
22	10 35·5	10 37·2	10 06·5	2·2	1·6	8·2	5·8	14·2	10·1
23	10 35·8	10 37·5	10 06·8	2·3	1·6	8·3	5·9	14·3	10·1
24	10 36·0	10 37·7	10 07·0	2·4	1·7	8·4	6·0	14·4	10·2
25	10 36·3	10 38·0	10 07·3	2·5	1·8	8·5	6·0	14·5	10·3
26	10 36·5	10 38·2	10 07·5	2·6	1·8	8·6	6·1	14·6	10·3
27	10 36·8	10 38·5	10 07·7	2·7	1·9	8·7	6·2	14·7	10·4
28	10 37·0	10 38·7	10 08·0	2·8	2·0	8·8	6·2	14·8	10·5
29	10 37·3	10 39·0	10 08·2	2·9	2·1	8·9	6·3	14·9	10·6
30	10 37·5	10 39·2	10 08·5	3·0	2·1	9·0	6·4	15·0	10·6
31	10 37·8	10 39·5	10 08·7	3·1	2·2	9·1	6·4	15·1	10·7
32	10 38·0	10 39·7	10 08·9	3·2	2·3	9·2	6·5	15·2	10·8
33	10 38·3	10 40·0	10 09·2	3·3	2·3	9·3	6·6	15·3	10·8
34	10 38·5	10 40·2	10 09·4	3·4	2·4	9·4	6·7	15·4	10·9
35	10 38·8	10 40·5	10 09·7	3·5	2·5	9·5	6·7	15·5	11·0
36	10 39·0	10 40·7	10 09·9	3·6	2·6	9·6	6·8	15·6	11·1
37	10 39·3	10 41·0	10 10·1	3·7	2·6	9·7	6·9	15·7	11·1
38	10 39·5	10 41·3	10 10·4	3·8	2·7	9·8	6·9	15·8	11·2
39	10 39·8	10 41·5	10 10·6	3·9	2·8	9·9	7·0	15·9	11·3
40	10 40·0	10 41·8	10 10·8	4·0	2·8	10·0	7·1	16·0	11·3
41	10 40·3	10 42·0	10 11·1	4·1	2·9	10·1	7·2	16·1	11·4
42	10 40·5	10 42·3	10 11·3	4·2	3·0	10·2	7·2	16·2	11·5
43	10 40·8	10 42·5	10 11·6	4·3	3·0	10·3	7·3	16·3	11·5
44	10 41·0	10 42·8	10 11·8	4·4	3·1	10·4	7·4	16·4	11·6
45	10 41·3	10 43·0	10 12·0	4·5	3·2	10·5	7·4	16·5	11·7
46	10 41·5	10 43·3	10 12·3	4·6	3·3	10·6	7·5	16·6	11·8
47	10 41·8	10 43·5	10 12·5	4·7	3·3	10·7	7·6	16·7	11·8
48	10 42·0	10 43·8	10 12·8	4·8	3·4	10·8	7·7	16·8	11·9
49	10 42·3	10 44·0	10 13·0	4·9	3·5	10·9	7·7	16·9	12·0
50	10 42·5	10 44·3	10 13·2	5·0	3·5	11·0	7·8	17·0	12·0
51	10 42·8	10 44·5	10 13·5	5·1	3·6	11·1	7·9	17·1	12·1
52	10 43·0	10 44·8	10 13·7	5·2	3·7	11·2	7·9	17·2	12·2
53	10 43·3	10 45·0	10 13·9	5·3	3·8	11·3	8·0	17·3	12·3
54	10 43·5	10 45·3	10 14·2	5·4	3·8	11·4	8·1	17·4	12·3
55	10 43·8	10 45·5	10 14·4	5·5	3·9	11·5	8·1	17·5	12·4
56	10 44·0	10 45·8	10 14·7	5·6	4·0	11·6	8·2	17·6	12·5
57	10 44·3	10 46·0	10 14·9	5·7	4·0	11·7	8·3	17·7	12·5
58	10 44·5	10 46·3	10 15·1	5·8	4·1	11·8	8·4	17·8	12·6
59	10 44·8	10 46·5	10 15·4	5·9	4·2	11·9	8·4	17·9	12·7
60	10 45·0	10 46·8	10 15·6	6·0	4·3	12·0	8·5	18·0	12·8

43m s	SUN PLANETS ° ′	ARIES ° ′	MOON ° ′	v or d ′	Corrⁿ ′	v or d ′	Corrⁿ ′	v or d ′	Corrⁿ ′
00	10 45·0	10 46·8	10 15·6	0·0	0·0	6·0	4·4	12·0	8·7
01	10 45·3	10 47·0	10 15·9	0·1	0·1	6·1	4·4	12·1	8·8
02	10 45·5	10 47·3	10 16·1	0·2	0·1	6·2	4·5	12·2	8·8
03	10 45·8	10 47·5	10 16·3	0·3	0·2	6·3	4·6	12·3	8·9
04	10 46·0	10 47·8	10 16·6	0·4	0·3	6·4	4·6	12·4	9·0
05	10 46·3	10 48·0	10 16·8	0·5	0·4	6·5	4·7	12·5	9·1
06	10 46·5	10 48·3	10 17·0	0·6	0·4	6·6	4·8	12·6	9·1
07	10 46·8	10 48·5	10 17·3	0·7	0·5	6·7	4·9	12·7	9·2
08	10 47·0	10 48·8	10 17·5	0·8	0·6	6·8	4·9	12·8	9·3
09	10 47·3	10 49·0	10 17·8	0·9	0·7	6·9	5·0	12·9	9·4
10	10 47·5	10 49·3	10 18·0	1·0	0·7	7·0	5·1	13·0	9·4
11	10 47·8	10 49·5	10 18·2	1·1	0·8	7·1	5·1	13·1	9·5
12	10 48·0	10 49·8	10 18·5	1·2	0·9	7·2	5·2	13·2	9·6
13	10 48·3	10 50·0	10 18·7	1·3	0·9	7·3	5·3	13·3	9·6
14	10 48·5	10 50·3	10 19·0	1·4	1·0	7·4	5·4	13·4	9·7
15	10 48·8	10 50·5	10 19·2	1·5	1·1	7·5	5·4	13·5	9·8
16	10 49·0	10 50·8	10 19·4	1·6	1·2	7·6	5·5	13·6	9·9
17	10 49·3	10 51·0	10 19·7	1·7	1·2	7·7	5·6	13·7	9·9
18	10 49·5	10 51·3	10 19·9	1·8	1·3	7·8	5·7	13·8	10·0
19	10 49·8	10 51·5	10 20·2	1·9	1·4	7·9	5·7	13·9	10·1
20	10 50·0	10 51·8	10 20·4	2·0	1·5	8·0	5·8	14·0	10·2
21	10 50·3	10 52·0	10 20·6	2·1	1·5	8·1	5·9	14·1	10·2
22	10 50·5	10 52·3	10 20·9	2·2	1·6	8·2	5·9	14·2	10·3
23	10 50·8	10 52·5	10 21·1	2·3	1·7	8·3	6·0	14·3	10·4
24	10 51·0	10 52·8	10 21·3	2·4	1·7	8·4	6·1	14·4	10·4
25	10 51·3	10 53·0	10 21·6	2·5	1·8	8·5	6·2	14·5	10·5
26	10 51·5	10 53·3	10 21·8	2·6	1·9	8·6	6·2	14·6	10·6
27	10 51·8	10 53·5	10 22·1	2·7	2·0	8·7	6·3	14·7	10·7
28	10 52·0	10 53·8	10 22·3	2·8	2·0	8·8	6·4	14·8	10·7
29	10 52·3	10 54·0	10 22·5	2·9	2·1	8·9	6·5	14·9	10·8
30	10 52·5	10 54·3	10 22·8	3·0	2·2	9·0	6·5	15·0	10·9
31	10 52·8	10 54·5	10 23·0	3·1	2·2	9·1	6·6	15·1	10·9
32	10 53·0	10 54·8	10 23·3	3·2	2·3	9·2	6·7	15·2	11·0
33	10 53·3	10 55·0	10 23·5	3·3	2·4	9·3	6·7	15·3	11·1
34	10 53·5	10 55·3	10 23·7	3·4	2·5	9·4	6·8	15·4	11·2
35	10 53·8	10 55·5	10 24·0	3·5	2·5	9·5	6·9	15·5	11·2
36	10 54·0	10 55·8	10 24·2	3·6	2·6	9·6	7·0	15·6	11·3
37	10 54·3	10 56·0	10 24·4	3·7	2·7	9·7	7·0	15·7	11·4
38	10 54·5	10 56·3	10 24·7	3·8	2·8	9·8	7·1	15·8	11·5
39	10 54·8	10 56·5	10 24·9	3·9	2·8	9·9	7·2	15·9	11·5
40	10 55·0	10 56·8	10 25·2	4·0	2·9	10·0	7·3	16·0	11·6
41	10 55·3	10 57·0	10 25·4	4·1	3·0	10·1	7·3	16·1	11·7
42	10 55·5	10 57·3	10 25·6	4·2	3·0	10·2	7·4	16·2	11·7
43	10 55·8	10 57·5	10 25·9	4·3	3·1	10·3	7·5	16·3	11·8
44	10 56·0	10 57·8	10 26·1	4·4	3·2	10·4	7·5	16·4	11·9
45	10 56·3	10 58·0	10 26·4	4·5	3·3	10·5	7·6	16·5	12·0
46	10 56·5	10 58·3	10 26·6	4·6	3·3	10·6	7·7	16·6	12·0
47	10 56·8	10 58·5	10 26·8	4·7	3·4	10·7	7·8	16·7	12·1
48	10 57·0	10 58·8	10 27·1	4·8	3·5	10·8	7·8	16·8	12·2
49	10 57·3	10 59·0	10 27·3	4·9	3·6	10·9	7·9	16·9	12·3
50	10 57·5	10 59·3	10 27·5	5·0	3·6	11·0	8·0	17·0	12·3
51	10 57·8	10 59·6	10 27·8	5·1	3·7	11·1	8·0	17·1	12·4
52	10 58·0	10 59·8	10 28·0	5·2	3·8	11·2	8·1	17·2	12·5
53	10 58·3	11 00·1	10 28·3	5·3	3·8	11·3	8·2	17·3	12·5
54	10 58·5	11 00·3	10 28·5	5·4	3·9	11·4	8·3	17·4	12·6
55	10 58·8	11 00·6	10 28·7	5·5	4·0	11·5	8·3	17·5	12·7
56	10 59·0	11 00·8	10 29·0	5·6	4·1	11·6	8·4	17·6	12·8
57	10 59·3	11 01·1	10 29·2	5·7	4·1	11·7	8·5	17·7	12·8
58	10 59·5	11 01·3	10 29·5	5·8	4·2	11·8	8·6	17·8	12·9
59	10 59·8	11 01·6	10 29·7	5·9	4·3	11·9	8·6	17·9	13·0
60	11 00·0	11 01·8	10 29·9	6·0	4·4	12·0	8·7	18·0	13·1

44ᵐ	SUN PLANETS	ARIES	MOON	v or d	Corrⁿ	v or d	Corrⁿ	v or d	Corrⁿ
s	° ′	° ′	° ′	′	′	′	′	′	′
00	11 00·0	11 01·8	10 29·9	0·0	0·0	6·0	4·5	12·0	8·9
01	11 00·3	11 02·1	10 30·2	0·1	0·1	6·1	4·5	12·1	9·0
02	11 00·5	11 02·3	10 30·4	0·2	0·1	6·2	4·6	12·2	9·0
03	11 00·8	11 02·6	10 30·6	0·3	0·2	6·3	4·7	12·3	9·1
04	11 01·0	11 02·8	10 30·9	0·4	0·3	6·4	4·7	12·4	9·2
05	11 01·3	11 03·1	10 31·1	0·5	0·4	6·5	4·8	12·5	9·3
06	11 01·5	11 03·3	10 31·4	0·6	0·4	6·6	4·9	12·6	9·3
07	11 01·8	11 03·6	10 31·6	0·7	0·5	6·7	5·0	12·7	9·4
08	11 02·0	11 03·8	10 31·8	0·8	0·6	6·8	5·0	12·8	9·5
09	11 02·3	11 04·1	10 32·1	0·9	0·7	6·9	5·1	12·9	9·6
10	11 02·5	11 04·3	10 32·3	1·0	0·7	7·0	5·2	13·0	9·6
11	11 02·8	11 04·6	10 32·6	1·1	0·8	7·1	5·3	13·1	9·7
12	11 03·0	11 04·8	10 32·8	1·2	0·9	7·2	5·3	13·2	9·8
13	11 03·3	11 05·1	10 33·0	1·3	1·0	7·3	5·4	13·3	9·9
14	11 03·5	11 05·3	10 33·3	1·4	1·0	7·4	5·5	13·4	9·9
15	11 03·8	11 05·6	10 33·5	1·5	1·1	7·5	5·6	13·5	10·0
16	11 04·0	11 05·8	10 33·8	1·6	1·2	7·6	5·6	13·6	10·1
17	11 04·3	11 06·1	10 34·0	1·7	1·3	7·7	5·7	13·7	10·2
18	11 04·5	11 06·3	10 34·2	1·8	1·3	7·8	5·8	13·8	10·2
19	11 04·8	11 06·6	10 34·5	1·9	1·4	7·9	5·9	13·9	10·3
20	11 05·0	11 06·8	10 34·7	2·0	1·5	8·0	5·9	14·0	10·4
21	11 05·3	11 07·1	10 34·9	2·1	1·6	8·1	6·0	14·1	10·5
22	11 05·5	11 07·3	10 35·2	2·2	1·6	8·2	6·1	14·2	10·5
23	11 05·8	11 07·6	10 35·4	2·3	1·7	8·3	6·2	14·3	10·6
24	11 06·0	11 07·8	10 35·7	2·4	1·8	8·4	6·2	14·4	10·7
25	11 06·3	11 08·1	10 35·9	2·5	1·9	8·5	6·3	14·5	10·8
26	11 06·5	11 08·3	10 36·1	2·6	1·9	8·6	6·4	14·6	10·8
27	11 06·8	11 08·6	10 36·4	2·7	2·0	8·7	6·5	14·7	10·9
28	11 07·0	11 08·8	10 36·6	2·8	2·1	8·8	6·5	14·8	11·0
29	11 07·3	11 09·1	10 36·9	2·9	2·2	8·9	6·6	14·9	11·1
30	11 07·5	11 09·3	10 37·1	3·0	2·2	9·0	6·7	15·0	11·1
31	11 07·8	11 09·6	10 37·3	3·1	2·3	9·1	6·7	15·1	11·2
32	11 08·0	11 09·8	10 37·6	3·2	2·4	9·2	6·8	15·2	11·3
33	11 08·3	11 10·1	10 37·8	3·3	2·4	9·3	6·9	15·3	11·3
34	11 08·5	11 10·3	10 38·0	3·4	2·5	9·4	7·0	15·4	11·4
35	11 08·8	11 10·6	10 38·3	3·5	2·6	9·5	7·0	15·5	11·5
36	11 09·0	11 10·8	10 38·5	3·6	2·7	9·6	7·1	15·6	11·6
37	11 09·3	11 11·1	10 38·8	3·7	2·7	9·7	7·2	15·7	11·6
38	11 09·5	11 11·3	10 39·0	3·8	2·8	9·8	7·3	15·8	11·7
39	11 09·8	11 11·6	10 39·2	3·9	2·9	9·9	7·3	15·9	11·8
40	11 10·0	11 11·8	10 39·5	4·0	3·0	10·0	7·4	16·0	11·9
41	11 10·3	11 12·1	10 39·7	4·1	3·0	10·1	7·5	16·1	11·9
42	11 10·5	11 12·3	10 40·0	4·2	3·1	10·2	7·6	16·2	12·0
43	11 10·8	11 12·6	10 40·2	4·3	3·2	10·3	7·6	16·3	12·1
44	11 11·0	11 12·8	10 40·4	4·4	3·3	10·4	7·7	16·4	12·2
45	11 11·3	11 13·1	10 40·7	4·5	3·3	10·5	7·8	16·5	12·2
46	11 11·5	11 13·3	10 40·9	4·6	3·4	10·6	7·9	16·6	12·3
47	11 11·8	11 13·6	10 41·1	4·7	3·5	10·7	7·9	16·7	12·4
48	11 12·0	11 13·8	10 41·4	4·8	3·6	10·8	8·0	16·8	12·5
49	11 12·3	11 14·1	10 41·6	4·9	3·6	10·9	8·1	16·9	12·5
50	11 12·5	11 14·3	10 41·9	5·0	3·7	11·0	8·2	17·0	12·6
51	11 12·8	11 14·6	10 42·1	5·1	3·8	11·1	8·2	17·1	12·7
52	11 13·0	11 14·8	10 42·3	5·2	3·9	11·2	8·3	17·2	12·8
53	11 13·3	11 15·1	10 42·6	5·3	3·9	11·3	8·4	17·3	12·8
54	11 13·5	11 15·3	10 42·8	5·4	4·0	11·4	8·5	17·4	12·9
55	11 13·8	11 15·6	10 43·1	5·5	4·1	11·5	8·5	17·5	13·0
56	11 14·0	11 15·8	10 43·3	5·6	4·2	11·6	8·6	17·6	13·1
57	11 14·3	11 16·1	10 43·5	5·7	4·2	11·7	8·7	17·7	13·1
58	11 14·5	11 16·3	10 43·8	5·8	4·3	11·8	8·8	17·8	13·2
59	11 14·8	11 16·6	10 44·0	5·9	4·4	11·9	8·8	17·9	13·3
60	11 15·0	11 16·8	10 44·3	6·0	4·5	12·0	8·9	18·0	13·4

45ᵐ	SUN PLANETS	ARIES	MOON	v or d	Corrⁿ	v or d	Corrⁿ	v or d	Corrⁿ
s	° ′	° ′	° ′	′	′	′	′	′	′
00	11 15·0	11 16·8	10 44·3	0·0	0·0	6·0	4·6	12·0	9·1
01	11 15·3	11 17·1	10 44·5	0·1	0·1	6·1	4·6	12·1	9·2
02	11 15·5	11 17·3	10 44·7	0·2	0·2	6·2	4·7	12·2	9·3
03	11 15·8	11 17·6	10 45·0	0·3	0·2	6·3	4·8	12·3	9·3
04	11 16·0	11 17·9	10 45·2	0·4	0·3	6·4	4·9	12·4	9·4
05	11 16·3	11 18·1	10 45·4	0·5	0·4	6·5	4·9	12·5	9·5
06	11 16·5	11 18·4	10 45·7	0·6	0·5	6·6	5·0	12·6	9·6
07	11 16·8	11 18·6	10 45·9	0·7	0·5	6·7	5·1	12·7	9·6
08	11 17·0	11 18·9	10 46·2	0·8	0·6	6·8	5·2	12·8	9·7
09	11 17·3	11 19·1	10 46·4	0·9	0·7	6·9	5·2	12·9	9·8
10	11 17·5	11 19·4	10 46·6	1·0	0·8	7·0	5·3	13·0	9·9
11	11 17·8	11 19·6	10 46·9	1·1	0·8	7·1	5·4	13·1	9·9
12	11 18·0	11 19·9	10 47·1	1·2	0·9	7·2	5·5	13·2	10·0
13	11 18·3	11 20·1	10 47·4	1·3	1·0	7·3	5·5	13·3	10·1
14	11 18·5	11 20·4	10 47·6	1·4	1·1	7·4	5·6	13·4	10·2
15	11 18·8	11 20·6	10 47·8	1·5	1·1	7·5	5·7	13·5	10·2
16	11 19·0	11 20·9	10 48·1	1·6	1·2	7·6	5·8	13·6	10·3
17	11 19·3	11 21·1	10 48·3	1·7	1·3	7·7	5·8	13·7	10·4
18	11 19·5	11 21·4	10 48·5	1·8	1·4	7·8	5·9	13·8	10·5
19	11 19·8	11 21·6	10 48·8	1·9	1·4	7·9	6·0	13·9	10·5
20	11 20·0	11 21·9	10 49·0	2·0	1·5	8·0	6·1	14·0	10·6
21	11 20·3	11 22·1	10 49·3	2·1	1·6	8·1	6·1	14·1	10·7
22	11 20·5	11 22·4	10 49·5	2·2	1·7	8·2	6·2	14·2	10·8
23	11 20·8	11 22·6	10 49·7	2·3	1·7	8·3	6·3	14·3	10·8
24	11 21·0	11 22·9	10 50·0	2·4	1·8	8·4	6·4	14·4	10·9
25	11 21·3	11 23·1	10 50·2	2·5	1·9	8·5	6·4	14·5	11·0
26	11 21·5	11 23·4	10 50·5	2·6	2·0	8·6	6·5	14·6	11·1
27	11 21·8	11 23·6	10 50·7	2·7	2·0	8·7	6·6	14·7	11·1
28	11 22·0	11 23·9	10 50·9	2·8	2·1	8·8	6·7	14·8	11·2
29	11 22·3	11 24·1	10 51·2	2·9	2·2	8·9	6·7	14·9	11·3
30	11 22·5	11 24·4	10 51·4	3·0	2·3	9·0	6·8	15·0	11·4
31	11 22·8	11 24·6	10 51·6	3·1	2·4	9·1	6·9	15·1	11·5
32	11 23·0	11 24·9	10 51·9	3·2	2·4	9·2	7·0	15·2	11·5
33	11 23·3	11 25·1	10 52·1	3·3	2·5	9·3	7·1	15·3	11·6
34	11 23·5	11 25·4	10 52·4	3·4	2·6	9·4	7·1	15·4	11·7
35	11 23·8	11 25·6	10 52·6	3·5	2·7	9·5	7·2	15·5	11·8
36	11 24·0	11 25·9	10 52·8	3·6	2·7	9·6	7·3	15·6	11·8
37	11 24·3	11 26·1	10 53·1	3·7	2·8	9·7	7·4	15·7	11·9
38	11 24·5	11 26·4	10 53·3	3·8	2·9	9·8	7·4	15·8	12·0
39	11 24·8	11 26·6	10 53·6	3·9	3·0	9·9	7·5	15·9	12·1
40	11 25·0	11 26·9	10 53·8	4·0	3·0	10·0	7·6	16·0	12·1
41	11 25·3	11 27·1	10 54·0	4·1	3·1	10·1	7·7	16·1	12·2
42	11 25·5	11 27·4	10 54·3	4·2	3·2	10·2	7·7	16·2	12·3
43	11 25·8	11 27·6	10 54·5	4·3	3·3	10·3	7·8	16·3	12·4
44	11 26·0	11 27·9	10 54·7	4·4	3·3	10·4	7·9	16·4	12·4
45	11 26·3	11 28·1	10 55·0	4·5	3·4	10·5	8·0	16·5	12·5
46	11 26·5	11 28·4	10 55·2	4·6	3·5	10·6	8·0	16·6	12·6
47	11 26·8	11 28·6	10 55·5	4·7	3·6	10·7	8·1	16·7	12·7
48	11 27·0	11 28·9	10 55·7	4·8	3·6	10·8	8·2	16·8	12·7
49	11 27·3	11 29·1	10 55·9	4·9	3·7	10·9	8·3	16·9	12·8
50	11 27·5	11 29·4	10 56·2	5·0	3·8	11·0	8·3	17·0	12·9
51	11 27·8	11 29·6	10 56·4	5·1	3·9	11·1	8·4	17·1	13·0
52	11 28·0	11 29·9	10 56·7	5·2	3·9	11·2	8·5	17·2	13·0
53	11 28·3	11 30·1	10 56·9	5·3	4·0	11·3	8·6	17·3	13·1
54	11 28·5	11 30·4	10 57·1	5·4	4·1	11·4	8·6	17·4	13·2
55	11 28·8	11 30·6	10 57·4	5·5	4·2	11·5	8·7	17·5	13·3
56	11 29·0	11 30·9	10 57·6	5·6	4·2	11·6	8·8	17·6	13·3
57	11 29·3	11 31·1	10 57·9	5·7	4·3	11·7	8·9	17·7	13·4
58	11 29·5	11 31·4	10 58·1	5·8	4·4	11·8	8·9	17·8	13·5
59	11 29·8	11 31·6	10 58·3	5·9	4·5	11·9	9·0	17·9	13·6
60	11 30·0	11 31·9	10 58·6	6·0	4·6	12·0	9·1	18·0	13·7

46^m	SUN PLANETS	ARIES	MOON	v or d	Corr^n	v or d	Corr^n	v or d	Corr^n
s	° ′	° ′	° ′	′	′	′	′	′	′
00	11 30·0	11 31·9	10 58·6	0·0	0·0	6·0	4·7	12·0	9·3
01	11 30·3	11 32·1	10 58·8	0·1	0·1	6·1	4·7	12·1	9·4
02	11 30·5	11 32·4	10 59·0	0·2	0·2	6·2	4·8	12·2	9·5
03	11 30·8	11 32·6	10 59·3	0·3	0·2	6·3	4·9	12·3	9·5
04	11 31·0	11 32·9	10 59·5	0·4	0·3	6·4	5·0	12·4	9·6
05	11 31·3	11 33·1	10 59·8	0·5	0·4	6·5	5·0	12·5	9·7
06	11 31·5	11 33·4	11 00·0	0·6	0·5	6·6	5·1	12·6	9·8
07	11 31·8	11 33·6	11 00·2	0·7	0·5	6·7	5·2	12·7	9·8
08	11 32·0	11 33·9	11 00·5	0·8	0·6	6·8	5·3	12·8	9·9
09	11 32·3	11 34·1	11 00·7	0·9	0·7	6·9	5·3	12·9	10·0
10	11 32·5	11 34·4	11 01·0	1·0	0·8	7·0	5·4	13·0	10·1
11	11 32·8	11 34·6	11 01·2	1·1	0·9	7·1	5·5	13·1	10·2
12	11 33·0	11 34·9	11 01·4	1·2	0·9	7·2	5·6	13·2	10·2
13	11 33·3	11 35·1	11 01·7	1·3	1·0	7·3	5·7	13·3	10·3
14	11 33·5	11 35·4	11 01·9	1·4	1·1	7·4	5·7	13·4	10·4
15	11 33·8	11 35·6	11 02·1	1·5	1·2	7·5	5·8	13·5	10·5
16	11 34·0	11 35·9	11 02·4	1·6	1·2	7·6	5·9	13·6	10·5
17	11 34·3	11 36·2	11 02·6	1·7	1·3	7·7	6·0	13·7	10·6
18	11 34·5	11 36·4	11 02·9	1·8	1·4	7·8	6·0	13·8	10·7
19	11 34·8	11 36·7	11 03·1	1·9	1·5	7·9	6·1	13·9	10·8
20	11 35·0	11 36·9	11 03·3	2·0	1·6	8·0	6·2	14·0	10·9
21	11 35·3	11 37·2	11 03·6	2·1	1·6	8·1	6·3	14·1	10·9
22	11 35·5	11 37·4	11 03·8	2·2	1·7	8·2	6·4	14·2	11·0
23	11 35·8	11 37·7	11 04·1	2·3	1·8	8·3	6·4	14·3	11·1
24	11 36·0	11 37·9	11 04·3	2·4	1·9	8·4	6·5	14·4	11·2
25	11 36·3	11 38·2	11 04·5	2·5	1·9	8·5	6·6	14·5	11·2
26	11 36·5	11 38·4	11 04·8	2·6	2·0	8·6	6·7	14·6	11·3
27	11 36·8	11 38·7	11 05·0	2·7	2·1	8·7	6·7	14·7	11·4
28	11 37·0	11 38·9	11 05·2	2·8	2·2	8·8	6·8	14·8	11·5
29	11 37·3	11 39·2	11 05·5	2·9	2·2	8·9	6·9	14·9	11·5
30	11 37·5	11 39·4	11 05·7	3·0	2·3	9·0	7·0	15·0	11·6
31	11 37·8	11 39·7	11 06·0	3·1	2·4	9·1	7·1	15·1	11·7
32	11 38·0	11 39·9	11 06·2	3·2	2·5	9·2	7·1	15·2	11·8
33	11 38·3	11 40·2	11 06·4	3·3	2·6	9·3	7·2	15·3	11·9
34	11 38·5	11 40·4	11 06·7	3·4	2·6	9·4	7·3	15·4	11·9
35	11 38·8	11 40·7	11 06·9	3·5	2·7	9·5	7·4	15·5	12·0
36	11 39·0	11 40·9	11 07·2	3·6	2·8	9·6	7·4	15·6	12·1
37	11 39·3	11 41·2	11 07·4	3·7	2·9	9·7	7·5	15·7	12·2
38	11 39·5	11 41·4	11 07·6	3·8	2·9	9·8	7·6	15·8	12·2
39	11 39·8	11 41·7	11 07·9	3·9	3·0	9·9	7·7	15·9	12·3
40	11 40·0	11 41·9	11 08·1	4·0	3·1	10·0	7·8	16·0	12·4
41	11 40·3	11 42·2	11 08·3	4·1	3·2	10·1	7·8	16·1	12·5
42	11 40·5	11 42·4	11 08·6	4·2	3·3	10·2	7·9	16·2	12·6
43	11 40·8	11 42·7	11 08·8	4·3	3·3	10·3	8·0	16·3	12·6
44	11 41·0	11 42·9	11 09·1	4·4	3·4	10·4	8·1	16·4	12·7
45	11 41·3	11 43·2	11 09·3	4·5	3·5	10·5	8·1	16·5	12·8
46	11 41·5	11 43·4	11 09·5	4·6	3·6	10·6	8·2	16·6	12·9
47	11 41·8	11 43·7	11 09·8	4·7	3·6	10·7	8·3	16·7	12·9
48	11 42·0	11 43·9	11 10·0	4·8	3·7	10·8	8·4	16·8	13·0
49	11 42·3	11 44·2	11 10·3	4·9	3·8	10·9	8·4	16·9	13·1
50	11 42·5	11 44·4	11 10·5	5·0	3·9	11·0	8·5	17·0	13·2
51	11 42·8	11 44·7	11 10·7	5·1	4·0	11·1	8·6	17·1	13·3
52	11 43·0	11 44·9	11 11·0	5·2	4·0	11·2	8·7	17·2	13·3
53	11 43·3	11 45·2	11 11·2	5·3	4·1	11·3	8·8	17·3	13·4
54	11 43·5	11 45·4	11 11·5	5·4	4·2	11·4	8·8	17·4	13·5
55	11 43·8	11 45·7	11 11·7	5·5	4·3	11·5	8·9	17·5	13·6
56	11 44·0	11 45·9	11 11·9	5·6	4·3	11·6	9·0	17·6	13·6
57	11 44·3	11 46·2	11 12·2	5·7	4·4	11·7	9·1	17·7	13·7
58	11 44·5	11 46·4	11 12·4	5·8	4·5	11·8	9·1	17·8	13·8
59	11 44·8	11 46·7	11 12·6	5·9	4·6	11·9	9·2	17·9	13·9
60	11 45·0	11 46·9	11 12·9	6·0	4·7	12·0	9·3	18·0	14·0

47^m	SUN PLANETS	ARIES	MOON	v or d	Corr^n	v or d	Corr^n	v or d	Corr^n
s	° ′	° ′	° ′	′	′	′	′	′	′
00	11 45·0	11 46·9	11 12·9	0·0	0·0	6·0	4·8	12·0	9·5
01	11 45·3	11 47·2	11 13·1	0·1	0·1	6·1	4·8	12·1	9·6
02	11 45·5	11 47·4	11 13·4	0·2	0·2	6·2	4·9	12·2	9·7
03	11 45·8	11 47·7	11 13·6	0·3	0·2	6·3	5·0	12·3	9·7
04	11 46·0	11 47·9	11 13·8	0·4	0·3	6·4	5·1	12·4	9·8
05	11 46·3	11 48·2	11 14·1	0·5	0·4	6·5	5·1	12·5	9·9
06	11 46·5	11 48·4	11 14·3	0·6	0·5	6·6	5·2	12·6	10·0
07	11 46·8	11 48·7	11 14·6	0·7	0·6	6·7	5·3	12·7	10·1
08	11 47·0	11 48·9	11 14·8	0·8	0·6	6·8	5·4	12·8	10·1
09	11 47·3	11 49·2	11 15·0	0·9	0·7	6·9	5·5	12·9	10·2
10	11 47·5	11 49·4	11 15·3	1·0	0·8	7·0	5·5	13·0	10·3
11	11 47·8	11 49·7	11 15·5	1·1	0·9	7·1	5·6	13·1	10·4
12	11 48·0	11 49·9	11 15·7	1·2	1·0	7·2	5·7	13·2	10·5
13	11 48·3	11 50·2	11 16·0	1·3	1·0	7·3	5·8	13·3	10·5
14	11 48·5	11 50·4	11 16·2	1·4	1·1	7·4	5·9	13·4	10·6
15	11 48·8	11 50·7	11 16·5	1·5	1·2	7·5	5·9	13·5	10·7
16	11 49·0	11 50·9	11 16·7	1·6	1·3	7·6	6·0	13·6	10·8
17	11 49·3	11 51·2	11 16·9	1·7	1·3	7·7	6·1	13·7	10·8
18	11 49·5	11 51·4	11 17·2	1·8	1·4	7·8	6·2	13·8	10·9
19	11 49·8	11 51·7	11 17·4	1·9	1·5	7·9	6·3	13·9	11·0
20	11 50·0	11 51·9	11 17·7	2·0	1·6	8·0	6·3	14·0	11·1
21	11 50·3	11 52·2	11 17·9	2·1	1·7	8·1	6·4	14·1	11·2
22	11 50·5	11 52·4	11 18·1	2·2	1·7	8·2	6·5	14·2	11·2
23	11 50·8	11 52·7	11 18·4	2·3	1·8	8·3	6·6	14·3	11·3
24	11 51·0	11 52·9	11 18·6	2·4	1·9	8·4	6·7	14·4	11·4
25	11 51·3	11 53·2	11 18·8	2·5	2·0	8·5	6·7	14·5	11·5
26	11 51·5	11 53·4	11 19·1	2·6	2·1	8·6	6·8	14·6	11·6
27	11 51·8	11 53·7	11 19·3	2·7	2·1	8·7	6·9	14·7	11·6
28	11 52·0	11 53·9	11 19·6	2·8	2·2	8·8	7·0	14·8	11·7
29	11 52·3	11 54·2	11 19·8	2·9	2·3	8·9	7·0	14·9	11·8
30	11 52·5	11 54·5	11 20·0	3·0	2·4	9·0	7·1	15·0	11·9
31	11 52·8	11 54·7	11 20·3	3·1	2·5	9·1	7·2	15·1	12·0
32	11 53·0	11 55·0	11 20·5	3·2	2·5	9·2	7·3	15·2	12·0
33	11 53·3	11 55·2	11 20·8	3·3	2·6	9·3	7·4	15·3	12·1
34	11 53·5	11 55·5	11 21·0	3·4	2·7	9·4	7·4	15·4	12·2
35	11 53·8	11 55·7	11 21·2	3·5	2·8	9·5	7·5	15·5	12·3
36	11 54·0	11 56·0	11 21·5	3·6	2·9	9·6	7·6	15·6	12·4
37	11 54·3	11 56·2	11 21·7	3·7	2·9	9·7	7·7	15·7	12·4
38	11 54·5	11 56·5	11 22·0	3·8	3·0	9·8	7·8	15·8	12·5
39	11 54·8	11 56·7	11 22·2	3·9	3·1	9·9	7·8	15·9	12·6
40	11 55·0	11 57·0	11 22·4	4·0	3·2	10·0	7·9	16·0	12·7
41	11 55·3	11 57·2	11 22·7	4·1	3·2	10·1	8·0	16·1	12·7
42	11 55·5	11 57·5	11 22·9	4·2	3·3	10·2	8·1	16·2	12·8
43	11 55·8	11 57·7	11 23·1	4·3	3·4	10·3	8·2	16·3	12·9
44	11 56·0	11 58·0	11 23·4	4·4	3·5	10·4	8·2	16·4	13·0
45	11 56·3	11 58·2	11 23·6	4·5	3·6	10·5	8·3	16·5	13·1
46	11 56·5	11 58·5	11 23·9	4·6	3·6	10·6	8·4	16·6	13·1
47	11 56·8	11 58·7	11 24·1	4·7	3·7	10·7	8·5	16·7	13·2
48	11 57·0	11 59·0	11 24·3	4·8	3·8	10·8	8·6	16·8	13·3
49	11 57·3	11 59·2	11 24·6	4·9	3·9	10·9	8·6	16·9	13·4
50	11 57·5	11 59·5	11 24·8	5·0	4·0	11·0	8·7	17·0	13·5
51	11 57·8	11 59·7	11 25·1	5·1	4·0	11·1	8·8	17·1	13·5
52	11 58·0	12 00·0	11 25·3	5·2	4·1	11·2	8·9	17·2	13·6
53	11 58·3	12 00·2	11 25·5	5·3	4·2	11·3	8·9	17·3	13·7
54	11 58·5	12 00·5	11 25·8	5·4	4·3	11·4	9·0	17·4	13·8
55	11 58·8	12 00·7	11 26·0	5·5	4·4	11·5	9·1	17·5	13·9
56	11 59·0	12 01·0	11 26·2	5·6	4·4	11·6	9·2	17·6	13·9
57	11 59·3	12 01·2	11 26·5	5·7	4·5	11·7	9·3	17·7	14·0
58	11 59·5	12 01·5	11 26·7	5·8	4·6	11·8	9·3	17·8	14·1
59	11 59·8	12 01·7	11 27·0	5·9	4·7	11·9	9·4	17·9	14·2
60	12 00·0	12 02·0	11 27·2	6·0	4·8	12·0	9·5	18·0	14·3

48m s	SUN PLANETS ° ′	ARIES ° ′	MOON ° ′	v or d ′	Corrn ′	v or d ′	Corrn ′	v or d ′	Corrn ′
00	12 00·0	12 02·0	11 27·2	0·0	0·0	6·0	4·9	12·0	9·7
01	12 00·3	12 02·2	11 27·4	0·1	0·1	6·1	4·9	12·1	9·8
02	12 00·5	12 02·5	11 27·7	0·2	0·2	6·2	5·0	12·2	9·9
03	12 00·8	12 02·7	11 27·9	0·3	0·2	6·3	5·1	12·3	9·9
04	12 01·0	12 03·0	11 28·2	0·4	0·3	6·4	5·2	12·4	10·0
05	12 01·3	12 03·2	11 28·4	0·5	0·4	6·5	5·3	12·5	10·1
06	12 01·5	12 03·5	11 28·6	0·6	0·5	6·6	5·3	12·6	10·2
07	12 01·8	12 03·7	11 28·9	0·7	0·6	6·7	5·4	12·7	10·3
08	12 02·0	12 04·0	11 29·1	0·8	0·6	6·8	5·5	12·8	10·3
09	12 02·3	12 04·2	11 29·3	0·9	0·7	6·9	5·6	12·9	10·4
10	12 02·5	12 04·5	11 29·6	1·0	0·8	7·0	5·7	13·0	10·5
11	12 02·8	12 04·7	11 29·8	1·1	0·9	7·1	5·7	13·1	10·6
12	12 03·0	12 05·0	11 30·1	1·2	1·0	7·2	5·8	13·2	10·7
13	12 03·3	12 05·2	11 30·3	1·3	1·1	7·3	5·9	13·3	10·8
14	12 03·5	12 05·5	11 30·5	1·4	1·1	7·4	6·0	13·4	10·8
15	12 03·8	12 05·7	11 30·8	1·5	1·2	7·5	6·1	13·5	10·9
16	12 04·0	12 06·0	11 31·0	1·6	1·3	7·6	6·1	13·6	11·0
17	12 04·3	12 06·2	11 31·3	1·7	1·4	7·7	6·2	13·7	11·1
18	12 04·5	12 06·5	11 31·5	1·8	1·5	7·8	6·3	13·8	11·2
19	12 04·8	12 06·7	11 31·7	1·9	1·5	7·9	6·4	13·9	11·2
20	12 05·0	12 07·0	11 32·0	2·0	1·6	8·0	6·5	14·0	11·3
21	12 05·3	12 07·2	11 32·2	2·1	1·7	8·1	6·5	14·1	11·4
22	12 05·5	12 07·5	11 32·4	2·2	1·8	8·2	6·6	14·2	11·5
23	12 05·8	12 07·7	11 32·7	2·3	1·9	8·3	6·7	14·3	11·6
24	12 06·0	12 08·0	11 32·9	2·4	1·9	8·4	6·8	14·4	11·6
25	12 06·3	12 08·2	11 33·2	2·5	2·0	8·5	6·9	14·5	11·7
26	12 06·5	12 08·5	11 33·4	2·6	2·1	8·6	7·0	14·6	11·8
27	12 06·8	12 08·7	11 33·6	2·7	2·2	8·7	7·0	14·7	11·9
28	12 07·0	12 09·0	11 33·9	2·8	2·3	8·8	7·1	14·8	12·0
29	12 07·3	12 09·2	11 34·1	2·9	2·3	8·9	7·2	14·9	12·0
30	12 07·5	12 09·5	11 34·4	3·0	2·4	9·0	7·3	15·0	12·1
31	12 07·8	12 09·7	11 34·6	3·1	2·5	9·1	7·4	15·1	12·2
32	12 08·0	12 10·0	11 34·8	3·2	2·6	9·2	7·4	15·2	12·3
33	12 08·3	12 10·2	11 35·1	3·3	2·7	9·3	7·5	15·3	12·4
34	12 08·5	12 10·5	11 35·3	3·4	2·7	9·4	7·6	15·4	12·4
35	12 08·8	12 10·7	11 35·6	3·5	2·8	9·5	7·7	15·5	12·5
36	12 09·0	12 11·0	11 35·8	3·6	2·9	9·6	7·8	15·6	12·6
37	12 09·3	12 11·2	11 36·0	3·7	3·0	9·7	7·8	15·7	12·7
38	12 09·5	12 11·5	11 36·3	3·8	3·1	9·8	7·9	15·8	12·8
39	12 09·8	12 11·7	11 36·5	3·9	3·2	9·9	8·0	15·9	12·9
40	12 10·0	12 12·0	11 36·7	4·0	3·2	10·0	8·1	16·0	12·9
41	12 10·3	12 12·2	11 37·0	4·1	3·3	10·1	8·2	16·1	13·0
42	12 10·5	12 12·5	11 37·2	4·2	3·4	10·2	8·2	16·2	13·1
43	12 10·8	12 12·8	11 37·5	4·3	3·5	10·3	8·3	16·3	13·2
44	12 11·0	12 13·0	11 37·7	4·4	3·6	10·4	8·4	16·4	13·3
45	12 11·3	12 13·3	11 37·9	4·5	3·6	10·5	8·5	16·5	13·3
46	12 11·5	12 13·5	11 38·2	4·6	3·7	10·6	8·6	16·6	13·4
47	12 11·8	12 13·8	11 38·4	4·7	3·8	10·7	8·6	16·7	13·5
48	12 12·0	12 14·0	11 38·7	4·8	3·9	10·8	8·7	16·8	13·6
49	12 12·3	12 14·3	11 38·9	4·9	4·0	10·9	8·8	16·9	13·7
50	12 12·5	12 14·5	11 39·1	5·0	4·0	11·0	8·9	17·0	13·7
51	12 12·8	12 14·8	11 39·4	5·1	4·1	11·1	9·0	17·1	13·8
52	12 13·0	12 15·0	11 39·6	5·2	4·2	11·2	9·1	17·2	13·9
53	12 13·3	12 15·3	11 39·8	5·3	4·3	11·3	9·1	17·3	14·0
54	12 13·5	12 15·5	11 40·1	5·4	4·4	11·4	9·2	17·4	14·1
55	12 13·8	12 15·8	11 40·3	5·5	4·4	11·5	9·3	17·5	14·1
56	12 14·0	12 16·0	11 40·6	5·6	4·5	11·6	9·4	17·6	14·2
57	12 14·3	12 16·3	11 40·8	5·7	4·6	11·7	9·5	17·7	14·3
58	12 14·5	12 16·5	11 41·0	5·8	4·7	11·8	9·5	17·8	14·4
59	12 14·8	12 16·8	11 41·3	5·9	4·8	11·9	9·6	17·9	14·5
60	12 15·0	12 17·0	11 41·5	6·0	4·9	12·0	9·7	18·0	14·6

49m s	SUN PLANETS ° ′	ARIES ° ′	MOON ° ′	v or d ′	Corrn ′	v or d ′	Corrn ′	v or d ′	Corrn ′
00	12 15·0	12 17·0	11 41·5	0·0	0·0	6·0	5·0	12·0	9·9
01	12 15·3	12 17·3	11 41·8	0·1	0·1	6·1	5·0	12·1	10·0
02	12 15·5	12 17·5	11 42·0	0·2	0·2	6·2	5·1	12·2	10·1
03	12 15·8	12 17·8	11 42·2	0·3	0·2	6·3	5·2	12·3	10·1
04	12 16·0	12 18·0	11 42·5	0·4	0·3	6·4	5·3	12·4	10·2
05	12 16·3	12 18·3	11 42·7	0·5	0·4	6·5	5·4	12·5	10·3
06	12 16·5	12 18·5	11 42·9	0·6	0·5	6·6	5·4	12·6	10·4
07	12 16·8	12 18·8	11 43·2	0·7	0·6	6·7	5·5	12·7	10·5
08	12 17·0	12 19·0	11 43·4	0·8	0·7	6·8	5·6	12·8	10·6
09	12 17·3	12 19·3	11 43·7	0·9	0·7	6·9	5·7	12·9	10·6
10	12 17·5	12 19·5	11 43·9	1·0	0·8	7·0	5·8	13·0	10·7
11	12 17·8	12 19·8	11 44·1	1·1	0·9	7·1	5·9	13·1	10·8
12	12 18·0	12 20·0	11 44·4	1·2	1·0	7·2	5·9	13·2	10·9
13	12 18·3	12 20·3	11 44·6	1·3	1·1	7·3	6·0	13·3	11·0
14	12 18·5	12 20·5	11 44·9	1·4	1·2	7·4	6·1	13·4	11·1
15	12 18·8	12 20·8	11 45·1	1·5	1·2	7·5	6·2	13·5	11·1
16	12 19·0	12 21·0	11 45·3	1·6	1·3	7·6	6·3	13·6	11·2
17	12 19·3	12 21·3	11 45·6	1·7	1·4	7·7	6·4	13·7	11·3
18	12 19·5	12 21·5	11 45·8	1·8	1·5	7·8	6·4	13·8	11·4
19	12 19·8	12 21·8	11 46·1	1·9	1·6	7·9	6·5	13·9	11·5
20	12 20·0	12 22·0	11 46·3	2·0	1·7	8·0	6·6	14·0	11·6
21	12 20·3	12 22·3	11 46·5	2·1	1·7	8·1	6·7	14·1	11·6
22	12 20·5	12 22·5	11 46·8	2·2	1·8	8·2	6·8	14·2	11·7
23	12 20·8	12 22·8	11 47·0	2·3	1·9	8·3	6·8	14·3	11·8
24	12 21·0	12 23·0	11 47·2	2·4	2·0	8·4	6·9	14·4	11·9
25	12 21·3	12 23·3	11 47·5	2·5	2·1	8·5	7·0	14·5	12·0
26	12 21·5	12 23·5	11 47·7	2·6	2·1	8·6	7·1	14·6	12·0
27	12 21·8	12 23·8	11 48·0	2·7	2·2	8·7	7·2	14·7	12·1
28	12 22·0	12 24·0	11 48·2	2·8	2·3	8·8	7·3	14·8	12·2
29	12 22·3	12 24·3	11 48·4	2·9	2·4	8·9	7·3	14·9	12·3
30	12 22·5	12 24·5	11 48·7	3·0	2·5	9·0	7·4	15·0	12·4
31	12 22·8	12 24·8	11 48·9	3·1	2·6	9·1	7·5	15·1	12·5
32	12 23·0	12 25·0	11 49·2	3·2	2·6	9·2	7·6	15·2	12·5
33	12 23·3	12 25·3	11 49·4	3·3	2·7	9·3	7·7	15·3	12·6
34	12 23·5	12 25·5	11 49·6	3·4	2·8	9·4	7·8	15·4	12·7
35	12 23·8	12 25·8	11 49·9	3·5	2·9	9·5	7·8	15·5	12·8
36	12 24·0	12 26·0	11 50·1	3·6	3·0	9·6	7·9	15·6	12·9
37	12 24·3	12 26·3	11 50·3	3·7	3·1	9·7	8·0	15·7	13·0
38	12 24·5	12 26·5	11 50·6	3·8	3·1	9·8	8·1	15·8	13·0
39	12 24·8	12 26·8	11 50·8	3·9	3·2	9·9	8·2	15·9	13·1
40	12 25·0	12 27·0	11 51·1	4·0	3·3	10·0	8·3	16·0	13·2
41	12 25·3	12 27·3	11 51·3	4·1	3·4	10·1	8·3	16·1	13·3
42	12 25·5	12 27·5	11 51·5	4·2	3·5	10·2	8·4	16·2	13·4
43	12 25·8	12 27·8	11 51·8	4·3	3·5	10·3	8·5	16·3	13·4
44	12 26·0	12 28·0	11 52·0	4·4	3·6	10·4	8·6	16·4	13·5
45	12 26·3	12 28·3	11 52·3	4·5	3·7	10·5	8·7	16·5	13·6
46	12 26·5	12 28·5	11 52·5	4·6	3·8	10·6	8·7	16·6	13·7
47	12 26·8	12 28·8	11 52·7	4·7	3·9	10·7	8·8	16·7	13·8
48	12 27·0	12 29·0	11 53·0	4·8	4·0	10·8	8·9	16·8	13·9
49	12 27·3	12 29·3	11 53·2	4·9	4·0	10·9	9·0	16·9	13·9
50	12 27·5	12 29·5	11 53·4	5·0	4·1	11·0	9·1	17·0	14·0
51	12 27·8	12 29·8	11 53·7	5·1	4·2	11·1	9·2	17·1	14·1
52	12 28·0	12 30·0	11 53·9	5·2	4·3	11·2	9·2	17·2	14·2
53	12 28·3	12 30·3	11 54·2	5·3	4·4	11·3	9·3	17·3	14·3
54	12 28·5	12 30·5	11 54·4	5·4	4·5	11·4	9·4	17·4	14·4
55	12 28·8	12 30·8	11 54·6	5·5	4·5	11·5	9·5	17·5	14·4
56	12 29·0	12 31·1	11 54·9	5·6	4·6	11·6	9·6	17·6	14·5
57	12 29·3	12 31·3	11 55·1	5·7	4·7	11·7	9·7	17·7	14·6
58	12 29·5	12 31·6	11 55·4	5·8	4·8	11·8	9·7	17·8	14·7
59	12 29·8	12 31·8	11 55·6	5·9	4·9	11·9	9·8	17·9	14·8
60	12 30·0	12 32·1	11 55·8	6·0	5·0	12·0	9·9	18·0	14·9

50m	SUN PLANETS	ARIES	MOON	v or d	Corrn	v or d	Corrn	v or d	Corrn
s	° ′	° ′	° ′	′	′	′	′	′	′
00	12 30·0	12 32·1	11 55·8	0·0	0·0	6·0	5·1	12·0	10·1
01	12 30·3	12 32·3	11 56·1	0·1	0·1	6·1	5·1	12·1	10·2
02	12 30·5	12 32·6	11 56·3	0·2	0·2	6·2	5·2	12·2	10·3
03	12 30·8	12 32·8	11 56·5	0·3	0·3	6·3	5·3	12·3	10·4
04	12 31·0	12 33·1	11 56·8	0·4	0·3	6·4	5·4	12·4	10·4
05	12 31·3	12 33·3	11 57·0	0·5	0·4	6·5	5·5	12·5	10·5
06	12 31·5	12 33·6	11 57·3	0·6	0·5	6·6	5·6	12·6	10·6
07	12 31·8	12 33·8	11 57·5	0·7	0·6	6·7	5·6	12·7	10·7
08	12 32·0	12 34·1	11 57·7	0·8	0·7	6·8	5·7	12·8	10·8
09	12 32·3	12 34·3	11 58·0	0·9	0·8	6·9	5·8	12·9	10·9
10	12 32·5	12 34·6	11 58·2	1·0	0·8	7·0	5·9	13·0	10·9
11	12 32·8	12 34·8	11 58·5	1·1	0·9	7·1	6·0	13·1	11·0
12	12 33·0	12 35·1	11 58·7	1·2	1·0	7·2	6·1	13·2	11·1
13	12 33·3	12 35·3	11 58·9	1·3	1·1	7·3	6·1	13·3	11·2
14	12 33·5	12 35·6	11 59·2	1·4	1·2	7·4	6·2	13·4	11·3
15	12 33·8	12 35·8	11 59·4	1·5	1·3	7·5	6·3	13·5	11·4
16	12 34·0	12 36·1	11 59·7	1·6	1·3	7·6	6·4	13·6	11·4
17	12 34·3	12 36·3	11 59·9	1·7	1·4	7·7	6·5	13·7	11·5
18	12 34·5	12 36·6	12 00·1	1·8	1·5	7·8	6·6	13·8	11·6
19	12 34·8	12 36·8	12 00·4	1·9	1·6	7·9	6·6	13·9	11·7
20	12 35·0	12 37·1	12 00·6	2·0	1·7	8·0	6·7	14·0	11·8
21	12 35·3	12 37·3	12 00·8	2·1	1·8	8·1	6·8	14·1	11·9
22	12 35·5	12 37·6	12 01·1	2·2	1·9	8·2	6·9	14·2	12·0
23	12 35·8	12 37·8	12 01·3	2·3	1·9	8·3	7·0	14·3	12·0
24	12 36·0	12 38·1	12 01·6	2·4	2·0	8·4	7·1	14·4	12·1
25	12 36·3	12 38·3	12 01·8	2·5	2·1	8·5	7·2	14·5	12·2
26	12 36·5	12 38·6	12 02·0	2·6	2·2	8·6	7·2	14·6	12·3
27	12 36·8	12 38·8	12 02·3	2·7	2·3	8·7	7·3	14·7	12·4
28	12 37·0	12 39·1	12 02·5	2·8	2·4	8·8	7·4	14·8	12·5
29	12 37·3	12 39·3	12 02·8	2·9	2·4	8·9	7·5	14·9	12·5
30	12 37·5	12 39·6	12 03·0	3·0	2·5	9·0	7·6	15·0	12·6
31	12 37·8	12 39·8	12 03·2	3·1	2·6	9·1	7·7	15·1	12·7
32	12 38·0	12 40·1	12 03·5	3·2	2·7	9·2	7·7	15·2	12·8
33	12 38·3	12 40·3	12 03·7	3·3	2·8	9·3	7·8	15·3	12·9
34	12 38·5	12 40·6	12 03·9	3·4	2·9	9·4	7·9	15·4	13·0
35	12 38·8	12 40·8	12 04·2	3·5	2·9	9·5	8·0	15·5	13·0
36	12 39·0	12 41·1	12 04·4	3·6	3·0	9·6	8·1	15·6	13·1
37	12 39·3	12 41·3	12 04·7	3·7	3·1	9·7	8·2	15·7	13·2
38	12 39·5	12 41·6	12 04·9	3·8	3·2	9·8	8·2	15·8	13·3
39	12 39·8	12 41·8	12 05·1	3·9	3·3	9·9	8·3	15·9	13·4
40	12 40·0	12 42·1	12 05·4	4·0	3·4	10·0	8·4	16·0	13·5
41	12 40·3	12 42·3	12 05·6	4·1	3·5	10·1	8·5	16·1	13·6
42	12 40·5	12 42·6	12 05·9	4·2	3·5	10·2	8·6	16·2	13·6
43	12 40·8	12 42·8	12 06·1	4·3	3·6	10·3	8·7	16·3	13·7
44	12 41·0	12 43·1	12 06·3	4·4	3·7	10·4	8·8	16·4	13·8
45	12 41·3	12 43·3	12 06·6	4·5	3·8	10·5	8·8	16·5	13·9
46	12 41·5	12 43·6	12 06·8	4·6	3·9	10·6	8·9	16·6	14·0
47	12 41·8	12 43·8	12 07·0	4·7	4·0	10·7	9·0	16·7	14·1
48	12 42·0	12 44·1	12 07·3	4·8	4·0	10·8	9·1	16·8	14·1
49	12 42·3	12 44·3	12 07·5	4·9	4·1	10·9	9·2	16·9	14·2
50	12 42·5	12 44·6	12 07·8	5·0	4·2	11·0	9·3	17·0	14·3
51	12 42·8	12 44·8	12 08·0	5·1	4·3	11·1	9·3	17·1	14·4
52	12 43·0	12 45·1	12 08·2	5·2	4·4	11·2	9·4	17·2	14·5
53	12 43·3	12 45·3	12 08·5	5·3	4·5	11·3	9·5	17·3	14·6
54	12 43·5	12 45·6	12 08·7	5·4	4·5	11·4	9·6	17·4	14·6
55	12 43·8	12 45·8	12 09·0	5·5	4·6	11·5	9·7	17·5	14·7
56	12 44·0	12 46·1	12 09·2	5·6	4·7	11·6	9·8	17·6	14·8
57	12 44·3	12 46·3	12 09·4	5·7	4·8	11·7	9·8	17·7	14·9
58	12 44·5	12 46·6	12 09·7	5·8	4·9	11·8	9·9	17·8	15·0
59	12 44·8	12 46·8	12 09·9	5·9	5·0	11·9	10·0	17·9	15·1
60	12 45·0	12 47·1	12 10·2	6·0	5·1	12·0	10·1	18·0	15·2

51m	SUN PLANETS	ARIES	MOON	v or d	Corrn	v or d	Corrn	v or d	Corrn
s	° ′	° ′	° ′	′	′	′	′	′	′
00	12 45·0	12 47·1	12 10·2	0·0	0·0	6·0	5·2	12·0	10·3
01	12 45·3	12 47·3	12 10·4	0·1	0·1	6·1	5·2	12·1	10·4
02	12 45·5	12 47·6	12 10·6	0·2	0·2	6·2	5·3	12·2	10·5
03	12 45·8	12 47·8	12 10·9	0·3	0·3	6·3	5·4	12·3	10·6
04	12 46·0	12 48·1	12 11·1	0·4	0·3	6·4	5·5	12·4	10·6
05	12 46·3	12 48·3	12 11·3	0·5	0·4	6·5	5·6	12·5	10·7
06	12 46·5	12 48·6	12 11·6	0·6	0·5	6·6	5·7	12·6	10·8
07	12 46·8	12 48·8	12 11·8	0·7	0·6	6·7	5·8	12·7	10·9
08	12 47·0	12 49·1	12 12·1	0·8	0·7	6·8	5·8	12·8	11·0
09	12 47·3	12 49·4	12 12·3	0·9	0·8	6·9	5·9	12·9	11·1
10	12 47·5	12 49·6	12 12·5	1·0	0·9	7·0	6·0	13·0	11·2
11	12 47·8	12 49·9	12 12·8	1·1	0·9	7·1	6·1	13·1	11·2
12	12 48·0	12 50·1	12 13·0	1·2	1·0	7·2	6·2	13·2	11·3
13	12 48·3	12 50·4	12 13·3	1·3	1·1	7·3	6·3	13·3	11·4
14	12 48·5	12 50·6	12 13·5	1·4	1·2	7·4	6·4	13·4	11·5
15	12 48·8	12 50·9	12 13·7	1·5	1·3	7·5	6·4	13·5	11·6
16	12 49·0	12 51·1	12 14·0	1·6	1·4	7·6	6·5	13·6	11·7
17	12 49·3	12 51·4	12 14·2	1·7	1·5	7·7	6·6	13·7	11·8
18	12 49·5	12 51·6	12 14·4	1·8	1·5	7·8	6·7	13·8	11·8
19	12 49·8	12 51·9	12 14·7	1·9	1·6	7·9	6·8	13·9	11·9
20	12 50·0	12 52·1	12 14·9	2·0	1·7	8·0	6·9	14·0	12·0
21	12 50·3	12 52·4	12 15·2	2·1	1·8	8·1	7·0	14·1	12·1
22	12 50·5	12 52·6	12 15·4	2·2	1·9	8·2	7·0	14·2	12·2
23	12 50·8	12 52·9	12 15·6	2·3	2·0	8·3	7·1	14·3	12·3
24	12 51·0	12 53·1	12 15·9	2·4	2·1	8·4	7·2	14·4	12·4
25	12 51·3	12 53·4	12 16·1	2·5	2·1	8·5	7·3	14·5	12·4
26	12 51·5	12 53·6	12 16·4	2·6	2·2	8·6	7·4	14·6	12·5
27	12 51·8	12 53·9	12 16·6	2·7	2·3	8·7	7·5	14·7	12·6
28	12 52·0	12 54·1	12 16·8	2·8	2·4	8·8	7·6	14·8	12·7
29	12 52·3	12 54·4	12 17·1	2·9	2·5	8·9	7·6	14·9	12·8
30	12 52·5	12 54·6	12 17·3	3·0	2·6	9·0	7·7	15·0	12·9
31	12 52·8	12 54·9	12 17·5	3·1	2·7	9·1	7·8	15·1	13·0
32	12 53·0	12 55·1	12 17·8	3·2	2·7	9·2	7·9	15·2	13·0
33	12 53·3	12 55·4	12 18·0	3·3	2·8	9·3	8·0	15·3	13·1
34	12 53·5	12 55·6	12 18·3	3·4	2·9	9·4	8·1	15·4	13·2
35	12 53·8	12 55·9	12 18·5	3·5	3·0	9·5	8·2	15·5	13·3
36	12 54·0	12 56·1	12 18·7	3·6	3·1	9·6	8·2	15·6	13·4
37	12 54·3	12 56·4	12 19·0	3·7	3·2	9·7	8·3	15·7	13·5
38	12 54·5	12 56·6	12 19·2	3·8	3·3	9·8	8·4	15·8	13·6
39	12 54·8	12 56·9	12 19·5	3·9	3·3	9·9	8·5	15·9	13·6
40	12 55·0	12 57·1	12 19·7	4·0	3·4	10·0	8·6	16·0	13·7
41	12 55·3	12 57·4	12 19·9	4·1	3·5	10·1	8·7	16·1	13·8
42	12 55·5	12 57·6	12 20·2	4·2	3·6	10·2	8·8	16·2	13·9
43	12 55·8	12 57·9	12 20·4	4·3	3·7	10·3	8·8	16·3	14·0
44	12 56·0	12 58·1	12 20·6	4·4	3·8	10·4	8·9	16·4	14·1
45	12 56·3	12 58·4	12 20·9	4·5	3·9	10·5	9·0	16·5	14·2
46	12 56·5	12 58·6	12 21·1	4·6	3·9	10·6	9·1	16·6	14·2
47	12 56·8	12 58·9	12 21·4	4·7	4·0	10·7	9·2	16·7	14·3
48	12 57·0	12 59·1	12 21·6	4·8	4·1	10·8	9·3	16·8	14·4
49	12 57·3	12 59·4	12 21·8	4·9	4·2	10·9	9·4	16·9	14·5
50	12 57·5	12 59·6	12 22·1	5·0	4·3	11·0	9·4	17·0	14·6
51	12 57·8	12 59·9	12 22·3	5·1	4·4	11·1	9·5	17·1	14·7
52	12 58·0	13 00·1	12 22·6	5·2	4·5	11·2	9·6	17·2	14·8
53	12 58·3	13 00·4	12 22·8	5·3	4·5	11·3	9·7	17·3	14·8
54	12 58·5	13 00·6	12 23·0	5·4	4·6	11·4	9·8	17·4	14·9
55	12 58·8	13 00·9	12 23·3	5·5	4·7	11·5	9·9	17·5	15·0
56	12 59·0	13 01·1	12 23·5	5·6	4·8	11·6	10·0	17·6	15·1
57	12 59·3	13 01·4	12 23·8	5·7	4·9	11·7	10·0	17·7	15·2
58	12 59·5	13 01·6	12 24·0	5·8	5·0	11·8	10·1	17·8	15·3
59	12 59·8	13 01·9	12 24·2	5·9	5·1	11·9	10·2	17·9	15·4
60	13 00·0	13 02·1	12 24·5	6·0	5·2	12·0	10·3	18·0	15·5

52ᵐ	SUN PLANETS	ARIES	MOON	v or d	Corrⁿ	v or d	Corrⁿ	v or d	Corrⁿ
s	° ′	° ′	° ′	′	′	′	′	′	′
00	13 00·0	13 02·1	12 24·5	0·0	0·0	6·0	5·3	12·0	10·5
01	13 00·3	13 02·4	12 24·7	0·1	0·1	6·1	5·3	12·1	10·6
02	13 00·5	13 02·6	12 24·9	0·2	0·2	6·2	5·4	12·2	10·7
03	13 00·8	13 02·9	12 25·2	0·3	0·3	6·3	5·5	12·3	10·8
04	13 01·0	13 03·1	12 25·4	0·4	0·4	6·4	5·6	12·4	10·9
05	13 01·3	13 03·4	12 25·7	0·5	0·4	6·5	5·7	12·5	10·9
06	13 01·5	13 03·6	12 25·9	0·6	0·5	6·6	5·8	12·6	11·0
07	13 01·8	13 03·9	12 26·1	0·7	0·6	6·7	5·9	12·7	11·1
08	13 02·0	13 04·1	12 26·4	0·8	0·7	6·8	6·0	12·8	11·2
09	13 02·3	13 04·4	12 26·6	0·9	0·8	6·9	6·0	12·9	11·3
10	13 02·5	13 04·6	12 26·9	1·0	0·9	7·0	6·1	13·0	11·4
11	13 02·8	13 04·9	12 27·1	1·1	1·0	7·1	6·2	13·1	11·5
12	13 03·0	13 05·1	12 27·3	1·2	1·1	7·2	6·3	13·2	11·6
13	13 03·3	13 05·4	12 27·6	1·3	1·1	7·3	6·4	13·3	11·6
14	13 03·5	13 05·6	12 27·8	1·4	1·2	7·4	6·5	13·4	11·7
15	13 03·8	13 05·9	12 28·0	1·5	1·3	7·5	6·6	13·5	11·8
16	13 04·0	13 06·1	12 28·3	1·6	1·4	7·6	6·7	13·6	11·9
17	13 04·3	13 06·4	12 28·5	1·7	1·5	7·7	6·7	13·7	12·0
18	13 04·5	13 06·6	12 28·8	1·8	1·6	7·8	6·8	13·8	12·1
19	13 04·8	13 06·9	12 29·0	1·9	1·7	7·9	6·9	13·9	12·2
20	13 05·0	13 07·1	12 29·2	2·0	1·8	8·0	7·0	14·0	12·3
21	13 05·3	13 07·4	12 29·5	2·1	1·8	8·1	7·1	14·1	12·3
22	13 05·5	13 07·7	12 29·7	2·2	1·9	8·2	7·2	14·2	12·4
23	13 05·8	13 07·9	12 30·0	2·3	2·0	8·3	7·3	14·3	12·5
24	13 06·0	13 08·2	12 30·2	2·4	2·1	8·4	7·4	14·4	12·6
25	13 06·3	13 08·4	12 30·4	2·5	2·2	8·5	7·4	14·5	12·7
26	13 06·5	13 08·7	12 30·7	2·6	2·3	8·6	7·5	14·6	12·8
27	13 06·8	13 08·9	12 30·9	2·7	2·4	8·7	7·6	14·7	12·9
28	13 07·0	13 09·2	12 31·1	2·8	2·5	8·8	7·7	14·8	13·0
29	13 07·3	13 09·4	12 31·4	2·9	2·5	8·9	7·8	14·9	13·0
30	13 07·5	13 09·7	12 31·6	3·0	2·6	9·0	7·9	15·0	13·1
31	13 07·8	13 09·9	12 31·9	3·1	2·7	9·1	8·0	15·1	13·2
32	13 08·0	13 10·2	12 32·1	3·2	2·8	9·2	8·1	15·2	13·3
33	13 08·3	13 10·4	12 32·3	3·3	2·9	9·3	8·1	15·3	13·4
34	13 08·5	13 10·7	12 32·6	3·4	3·0	9·4	8·2	15·4	13·5
35	13 08·8	13 10·9	12 32·8	3·5	3·1	9·5	8·3	15·5	13·6
36	13 09·0	13 11·2	12 33·1	3·6	3·2	9·6	8·4	15·6	13·7
37	13 09·3	13 11·4	12 33·3	3·7	3·2	9·7	8·5	15·7	13·7
38	13 09·5	13 11·7	12 33·5	3·8	3·3	9·8	8·6	15·8	13·8
39	13 09·8	13 11·9	12 33·8	3·9	3·4	9·9	8·7	15·9	13·9
40	13 10·0	13 12·2	12 34·0	4·0	3·5	10·0	8·8	16·0	14·0
41	13 10·3	13 12·4	12 34·2	4·1	3·6	10·1	8·8	16·1	14·1
42	13 10·5	13 12·7	12 34·5	4·2	3·7	10·2	8·9	16·2	14·2
43	13 10·8	13 12·9	12 34·7	4·3	3·8	10·3	9·0	16·3	14·3
44	13 11·0	13 13·2	12 35·0	4·4	3·9	10·4	9·1	16·4	14·4
45	13 11·3	13 13·4	12 35·2	4·5	3·9	10·5	9·2	16·5	14·4
46	13 11·5	13 13·7	12 35·4	4·6	4·0	10·6	9·3	16·6	14·5
47	13 11·8	13 13·9	12 35·7	4·7	4·1	10·7	9·4	16·7	14·6
48	13 12·0	13 14·2	12 35·9	4·8	4·2	10·8	9·5	16·8	14·7
49	13 12·3	13 14·4	12 36·2	4·9	4·3	10·9	9·5	16·9	14·8
50	13 12·5	13 14·7	12 36·4	5·0	4·4	11·0	9·6	17·0	14·9
51	13 12·8	13 14·9	12 36·6	5·1	4·5	11·1	9·7	17·1	15·0
52	13 13·0	13 15·2	12 36·9	5·2	4·6	11·2	9·8	17·2	15·1
53	13 13·3	13 15·4	12 37·1	5·3	4·6	11·3	9·9	17·3	15·1
54	13 13·5	13 15·7	12 37·4	5·4	4·7	11·4	10·0	17·4	15·2
55	13 13·8	13 15·9	12 37·6	5·5	4·8	11·5	10·1	17·5	15·3
56	13 14·0	13 16·2	12 37·8	5·6	4·9	11·6	10·2	17·6	15·4
57	13 14·3	13 16·4	12 38·1	5·7	5·0	11·7	10·2	17·7	15·5
58	13 14·5	13 16·7	12 38·3	5·8	5·1	11·8	10·3	17·8	15·6
59	13 14·8	13 16·9	12 38·5	5·9	5·2	11·9	10·4	17·9	15·7
60	13 15·0	13 17·2	12 38·8	6·0	5·3	12·0	10·5	18·0	15·8

53ᵐ	SUN PLANETS	ARIES	MOON	v or d	Corrⁿ	v or d	Corrⁿ	v or d	Corrⁿ
s	° ′	° ′	° ′	′	′	′	′	′	′
00	13 15·0	13 17·2	12 38·8	0·0	0·0	6·0	5·4	12·0	10·7
01	13 15·3	13 17·4	12 39·0	0·1	0·1	6·1	5·4	12·1	10·8
02	13 15·5	13 17·7	12 39·3	0·2	0·2	6·2	5·5	12·2	10·9
03	13 15·8	13 17·9	12 39·5	0·3	0·3	6·3	5·6	12·3	11·0
04	13 16·0	13 18·2	12 39·7	0·4	0·4	6·4	5·7	12·4	11·1
05	13 16·3	13 18·4	12 40·0	0·5	0·4	6·5	5·8	12·5	11·1
06	13 16·5	13 18·7	12 40·2	0·6	0·5	6·6	5·9	12·6	11·2
07	13 16·8	13 18·9	12 40·5	0·7	0·6	6·7	6·0	12·7	11·3
08	13 17·0	13 19·2	12 40·7	0·8	0·7	6·8	6·1	12·8	11·4
09	13 17·3	13 19·4	12 40·9	0·9	0·8	6·9	6·2	12·9	11·5
10	13 17·5	13 19·7	12 41·2	1·0	0·9	7·0	6·2	13·0	11·6
11	13 17·8	13 19·9	12 41·4	1·1	1·0	7·1	6·3	13·1	11·7
12	13 18·0	13 20·2	12 41·6	1·2	1·1	7·2	6·4	13·2	11·8
13	13 18·3	13 20·4	12 41·9	1·3	1·2	7·3	6·5	13·3	11·9
14	13 18·5	13 20·7	12 42·1	1·4	1·2	7·4	6·6	13·4	11·9
15	13 18·8	13 20·9	12 42·4	1·5	1·3	7·5	6·7	13·5	12·0
16	13 19·0	13 21·2	12 42·6	1·6	1·4	7·6	6·8	13·6	12·1
17	13 19·3	13 21·4	12 42·8	1·7	1·5	7·7	6·9	13·7	12·2
18	13 19·5	13 21·7	12 43·1	1·8	1·6	7·8	7·0	13·8	12·3
19	13 19·8	13 21·9	12 43·3	1·9	1·7	7·9	7·0	13·9	12·4
20	13 20·0	13 22·2	12 43·6	2·0	1·8	8·0	7·1	14·0	12·5
21	13 20·3	13 22·4	12 43·8	2·1	1·9	8·1	7·2	14·1	12·6
22	13 20·5	13 22·7	12 44·0	2·2	2·0	8·2	7·3	14·2	12·7
23	13 20·8	13 22·9	12 44·3	2·3	2·1	8·3	7·4	14·3	12·8
24	13 21·0	13 23·2	12 44·5	2·4	2·1	8·4	7·5	14·4	12·8
25	13 21·3	13 23·4	12 44·7	2·5	2·2	8·5	7·6	14·5	12·9
26	13 21·5	13 23·7	12 45·0	2·6	2·3	8·6	7·7	14·6	13·0
27	13 21·8	13 23·9	12 45·2	2·7	2·4	8·7	7·8	14·7	13·1
28	13 22·0	13 24·2	12 45·5	2·8	2·5	8·8	7·8	14·8	13·2
29	13 22·3	13 24·4	12 45·7	2·9	2·6	8·9	7·9	14·9	13·3
30	13 22·5	13 24·7	12 45·9	3·0	2·7	9·0	8·0	15·0	13·4
31	13 22·8	13 24·9	12 46·2	3·1	2·8	9·1	8·1	15·1	13·5
32	13 23·0	13 25·2	12 46·4	3·2	2·9	9·2	8·2	15·2	13·6
33	13 23·3	13 25·4	12 46·7	3·3	2·9	9·3	8·3	15·3	13·6
34	13 23·5	13 25·7	12 46·9	3·4	3·0	9·4	8·4	15·4	13·7
35	13 23·8	13 26·0	12 47·1	3·5	3·1	9·5	8·5	15·5	13·8
36	13 24·0	13 26·2	12 47·4	3·6	3·2	9·6	8·6	15·6	13·9
37	13 24·3	13 26·5	12 47·6	3·7	3·3	9·7	8·6	15·7	14·0
38	13 24·5	13 26·7	12 47·9	3·8	3·4	9·8	8·7	15·8	14·1
39	13 24·8	13 27·0	12 48·1	3·9	3·5	9·9	8·8	15·9	14·2
40	13 25·0	13 27·2	12 48·3	4·0	3·6	10·0	8·9	16·0	14·3
41	13 25·3	13 27·5	12 48·6	4·1	3·7	10·1	9·0	16·1	14·4
42	13 25·5	13 27·7	12 48·8	4·2	3·7	10·2	9·1	16·2	14·4
43	13 25·8	13 28·0	12 49·0	4·3	3·8	10·3	9·2	16·3	14·5
44	13 26·0	13 28·2	12 49·3	4·4	3·9	10·4	9·3	16·4	14·6
45	13 26·3	13 28·5	12 49·5	4·5	4·0	10·5	9·4	16·5	14·7
46	13 26·5	13 28·7	12 49·8	4·6	4·1	10·6	9·5	16·6	14·8
47	13 26·8	13 29·0	12 50·0	4·7	4·2	10·7	9·5	16·7	14·9
48	13 27·0	13 29·2	12 50·2	4·8	4·3	10·8	9·6	16·8	15·0
49	13 27·3	13 29·5	12 50·5	4·9	4·4	10·9	9·7	16·9	15·1
50	13 27·5	13 29·7	12 50·7	5·0	4·5	11·0	9·8	17·0	15·2
51	13 27·8	13 30·0	12 51·0	5·1	4·5	11·1	9·9	17·1	15·2
52	13 28·0	13 30·2	12 51·2	5·2	4·6	11·2	10·0	17·2	15·3
53	13 28·3	13 30·5	12 51·4	5·3	4·7	11·3	10·1	17·3	15·4
54	13 28·5	13 30·7	12 51·7	5·4	4·8	11·4	10·2	17·4	15·5
55	13 28·8	13 31·0	12 51·9	5·5	4·9	11·5	10·3	17·5	15·6
56	13 29·0	13 31·2	12 52·1	5·6	5·0	11·6	10·3	17·6	15·7
57	13 29·3	13 31·5	12 52·4	5·7	5·1	11·7	10·4	17·7	15·8
58	13 29·5	13 31·7	12 52·6	5·8	5·2	11·8	10·5	17·8	15·9
59	13 29·8	13 32·0	12 52·9	5·9	5·3	11·9	10·6	17·9	16·0
60	13 30·0	13 32·2	12 53·1	6·0	5·4	12·0	10·7	18·0	16·1

54^m	SUN PLANETS	ARIES	MOON	v or d	Corr^n	v or d	Corr^n	v or d	Corr^n
s	° ′	° ′	° ′	′	′	′	′	′	′
00	13 30·0	13 32·2	12 53·1	0·0	0·0	6·0	5·5	12·0	10·9
01	13 30·3	13 32·5	12 53·3	0·1	0·1	6·1	5·5	12·1	11·0
02	13 30·5	13 32·7	12 53·6	0·2	0·2	6·2	5·6	12·2	11·1
03	13 30·8	13 33·0	12 53·8	0·3	0·3	6·3	5·7	12·3	11·2
04	13 31·0	13 33·2	12 54·1	0·4	0·4	6·4	5·8	12·4	11·3
05	13 31·3	13 33·5	12 54·3	0·5	0·5	6·5	5·9	12·5	11·4
06	13 31·5	13 33·7	12 54·5	0·6	0·5	6·6	6·0	12·6	11·4
07	13 31·8	13 34·0	12 54·8	0·7	0·6	6·7	6·1	12·7	11·5
08	13 32·0	13 34·2	12 55·0	0·8	0·7	6·8	6·2	12·8	11·6
09	13 32·3	13 34·5	12 55·2	0·9	0·8	6·9	6·3	12·9	11·7
10	13 32·5	13 34·7	12 55·5	1·0	0·9	7·0	6·4	13·0	11·8
11	13 32·8	13 35·0	12 55·7	1·1	1·0	7·1	6·4	13·1	11·9
12	13 33·0	13 35·2	12 56·0	1·2	1·1	7·2	6·5	13·2	12·0
13	13 33·3	13 35·5	12 56·2	1·3	1·2	7·3	6·6	13·3	12·1
14	13 33·5	13 35·7	12 56·4	1·4	1·3	7·4	6·7	13·4	12·2
15	13 33·8	13 36·0	12 56·7	1·5	1·4	7·5	6·8	13·5	12·3
16	13 34·0	13 36·2	12 56·9	1·6	1·5	7·6	6·9	13·6	12·4
17	13 34·3	13 36·5	12 57·2	1·7	1·5	7·7	7·0	13·7	12·4
18	13 34·5	13 36·7	12 57·4	1·8	1·6	7·8	7·1	13·8	12·5
19	13 34·8	13 37·0	12 57·6	1·9	1·7	7·9	7·2	13·9	12·6
20	13 35·0	13 37·2	12 57·9	2·0	1·8	8·0	7·3	14·0	12·7
21	13 35·3	13 37·5	12 58·1	2·1	1·9	8·1	7·4	14·1	12·8
22	13 35·5	13 37·7	12 58·3	2·2	2·0	8·2	7·4	14·2	12·9
23	13 35·8	13 38·0	12 58·6	2·3	2·1	8·3	7·5	14·3	13·0
24	13 36·0	13 38·2	12 58·8	2·4	2·2	8·4	7·6	14·4	13·1
25	13 36·3	13 38·5	12 59·1	2·5	2·3	8·5	7·7	14·5	13·2
26	13 36·5	13 38·7	12 59·3	2·6	2·4	8·6	7·8	14·6	13·3
27	13 36·8	13 39·0	12 59·5	2·7	2·5	8·7	7·9	14·7	13·4
28	13 37·0	13 39·2	12 59·8	2·8	2·5	8·8	8·0	14·8	13·4
29	13 37·3	13 39·5	13 00·0	2·9	2·6	8·9	8·1	14·9	13·5
30	13 37·5	13 39·7	13 00·3	3·0	2·7	9·0	8·2	15·0	13·6
31	13 37·8	13 40·0	13 00·5	3·1	2·8	9·1	8·3	15·1	13·7
32	13 38·0	13 40·2	13 00·7	3·2	2·9	9·2	8·4	15·2	13·8
33	13 38·3	13 40·5	13 01·0	3·3	3·0	9·3	8·4	15·3	13·9
34	13 38·5	13 40·7	13 01·2	3·4	3·1	9·4	8·5	15·4	14·0
35	13 38·8	13 41·0	13 01·5	3·5	3·2	9·5	8·6	15·5	14·1
36	13 39·0	13 41·2	13 01·7	3·6	3·3	9·6	8·7	15·6	14·2
37	13 39·3	13 41·5	13 01·9	3·7	3·4	9·7	8·8	15·7	14·3
38	13 39·5	13 41·7	13 02·2	3·8	3·5	9·8	8·9	15·8	14·4
39	13 39·8	13 42·0	13 02·4	3·9	3·5	9·9	9·0	15·9	14·4
40	13 40·0	13 42·2	13 02·6	4·0	3·6	10·0	9·1	16·0	14·5
41	13 40·3	13 42·5	13 02·9	4·1	3·7	10·1	9·2	16·1	14·6
42	13 40·5	13 42·7	13 03·1	4·2	3·8	10·2	9·3	16·2	14·7
43	13 40·8	13 43·0	13 03·4	4·3	3·9	10·3	9·4	16·3	14·8
44	13 41·0	13 43·2	13 03·6	4·4	4·0	10·4	9·4	16·4	14·9
45	13 41·3	13 43·5	13 03·8	4·5	4·1	10·5	9·5	16·5	15·0
46	13 41·5	13 43·7	13 04·1	4·6	4·2	10·6	9·6	16·6	15·1
47	13 41·8	13 44·0	13 04·3	4·7	4·3	10·7	9·7	16·7	15·2
48	13 42·0	13 44·3	13 04·6	4·8	4·4	10·8	9·8	16·8	15·3
49	13 42·3	13 44·5	13 04·8	4·9	4·5	10·9	9·9	16·9	15·4
50	13 42·5	13 44·8	13 05·0	5·0	4·5	11·0	10·0	17·0	15·4
51	13 42·8	13 45·0	13 05·3	5·1	4·6	11·1	10·1	17·1	15·5
52	13 43·0	13 45·3	13 05·5	5·2	4·7	11·2	10·2	17·2	15·6
53	13 43·3	13 45·5	13 05·7	5·3	4·8	11·3	10·3	17·3	15·7
54	13 43·5	13 45·8	13 06·0	5·4	4·9	11·4	10·4	17·4	15·8
55	13 43·8	13 46·0	13 06·2	5·5	5·0	11·5	10·4	17·5	15·9
56	13 44·0	13 46·3	13 06·5	5·6	5·1	11·6	10·5	17·6	16·0
57	13 44·3	13 46·5	13 06·7	5·7	5·2	11·7	10·6	17·7	16·1
58	13 44·5	13 46·8	13 06·9	5·8	5·3	11·8	10·7	17·8	16·2
59	13 44·8	13 47·0	13 07·2	5·9	5·4	11·9	10·8	17·9	16·3
60	13 45·0	13 47·3	13 07·4	6·0	5·5	12·0	10·9	18·0	16·4

55^m	SUN PLANETS	ARIES	MOON	v or d	Corr^n	v or d	Corr^n	v or d	Corr^n
s	° ′	° ′	° ′	′	′	′	′	′	′
00	13 45·0	13 47·3	13 07·4	0·0	0·0	6·0	5·6	12·0	11·1
01	13 45·3	13 47·5	13 07·7	0·1	0·1	6·1	5·6	12·1	11·2
02	13 45·5	13 47·8	13 07·9	0·2	0·2	6·2	5·7	12·2	11·3
03	13 45·8	13 48·0	13 08·1	0·3	0·3	6·3	5·8	12·3	11·4
04	13 46·0	13 48·3	13 08·4	0·4	0·4	6·4	5·9	12·4	11·5
05	13 46·3	13 48·5	13 08·6	0·5	0·5	6·5	6·0	12·5	11·6
06	13 46·5	13 48·8	13 08·8	0·6	0·6	6·6	6·1	12·6	11·7
07	13 46·8	13 49·0	13 09·1	0·7	0·6	6·7	6·2	12·7	11·7
08	13 47·0	13 49·3	13 09·3	0·8	0·7	6·8	6·3	12·8	11·8
09	13 47·3	13 49·5	13 09·6	0·9	0·8	6·9	6·4	12·9	11·9
10	13 47·5	13 49·8	13 09·8	1·0	0·9	7·0	6·5	13·0	12·0
11	13 47·8	13 50·0	13 10·0	1·1	1·0	7·1	6·6	13·1	12·1
12	13 48·0	13 50·3	13 10·3	1·2	1·1	7·2	6·7	13·2	12·2
13	13 48·3	13 50·5	13 10·5	1·3	1·2	7·3	6·8	13·3	12·3
14	13 48·5	13 50·8	13 10·8	1·4	1·3	7·4	6·8	13·4	12·4
15	13 48·8	13 51·0	13 11·0	1·5	1·4	7·5	6·9	13·5	12·5
16	13 49·0	13 51·3	13 11·2	1·6	1·5	7·6	7·0	13·6	12·6
17	13 49·3	13 51·5	13 11·5	1·7	1·6	7·7	7·1	13·7	12·7
18	13 49·5	13 51·8	13 11·7	1·8	1·7	7·8	7·2	13·8	12·8
19	13 49·8	13 52·0	13 12·0	1·9	1·8	7·9	7·3	13·9	12·9
20	13 50·0	13 52·3	13 12·2	2·0	1·9	8·0	7·4	14·0	13·0
21	13 50·3	13 52·5	13 12·4	2·1	1·9	8·1	7·5	14·1	13·0
22	13 50·5	13 52·8	13 12·7	2·2	2·0	8·2	7·6	14·2	13·1
23	13 50·8	13 53·0	13 12·9	2·3	2·1	8·3	7·7	14·3	13·2
24	13 51·0	13 53·3	13 13·1	2·4	2·2	8·4	7·8	14·4	13·3
25	13 51·3	13 53·5	13 13·4	2·5	2·3	8·5	7·9	14·5	13·4
26	13 51·5	13 53·8	13 13·6	2·6	2·4	8·6	8·0	14·6	13·5
27	13 51·8	13 54·0	13 13·9	2·7	2·5	8·7	8·0	14·7	13·6
28	13 52·0	13 54·3	13 14·1	2·8	2·6	8·8	8·1	14·8	13·7
29	13 52·3	13 54·5	13 14·3	2·9	2·7	8·9	8·2	14·9	13·8
30	13 52·5	13 54·8	13 14·6	3·0	2·8	9·0	8·3	15·0	13·9
31	13 52·8	13 55·0	13 14·8	3·1	2·9	9·1	8·4	15·1	14·0
32	13 53·0	13 55·3	13 15·1	3·2	3·0	9·2	8·5	15·2	14·1
33	13 53·3	13 55·5	13 15·3	3·3	3·1	9·3	8·6	15·3	14·2
34	13 53·5	13 55·8	13 15·5	3·4	3·1	9·4	8·7	15·4	14·2
35	13 53·8	13 56·0	13 15·8	3·5	3·2	9·5	8·8	15·5	14·3
36	13 54·0	13 56·3	13 16·0	3·6	3·3	9·6	8·9	15·6	14·4
37	13 54·3	13 56·5	13 16·2	3·7	3·4	9·7	9·0	15·7	14·5
38	13 54·5	13 56·8	13 16·5	3·8	3·5	9·8	9·1	15·8	14·6
39	13 54·8	13 57·0	13 16·7	3·9	3·6	9·9	9·2	15·9	14·7
40	13 55·0	13 57·3	13 17·0	4·0	3·7	10·0	9·3	16·0	14·8
41	13 55·3	13 57·5	13 17·2	4·1	3·8	10·1	9·3	16·1	14·9
42	13 55·5	13 57·8	13 17·4	4·2	3·9	10·2	9·4	16·2	15·0
43	13 55·8	13 58·0	13 17·7	4·3	4·0	10·3	9·5	16·3	15·1
44	13 56·0	13 58·3	13 17·9	4·4	4·1	10·4	9·6	16·4	15·2
45	13 56·3	13 58·5	13 18·2	4·5	4·2	10·5	9·7	16·5	15·3
46	13 56·5	13 58·8	13 18·4	4·6	4·3	10·6	9·8	16·6	15·4
47	13 56·8	13 59·0	13 18·6	4·7	4·3	10·7	9·9	16·7	15·4
48	13 57·0	13 59·3	13 18·9	4·8	4·4	10·8	10·0	16·8	15·5
49	13 57·3	13 59·5	13 19·1	4·9	4·5	10·9	10·1	16·9	15·6
50	13 57·5	13 59·8	13 19·3	5·0	4·6	11·0	10·2	17·0	15·7
51	13 57·8	14 00·0	13 19·6	5·1	4·7	11·1	10·3	17·1	15·8
52	13 58·0	14 00·3	13 19·8	5·2	4·8	11·2	10·4	17·2	15·9
53	13 58·3	14 00·5	13 20·1	5·3	4·9	11·3	10·5	17·3	16·0
54	13 58·5	14 00·8	13 20·3	5·4	5·0	11·4	10·5	17·4	16·1
55	13 58·8	14 01·0	13 20·5	5·5	5·1	11·5	10·6	17·5	16·2
56	13 59·0	14 01·3	13 20·8	5·6	5·2	11·6	10·7	17·6	16·3
57	13 59·3	14 01·5	13 21·0	5·7	5·3	11·7	10·8	17·7	16·4
58	13 59·5	14 01·8	13 21·3	5·8	5·4	11·8	10·9	17·8	16·5
59	13 59·8	14 02·0	13 21·5	5·9	5·5	11·9	11·0	17·9	16·6
60	14 00·0	14 02·3	13 21·7	6·0	5·6	12·0	11·1	18·0	16·7

56ᵐ	SUN PLANETS	ARIES	MOON	v or d	Corrⁿ	v or d	Corrⁿ	v or d	Corrⁿ
s	° ′	° ′	° ′	′	′	′	′	′	′
00	14 00·0	14 02·3	13 21·7	0·0	0·0	6·0	5·7	12·0	11·3
01	14 00·3	14 02·6	13 22·0	0·1	0·1	6·1	5·7	12·1	11·4
02	14 00·5	14 02·8	13 22·2	0·2	0·2	6·2	5·8	12·2	11·5
03	14 00·8	14 03·1	13 22·4	0·3	0·3	6·3	5·9	12·3	11·6
04	14 01·0	14 03·3	13 22·7	0·4	0·4	6·4	6·0	12·4	11·7
05	14 01·3	14 03·6	13 22·9	0·5	0·5	6·5	6·1	12·5	11·8
06	14 01·5	14 03·8	13 23·2	0·6	0·6	6·6	6·2	12·6	11·9
07	14 01·8	14 04·1	13 23·4	0·7	0·7	6·7	6·3	12·7	12·0
08	14 02·0	14 04·3	13 23·6	0·8	0·8	6·8	6·4	12·8	12·1
09	14 02·3	14 04·6	13 23·9	0·9	0·8	6·9	6·5	12·9	12·1
10	14 02·5	14 04·8	13 24·1	1·0	0·9	7·0	6·6	13·0	12·2
11	14 02·8	14 05·1	13 24·4	1·1	1·0	7·1	6·7	13·1	12·3
12	14 03·0	14 05·3	13 24·6	1·2	1·1	7·2	6·8	13·2	12·4
13	14 03·3	14 05·6	13 24·8	1·3	1·2	7·3	6·9	13·3	12·5
14	14 03·5	14 05·8	13 25·1	1·4	1·3	7·4	7·0	13·4	12·6
15	14 03·8	14 06·1	13 25·3	1·5	1·4	7·5	7·1	13·5	12·7
16	14 04·0	14 06·3	13 25·6	1·6	1·5	7·6	7·2	13·6	12·8
17	14 04·3	14 06·6	13 25·8	1·7	1·6	7·7	7·3	13·7	12·9
18	14 04·5	14 06·8	13 26·0	1·8	1·7	7·8	7·3	13·8	13·0
19	14 04·8	14 07·1	13 26·3	1·9	1·8	7·9	7·4	13·9	13·1
20	14 05·0	14 07·3	13 26·5	2·0	1·9	8·0	7·5	14·0	13·2
21	14 05·3	14 07·6	13 26·7	2·1	2·0	8·1	7·6	14·1	13·3
22	14 05·5	14 07·8	13 27·0	2·2	2·1	8·2	7·7	14·2	13·4
23	14 05·8	14 08·1	13 27·2	2·3	2·2	8·3	7·8	14·3	13·5
24	14 06·0	14 08·3	13 27·5	2·4	2·3	8·4	7·9	14·4	13·6
25	14 06·3	14 08·6	13 27·7	2·5	2·4	8·5	8·0	14·5	13·7
26	14 06·5	14 08·8	13 27·9	2·6	2·4	8·6	8·1	14·6	13·7
27	14 06·8	14 09·1	13 28·2	2·7	2·5	8·7	8·2	14·7	13·8
28	14 07·0	14 09·3	13 28·4	2·8	2·6	8·8	8·3	14·8	13·9
29	14 07·3	14 09·6	13 28·7	2·9	2·7	8·9	8·4	14·9	14·0
30	14 07·5	14 09·8	13 28·9	3·0	2·8	9·0	8·5	15·0	14·1
31	14 07·8	14 10·1	13 29·1	3·1	2·9	9·1	8·6	15·1	14·2
32	14 08·0	14 10·3	13 29·4	3·2	3·0	9·2	8·7	15·2	14·3
33	14 08·3	14 10·6	13 29·6	3·3	3·1	9·3	8·8	15·3	14·4
34	14 08·5	14 10·8	13 29·8	3·4	3·2	9·4	8·9	15·4	14·5
35	14 08·8	14 11·1	13 30·1	3·5	3·3	9·5	8·9	15·5	14·6
36	14 09·0	14 11·3	13 30·3	3·6	3·4	9·6	9·0	15·6	14·7
37	14 09·3	14 11·6	13 30·6	3·7	3·5	9·7	9·1	15·7	14·8
38	14 09·5	14 11·8	13 30·8	3·8	3·6	9·8	9·2	15·8	14·9
39	14 09·8	14 12·1	13 31·0	3·9	3·7	9·9	9·3	15·9	15·0
40	14 10·0	14 12·3	13 31·3	4·0	3·8	10·0	9·4	16·0	15·1
41	14 10·3	14 12·6	13 31·5	4·1	3·9	10·1	9·5	16·1	15·2
42	14 10·5	14 12·8	13 31·8	4·2	4·0	10·2	9·6	16·2	15·3
43	14 10·8	14 13·1	13 32·0	4·3	4·0	10·3	9·7	16·3	15·3
44	14 11·0	14 13·3	13 32·2	4·4	4·1	10·4	9·8	16·4	15·4
45	14 11·3	14 13·6	13 32·5	4·5	4·2	10·5	9·9	16·5	15·5
46	14 11·5	14 13·8	13 32·7	4·6	4·3	10·6	10·0	16·6	15·6
47	14 11·8	14 14·1	13 32·9	4·7	4·4	10·7	10·1	16·7	15·7
48	14 12·0	14 14·3	13 33·2	4·8	4·5	10·8	10·2	16·8	15·8
49	14 12·3	14 14·6	13 33·4	4·9	4·6	10·9	10·3	16·9	15·9
50	14 12·5	14 14·8	13 33·7	5·0	4·7	11·0	10·4	17·0	16·0
51	14 12·8	14 15·1	13 33·9	5·1	4·8	11·1	10·5	17·1	16·1
52	14 13·0	14 15·3	13 34·1	5·2	4·9	11·2	10·5	17·2	16·2
53	14 13·3	14 15·6	13 34·4	5·3	5·0	11·3	10·6	17·3	16·3
54	14 13·5	14 15·8	13 34·6	5·4	5·1	11·4	10·7	17·4	16·4
55	14 13·8	14 16·1	13 34·9	5·5	5·2	11·5	10·8	17·5	16·5
56	14 14·0	14 16·3	13 35·1	5·6	5·3	11·6	10·9	17·6	16·6
57	14 14·3	14 16·6	13 35·3	5·7	5·4	11·7	11·0	17·7	16·7
58	14 14·5	14 16·8	13 35·6	5·8	5·5	11·8	11·1	17·8	16·8
59	14 14·8	14 17·1	13 35·8	5·9	5·6	11·9	11·2	17·9	16·9
60	14 15·0	14 17·3	13 36·1	6·0	5·7	12·0	11·3	18·0	17·0

57ᵐ	SUN PLANETS	ARIES	MOON	v or d	Corrⁿ	v or d	Corrⁿ	v or d	Corrⁿ
s	° ′	° ′	° ′	′	′	′	′	′	′
00	14 15·0	14 17·3	13 36·1	0·0	0·0	6·0	5·8	12·0	11·5
01	14 15·3	14 17·6	13 36·3	0·1	0·1	6·1	5·8	12·1	11·6
02	14 15·5	14 17·8	13 36·5	0·2	0·2	6·2	5·9	12·2	11·7
03	14 15·8	14 18·1	13 36·8	0·3	0·3	6·3	6·0	12·3	11·8
04	14 16·0	14 18·3	13 37·0	0·4	0·4	6·4	6·1	12·4	11·9
05	14 16·3	14 18·6	13 37·2	0·5	0·5	6·5	6·2	12·5	12·0
06	14 16·5	14 18·8	13 37·5	0·6	0·6	6·6	6·3	12·6	12·1
07	14 16·8	14 19·1	13 37·7	0·7	0·7	6·7	6·4	12·7	12·2
08	14 17·0	14 19·3	13 38·0	0·8	0·8	6·8	6·5	12·8	12·3
09	14 17·3	14 19·6	13 38·2	0·9	0·9	6·9	6·6	12·9	12·4
10	14 17·5	14 19·8	13 38·4	1·0	1·0	7·0	6·7	13·0	12·5
11	14 17·8	14 20·1	13 38·7	1·1	1·1	7·1	6·8	13·1	12·6
12	14 18·0	14 20·3	13 38·9	1·2	1·2	7·2	6·9	13·2	12·7
13	14 18·3	14 20·6	13 39·2	1·3	1·2	7·3	7·0	13·3	12·7
14	14 18·5	14 20·9	13 39·4	1·4	1·3	7·4	7·1	13·4	12·8
15	14 18·8	14 21·1	13 39·6	1·5	1·4	7·5	7·2	13·5	12·9
16	14 19·0	14 21·4	13 39·9	1·6	1·5	7·6	7·3	13·6	13·0
17	14 19·3	14 21·6	13 40·1	1·7	1·6	7·7	7·4	13·7	13·1
18	14 19·5	14 21·9	13 40·3	1·8	1·7	7·8	7·5	13·8	13·2
19	14 19·8	14 22·1	13 40·6	1·9	1·8	7·9	7·6	13·9	13·3
20	14 20·0	14 22·4	13 40·8	2·0	1·9	8·0	7·7	14·0	13·4
21	14 20·3	14 22·6	13 41·1	2·1	2·0	8·1	7·8	14·1	13·5
22	14 20·5	14 22·9	13 41·3	2·2	2·1	8·2	7·9	14·2	13·6
23	14 20·8	14 23·1	13 41·5	2·3	2·2	8·3	8·0	14·3	13·7
24	14 21·0	14 23·4	13 41·8	2·4	2·3	8·4	8·1	14·4	13·8
25	14 21·3	14 23·6	13 42·0	2·5	2·4	8·5	8·1	14·5	13·9
26	14 21·5	14 23·9	13 42·3	2·6	2·5	8·6	8·2	14·6	14·0
27	14 21·8	14 24·1	13 42·5	2·7	2·6	8·7	8·3	14·7	14·1
28	14 22·0	14 24·4	13 42·7	2·8	2·7	8·8	8·4	14·8	14·2
29	14 22·3	14 24·6	13 43·0	2·9	2·8	8·9	8·5	14·9	14·3
30	14 22·5	14 24·9	13 43·2	3·0	2·9	9·0	8·6	15·0	14·4
31	14 22·8	14 25·1	13 43·4	3·1	3·0	9·1	8·7	15·1	14·5
32	14 23·0	14 25·4	13 43·7	3·2	3·1	9·2	8·8	15·2	14·6
33	14 23·3	14 25·6	13 43·9	3·3	3·2	9·3	8·9	15·3	14·7
34	14 23·5	14 25·9	13 44·2	3·4	3·3	9·4	9·0	15·4	14·8
35	14 23·8	14 26·1	13 44·4	3·5	3·4	9·5	9·1	15·5	14·9
36	14 24·0	14 26·4	13 44·6	3·6	3·5	9·6	9·2	15·6	15·0
37	14 24·3	14 26·6	13 44·9	3·7	3·5	9·7	9·3	15·7	15·0
38	14 24·5	14 26·9	13 45·1	3·8	3·6	9·8	9·4	15·8	15·1
39	14 24·8	14 27·1	13 45·4	3·9	3·7	9·9	9·5	15·9	15·2
40	14 25·0	14 27·4	13 45·6	4·0	3·8	10·0	9·6	16·0	15·3
41	14 25·3	14 27·6	13 45·8	4·1	3·9	10·1	9·7	16·1	15·4
42	14 25·5	14 27·9	13 46·1	4·2	4·0	10·2	9·8	16·2	15·5
43	14 25·8	14 28·1	13 46·3	4·3	4·1	10·3	9·9	16·3	15·6
44	14 26·0	14 28·4	13 46·5	4·4	4·2	10·4	10·0	16·4	15·7
45	14 26·3	14 28·6	13 46·8	4·5	4·3	10·5	10·1	16·5	15·8
46	14 26·5	14 28·9	13 47·0	4·6	4·4	10·6	10·2	16·6	15·9
47	14 26·8	14 29·1	13 47·3	4·7	4·5	10·7	10·3	16·7	16·0
48	14 27·0	14 29·4	13 47·5	4·8	4·6	10·8	10·4	16·8	16·1
49	14 27·3	14 29·6	13 47·7	4·9	4·7	10·9	10·4	16·9	16·2
50	14 27·5	14 29·9	13 48·0	5·0	4·8	11·0	10·5	17·0	16·3
51	14 27·8	14 30·1	13 48·2	5·1	4·9	11·1	10·6	17·1	16·4
52	14 28·0	14 30·4	13 48·5	5·2	5·0	11·2	10·7	17·2	16·5
53	14 28·3	14 30·6	13 48·7	5·3	5·1	11·3	10·8	17·3	16·6
54	14 28·5	14 30·9	13 48·9	5·4	5·2	11·4	10·9	17·4	16·7
55	14 28·8	14 31·1	13 49·2	5·5	5·3	11·5	11·0	17·5	16·8
56	14 29·0	14 31·4	13 49·4	5·6	5·4	11·6	11·1	17·6	16·9
57	14 29·3	14 31·6	13 49·7	5·7	5·5	11·7	11·2	17·7	17·0
58	14 29·5	14 31·9	13 49·9	5·8	5·6	11·8	11·3	17·8	17·1
59	14 29·8	14 32·1	13 50·1	5·9	5·7	11·9	11·4	17·9	17·2
60	14 30·0	14 32·4	13 50·4	6·0	5·8	12·0	11·5	18·0	17·3

58ᵐ	SUN PLANETS	ARIES	MOON	v or d	Corrⁿ	v or d	Corrⁿ	v or d	Corrⁿ
s	° ′	° ′	° ′	′	′	′	′	′	′
00	14 30·0	14 32·4	13 50·4	0·0	0·0	6·0	5·9	12·0	11·7
01	14 30·3	14 32·6	13 50·6	0·1	0·1	6·1	5·9	12·1	11·8
02	14 30·5	14 32·9	13 50·8	0·2	0·2	6·2	6·0	12·2	11·9
03	14 30·8	14 33·1	13 51·1	0·3	0·3	6·3	6·1	12·3	12·0
04	14 31·0	14 33·4	13 51·3	0·4	0·4	6·4	6·2	12·4	12·1
05	14 31·3	14 33·6	13 51·6	0·5	0·5	6·5	6·3	12·5	12·2
06	14 31·5	14 33·9	13 51·8	0·6	0·6	6·6	6·4	12·6	12·3
07	14 31·8	14 34·1	13 52·0	0·7	0·7	6·7	6·5	12·7	12·4
08	14 32·0	14 34·4	13 52·3	0·8	0·8	6·8	6·6	12·8	12·5
09	14 32·3	14 34·6	13 52·5	0·9	0·9	6·9	6·7	12·9	12·6
10	14 32·5	14 34·9	13 52·8	1·0	1·0	7·0	6·8	13·0	12·7
11	14 32·8	14 35·1	13 53·0	1·1	1·1	7·1	6·9	13·1	12·8
12	14 33·0	14 35·4	13 53·2	1·2	1·2	7·2	7·0	13·2	12·9
13	14 33·3	14 35·6	13 53·5	1·3	1·3	7·3	7·1	13·3	13·0
14	14 33·5	14 35·9	13 53·7	1·4	1·4	7·4	7·2	13·4	13·1
15	14 33·8	14 36·1	13 53·9	1·5	1·5	7·5	7·3	13·5	13·2
16	14 34·0	14 36·4	13 54·2	1·6	1·6	7·6	7·4	13·6	13·3
17	14 34·3	14 36·6	13 54·4	1·7	1·7	7·7	7·5	13·7	13·4
18	14 34·5	14 36·9	13 54·7	1·8	1·8	7·8	7·6	13·8	13·5
19	14 34·8	14 37·1	13 54·9	1·9	1·9	7·9	7·7	13·9	13·6
20	14 35·0	14 37·4	13 55·1	2·0	2·0	8·0	7·8	14·0	13·7
21	14 35·3	14 37·6	13 55·4	2·1	2·0	8·1	7·9	14·1	13·7
22	14 35·5	14 37·9	13 55·6	2·2	2·1	8·2	8·0	14·2	13·8
23	14 35·8	14 38·1	13 55·9	2·3	2·2	8·3	8·1	14·3	13·9
24	14 36·0	14 38·4	13 56·1	2·4	2·3	8·4	8·2	14·4	14·0
25	14 36·3	14 38·6	13 56·3	2·5	2·4	8·5	8·3	14·5	14·1
26	14 36·5	14 38·9	13 56·6	2·6	2·5	8·6	8·4	14·6	14·2
27	14 36·8	14 39·2	13 56·8	2·7	2·6	8·7	8·5	14·7	14·3
28	14 37·0	14 39·4	13 57·0	2·8	2·7	8·8	8·6	14·8	14·4
29	14 37·3	14 39·7	13 57·3	2·9	2·8	8·9	8·7	14·9	14·5
30	14 37·5	14 39·9	13 57·5	3·0	2·9	9·0	8·8	15·0	14·6
31	14 37·8	14 40·2	13 57·8	3·1	3·0	9·1	8·9	15·1	14·7
32	14 38·0	14 40·4	13 58·0	3·2	3·1	9·2	9·0	15·2	14·8
33	14 38·3	14 40·7	13 58·2	3·3	3·2	9·3	9·1	15·3	14·9
34	14 38·5	14 40·9	13 58·5	3·4	3·3	9·4	9·2	15·4	15·0
35	14 38·8	14 41·2	13 58·7	3·5	3·4	9·5	9·3	15·5	15·1
36	14 39·0	14 41·4	13 59·0	3·6	3·5	9·6	9·4	15·6	15·2
37	14 39·3	14 41·7	13 59·2	3·7	3·6	9·7	9·5	15·7	15·3
38	14 39·5	14 41·9	13 59·4	3·8	3·7	9·8	9·6	15·8	15·4
39	14 39·8	14 42·2	13 59·7	3·9	3·8	9·9	9·7	15·9	15·5
40	14 40·0	14 42·4	13 59·9	4·0	3·9	10·0	9·8	16·0	15·6
41	14 40·3	14 42·7	14 00·1	4·1	4·0	10·1	9·8	16·1	15·7
42	14 40·5	14 42·9	14 00·4	4·2	4·1	10·2	9·9	16·2	15·8
43	14 40·8	14 43·2	14 00·6	4·3	4·2	10·3	10·0	16·3	15·9
44	14 41·0	14 43·4	14 00·9	4·4	4·3	10·4	10·1	16·4	16·0
45	14 41·3	14 43·7	14 01·1	4·5	4·4	10·5	10·2	16·5	16·1
46	14 41·5	14 43·9	14 01·3	4·6	4·5	10·6	10·3	16·6	16·2
47	14 41·8	14 44·2	14 01·6	4·7	4·6	10·7	10·4	16·7	16·3
48	14 42·0	14 44·4	14 01·8	4·8	4·7	10·8	10·5	16·8	16·4
49	14 42·3	14 44·7	14 02·1	4·9	4·8	10·9	10·6	16·9	16·5
50	14 42·5	14 44·9	14 02·3	5·0	4·9	11·0	10·7	17·0	16·6
51	14 42·8	14 45·2	14 02·5	5·1	5·0	11·1	10·8	17·1	16·7
52	14 43·0	14 45·4	14 02·8	5·2	5·1	11·2	10·9	17·2	16·8
53	14 43·3	14 45·7	14 03·0	5·3	5·2	11·3	11·0	17·3	16·9
54	14 43·5	14 45·9	14 03·3	5·4	5·3	11·4	11·1	17·4	17·0
55	14 43·8	14 46·2	14 03·5	5·5	5·4	11·5	11·2	17·5	17·1
56	14 44·0	14 46·4	14 03·7	5·6	5·5	11·6	11·3	17·6	17·2
57	14 44·3	14 46·7	14 04·0	5·7	5·6	11·7	11·4	17·7	17·3
58	14 44·5	14 46·9	14 04·2	5·8	5·7	11·8	11·5	17·8	17·4
59	14 44·8	14 47·2	14 04·4	5·9	5·8	11·9	11·6	17·9	17·5
60	14 45·0	14 47·4	14 04·7	6·0	5·9	12·0	11·7	18·0	17·6

59ᵐ	SUN PLANETS	ARIES	MOON	v or d	Corrⁿ	v or d	Corrⁿ	v or d	Corrⁿ
s	° ′	° ′	° ′	′	′	′	′	′	′
00	14 45·0	14 47·4	14 04·7	0·0	0·0	6·0	6·0	12·0	11·9
01	14 45·3	14 47·7	14 04·9	0·1	0·1	6·1	6·0	12·1	12·0
02	14 45·5	14 47·9	14 05·2	0·2	0·2	6·2	6·1	12·2	12·1
03	14 45·8	14 48·2	14 05·4	0·3	0·3	6·3	6·2	12·3	12·2
04	14 46·0	14 48·4	14 05·6	0·4	0·4	6·4	6·3	12·4	12·3
05	14 46·3	14 48·7	14 05·9	0·5	0·5	6·5	6·4	12·5	12·4
06	14 46·5	14 48·9	14 06·1	0·6	0·6	6·6	6·5	12·6	12·5
07	14 46·8	14 49·2	14 06·4	0·7	0·7	6·7	6·6	12·7	12·6
08	14 47·0	14 49·4	14 06·6	0·8	0·8	6·8	6·7	12·8	12·7
09	14 47·3	14 49·7	14 06·8	0·9	0·9	6·9	6·8	12·9	12·8
10	14 47·5	14 49·9	14 07·1	1·0	1·0	7·0	6·9	13·0	12·9
11	14 47·8	14 50·2	14 07·3	1·1	1·1	7·1	7·0	13·1	13·0
12	14 48·0	14 50·4	14 07·5	1·2	1·2	7·2	7·1	13·2	13·1
13	14 48·3	14 50·7	14 07·8	1·3	1·3	7·3	7·2	13·3	13·2
14	14 48·5	14 50·9	14 08·0	1·4	1·4	7·4	7·3	13·4	13·3
15	14 48·8	14 51·2	14 08·3	1·5	1·5	7·5	7·4	13·5	13·4
16	14 49·0	14 51·4	14 08·5	1·6	1·6	7·6	7·5	13·6	13·5
17	14 49·3	14 51·7	14 08·7	1·7	1·7	7·7	7·6	13·7	13·6
18	14 49·5	14 51·9	14 09·0	1·8	1·8	7·8	7·7	13·8	13·7
19	14 49·8	14 52·2	14 09·2	1·9	1·9	7·9	7·8	13·9	13·8
20	14 50·0	14 52·4	14 09·5	2·0	2·0	8·0	7·9	14·0	13·9
21	14 50·3	14 52·7	14 09·7	2·1	2·1	8·1	8·0	14·1	14·0
22	14 50·5	14 52·9	14 09·9	2·2	2·2	8·2	8·1	14·2	14·1
23	14 50·8	14 53·2	14 10·2	2·3	2·3	8·3	8·2	14·3	14·2
24	14 51·0	14 53·4	14 10·4	2·4	2·4	8·4	8·3	14·4	14·3
25	14 51·3	14 53·7	14 10·6	2·5	2·5	8·5	8·4	14·5	14·4
26	14 51·5	14 53·9	14 10·9	2·6	2·6	8·6	8·5	14·6	14·5
27	14 51·8	14 54·2	14 11·1	2·7	2·7	8·7	8·6	14·7	14·6
28	14 52·0	14 54·4	14 11·4	2·8	2·8	8·8	8·7	14·8	14·7
29	14 52·3	14 54·7	14 11·6	2·9	2·9	8·9	8·8	14·9	14·8
30	14 52·5	14 54·9	14 11·8	3·0	3·0	9·0	8·9	15·0	14·9
31	14 52·8	14 55·2	14 12·1	3·1	3·1	9·1	9·0	15·1	15·0
32	14 53·0	14 55·4	14 12·3	3·2	3·2	9·2	9·1	15·2	15·1
33	14 53·3	14 55·7	14 12·6	3·3	3·3	9·3	9·2	15·3	15·2
34	14 53·5	14 55·9	14 12·8	3·4	3·4	9·4	9·3	15·4	15·3
35	14 53·8	14 56·2	14 13·0	3·5	3·5	9·5	9·4	15·5	15·4
36	14 54·0	14 56·4	14 13·3	3·6	3·6	9·6	9·5	15·6	15·5
37	14 54·3	14 56·7	14 13·5	3·7	3·7	9·7	9·6	15·7	15·6
38	14 54·5	14 56·9	14 13·8	3·8	3·8	9·8	9·7	15·8	15·7
39	14 54·8	14 57·2	14 14·0	3·9	3·9	9·9	9·8	15·9	15·8
40	14 55·0	14 57·5	14 14·2	4·0	4·0	10·0	9·9	16·0	15·9
41	14 55·3	14 57·7	14 14·5	4·1	4·1	10·1	10·0	16·1	16·0
42	14 55·5	14 58·0	14 14·7	4·2	4·2	10·2	10·1	16·2	16·1
43	14 55·8	14 58·2	14 14·9	4·3	4·3	10·3	10·2	16·3	16·2
44	14 56·0	14 58·5	14 15·2	4·4	4·4	10·4	10·3	16·4	16·3
45	14 56·3	14 58·7	14 15·4	4·5	4·5	10·5	10·4	16·5	16·4
46	14 56·5	14 59·0	14 15·7	4·6	4·6	10·6	10·5	16·6	16·5
47	14 56·8	14 59·2	14 15·9	4·7	4·7	10·7	10·6	16·7	16·6
48	14 57·0	14 59·5	14 16·1	4·8	4·8	10·8	10·7	16·8	16·7
49	14 57·3	14 59·7	14 16·4	4·9	4·9	10·9	10·8	16·9	16·8
50	14 57·5	15 00·0	14 16·6	5·0	5·0	11·0	10·9	17·0	16·9
51	14 57·8	15 00·2	14 16·9	5·1	5·1	11·1	11·0	17·1	17·0
52	14 58·0	15 00·5	14 17·1	5·2	5·2	11·2	11·1	17·2	17·1
53	14 58·3	15 00·7	14 17·3	5·3	5·3	11·3	11·2	17·3	17·2
54	14 58·5	15 01·0	14 17·6	5·4	5·4	11·4	11·3	17·4	17·3
55	14 58·8	15 01·2	14 17·8	5·5	5·5	11·5	11·4	17·5	17·4
56	14 59·0	15 01·5	14 18·0	5·6	5·6	11·6	11·5	17·6	17·5
57	14 59·3	15 01·7	14 18·3	5·7	5·7	11·7	11·6	17·7	17·6
58	14 59·5	15 02·0	14 18·5	5·8	5·8	11·8	11·7	17·8	17·7
59	14 59·8	15 02·2	14 18·8	5·9	5·9	11·9	11·8	17·9	17·8
60	15 00·0	15 02·5	14 19·0	6·0	6·0	12·0	11·9	18·0	17·9

TABLES FOR INTERPOLATING SUNRISE, MOONRISE, ETC.

TABLE I—FOR LATITUDE

Tabular Interval			Difference between the times for consecutive latitudes															
10°	5°	2°	5m	10m	15m	20m	25m	30m	35m	40m	45m	50m	55m	60m	1h 05m	1h 10m	1h 15m	1h 20m
° ′	° ′	° ′	m	m	m	m	m	m	m	m	m	m	m	m	h m	h m	h m	h m
0 30	0 15	0 06	0	0	1	1	1	1	1	2	2	2	2	2	0 02	0 02	0 02	0 02
1 00	0 30	0 12	0	1	1	2	2	3	3	3	4	4	4	5	05	05	05	05
1 30	0 45	0 18	1	1	2	3	3	4	4	5	5	6	7	7	07	07	07	07
2 00	1 00	0 24	1	2	3	4	5	5	6	7	7	8	9	10	10	10	10	10
2 30	1 15	0 30	1	2	4	5	6	7	8	9	9	10	11	12	12	13	13	13
3 00	1 30	0 36	1	3	4	6	7	8	9	10	11	12	13	14	0 15	0 15	0 16	0 16
3 30	1 45	0 42	2	3	5	7	8	10	11	12	13	14	16	17	18	18	19	19
4 00	2 00	0 48	2	4	6	8	9	11	13	14	15	16	18	19	20	21	22	22
4 30	2 15	0 54	2	4	7	9	11	13	15	16	18	19	21	22	23	24	25	26
5 00	2 30	1 00	2	5	7	10	12	14	16	18	20	22	23	25	26	27	28	29
5 30	2 45	1 06	3	5	8	11	13	16	18	20	22	24	26	28	0 29	0 30	0 31	0 32
6 00	3 00	1 12	3	6	9	12	14	17	20	22	24	26	29	31	32	33	34	36
6 30	3 15	1 18	3	6	10	13	16	19	22	24	26	29	31	34	36	37	38	40
7 00	3 30	1 24	3	7	10	14	17	20	23	26	29	31	34	37	39	41	42	44
7 30	3 45	1 30	4	7	11	15	18	22	25	28	31	34	37	40	43	44	46	48
8 00	4 00	1 36	4	8	12	16	20	23	27	30	34	37	41	44	0 47	0 48	0 51	0 53
8 30	4 15	1 42	4	8	13	17	21	25	29	33	36	40	44	48	0 51	0 53	0 56	0 58
9 00	4 30	1 48	4	9	13	18	22	27	31	35	39	43	47	52	0 55	0 58	1 01	1 04
9 30	4 45	1 54	5	9	14	19	24	28	33	38	42	47	51	56	1 00	1 04	1 08	1 12
10 00	5 00	2 00	5	10	15	20	25	30	35	40	45	50	55	60	1 05	1 10	1 15	1 20

Table I is for interpolating the L.M.T. of sunrise, twilight, moonrise, etc., for latitude. It is to be entered, in the appropriate column on the left, with the difference between true latitude and the nearest tabular latitude which is *less* than the true latitude; and with the argument at the top which is the nearest value of the difference between the times for the tabular latitude and the next higher one; the correction so obtained is applied to the time for the tabular latitude; the sign of the correction can be seen by inspection. It is to be noted that the interpolation is not linear, so that when using this table it is essential to take out the tabular phenomenon for the latitude *less* than the true latitude.

TABLE II—FOR LONGITUDE

Long. East or West	Difference between the times for given date and preceding date (for east longitude) or for given date and following date (for west longitude)																	
	10m	20m	30m	40m	50m	60m	1h + 10m	1h + 20m	1h + 30m	1h + 40m	1h + 50m	1h + 60m	2h 10m	2h 20m	2h 30m	2h 40m	2h 50m	3h 00m
°	m	m	m	m	m	m	m	m	m	m	m	m	h m	h m	h m	h m	h m	h m
0	0	0	0	0	0	0	0	0	0	0	0	0	0 00	0 00	0 00	0 00	0 00	0 00
10	0	1	1	1	1	2	2	2	2	3	3	3	04	04	04	04	05	05
20	1	1	2	2	3	3	4	4	5	6	6	7	07	08	08	09	09	10
30	1	2	2	3	4	5	6	7	7	8	9	10	11	12	12	13	14	15
40	1	2	3	4	6	7	8	9	10	11	12	13	14	16	17	18	19	20
50	1	3	4	6	7	8	10	11	12	14	15	17	0 18	0 19	0 21	0 22	0 24	0 25
60	2	3	5	7	8	10	12	13	15	17	18	20	22	23	25	27	28	30
70	2	4	6	8	10	12	14	16	17	19	21	23	25	27	29	31	33	35
80	2	4	7	9	11	13	16	18	20	22	24	27	29	31	33	36	38	40
90	2	5	7	10	12	15	17	20	22	25	27	30	32	35	37	40	42	45
100	3	6	8	11	14	17	19	22	25	28	31	33	0 36	0 39	0 42	0 44	0 47	0 50
110	3	6	9	12	15	18	21	24	27	31	34	37	40	43	46	49	0 52	0 55
120	3	7	10	13	17	20	23	27	30	33	37	40	43	47	50	53	0 57	1 00
130	4	7	11	14	18	22	25	29	32	36	40	43	47	51	54	0 58	1 01	1 05
140	4	8	12	16	19	23	27	31	35	39	43	47	51	54	0 58	1 02	1 06	1 10
150	4	8	13	17	21	25	29	33	38	42	46	50	0 54	0 58	1 03	1 07	1 11	1 15
160	4	9	13	18	22	27	31	36	40	44	49	53	0 58	1 02	1 07	1 11	1 16	1 20
170	5	9	14	19	24	28	33	38	42	47	52	57	1 01	1 06	1 11	1 16	1 20	1 25
180	5	10	15	20	25	30	35	40	45	50	55	60	1 05	1 10	1 15	1 20	1 25	1 30

Table II is for interpolating the L.M.T. of moonrise, moonset and the Moon's meridian passage for longitude. It is entered with longitude and with the difference between the times for the given date and for the preceding date (in east longitudes) or following date (in west longitudes). The correction is normally *added* for west longitudes and *subtracted* for east longitudes, but if, as occasionally happens, the times become earlier each day instead of later, the signs of the corrections must be reversed.

INDEX TO SELECTED STARS, 1993

Name	No	Mag	SHA	Dec
			°	°
Acamar	**7**	3·1	315	S 40
Achernar	**5**	0·6	336	S 57
Acrux	**30**	1·1	173	S 63
Adhara	**19**	1·6	255	S 29
Aldebaran	**10**	1·1	291	N 16
Alioth	**32**	1·7	167	N 56
Alkaid	**34**	1·9	153	N 49
Al Na'ir	**55**	2·2	28	S 47
Alnilam	**15**	1·8	276	S 1
Alphard	**25**	2·2	218	S 9
Alphecca	**41**	2·3	126	N 27
Alpheratz	**1**	2·2	358	N 29
Altair	**51**	0·9	62	N 9
Ankaa	**2**	2·4	354	S 42
Antares	**42**	1·2	113	S 26
Arcturus	**37**	0·2	146	N 19
Atria	**43**	1·9	108	S 69
Avior	**22**	1·7	234	S 59
Bellatrix	**13**	1·7	279	N 6
Betelgeuse	**16**	Var.*	271	N 7
Canopus	**17**	−0·9	264	S 53
Capella	**12**	0·2	281	N 46
Deneb	**53**	1·3	50	N 45
Denebola	**28**	2·2	183	N 15
Diphda	**4**	2·2	349	S 18
Dubhe	**27**	2·0	194	N 62
Elnath	**14**	1·8	279	N 29
Eltanin	**47**	2·4	91	N 51
Enif	**54**	2·5	34	N 10
Fomalhaut	**56**	1·3	16	S 30
Gacrux	**31**	1·6	172	S 57
Gienah	**29**	2·8	176	S 18
Hadar	**35**	0·9	149	S 60
Hamal	**6**	2·2	328	N 23
Kaus Australis	**48**	2·0	84	S 34
Kochab	**40**	2·2	137	N 74
Markab	**57**	2·6	14	N 15
Menkar	**8**	2·8	315	N 4
Menkent	**36**	2·3	148	S 36
Miaplacidus	**24**	1·8	222	S 70
Mirfak	**9**	1·9	309	N 50
Nunki	**50**	2·1	76	S 26
Peacock	**52**	2·1	54	S 57
Pollux	**21**	1·2	244	N 28
Procyon	**20**	0·5	245	N 5
Rasalhague	**46**	2·1	96	N 13
Regulus	**26**	1·3	208	N 12
Rigel	**11**	0·3	281	S 8
Rigil Kentaurus	**38**	0·1	140	S 61
Sabik	**44**	2·6	102	S 16
Schedar	**3**	2·5	350	N 57
Shaula	**45**	1·7	97	S 37
Sirius	**18**	−1·6	259	S 17
Spica	**33**	1·2	159	S 11
Suhail	**23**	2·2	223	S 43
Vega	**49**	0·1	81	N 39
Zubenelgenubi	**39**	2·9	137	S 16

No	Name	Mag	SHA	Dec
			°	°
1	*Alpheratz*	2·2	358	N 29
2	*Ankaa*	2·4	354	S 42
3	*Schedar*	2·5	350	N 57
4	*Diphda*	2·2	349	S 18
5	*Achernar*	0·6	336	S 57
6	*Hamal*	2·2	328	N 23
7	*Acamar*	3·1	315	S 40
8	*Menkar*	2·8	315	N 4
9	*Mirfak*	1·9	309	N 50
10	*Aldebaran*	1·1	291	N 16
11	*Rigel*	0·3	281	S 8
12	*Capella*	0·2	281	N 46
13	*Bellatrix*	1·7	279	N 6
14	*Elnath*	1·8	279	N 29
15	*Alnilam*	1·8	276	S 1
16	*Betelgeuse*	Var.*	271	N 7
17	*Canopus*	−0·9	264	S 53
18	*Sirius*	−1·6	259	S 17
19	*Adhara*	1·6	255	S 29
20	*Procyon*	0·5	245	N 5
21	*Pollux*	1·2	244	N 28
22	*Avior*	1·7	234	S 59
23	*Suhail*	2·2	223	S 43
24	*Miaplacidus*	1·8	222	S 70
25	*Alphard*	2·2	218	S 9
26	*Regulus*	1·3	208	N 12
27	*Dubhe*	2·0	194	N 62
28	*Denebola*	2·2	183	N 15
29	*Gienah*	2·8	176	S 18
30	*Acrux*	1·1	173	S 63
31	*Gacrux*	1·6	172	S 57
32	*Alioth*	1·7	167	N 56
33	*Spica*	1·2	159	S 11
34	*Alkaid*	1·9	153	N 49
35	*Hadar*	0·9	149	S 60
36	*Menkent*	2·3	148	S 36
37	*Arcturus*	0·2	146	N 19
38	*Rigil Kentaurus*	0·1	140	S 61
39	*Zubenelgenubi*	2·9	137	S 16
40	*Kochab*	2·2	137	N 74
41	*Alphecca*	2·3	126	N 27
42	*Antares*	1·2	113	S 26
43	*Atria*	1·9	108	S 69
44	*Sabik*	2·6	102	S 16
45	*Shaula*	1·7	97	S 37
46	*Rasalhague*	2·1	96	N 13
47	*Eltanin*	2·4	91	N 51
48	*Kaus Australis*	2·0	84	S 34
49	*Vega*	0·1	81	N 39
50	*Nunki*	2·1	76	S 26
51	*Altair*	0·9	62	N 9
52	*Peacock*	2·1	54	S 57
53	*Deneb*	1·3	50	N 45
54	*Enif*	2·5	34	N 10
55	*Al Na'ir*	2·2	28	S 47
56	*Fomalhaut*	1·3	16	S 30
57	*Markab*	2·6	14	N 15

*0·1 — 1·2

ALTITUDE CORRECTION TABLES 0°–35°—MOON

App. Alt.	0°–4° Corrⁿ	5°–9° Corrⁿ	10°–14° Corrⁿ	15°–19° Corrⁿ	20°–24° Corrⁿ	25°–29° Corrⁿ	30°–34° Corrⁿ	App. Alt.
′	° ′	° ′	° ′	° ′	° ′	° ′	° ′	′
00	0 33·8	5 58·2	10 62·1	15 62·8	20 62·2	25 60·8	30 58·9	00
10	35·9	58·5	62·2	62·8	62·1	60·8	58·8	10
20	37·8	58·7	62·2	62·8	62·1	60·7	58·8	20
30	39·6	58·9	62·3	62·8	62·1	60·7	58·7	30
40	41·2	59·1	62·3	62·8	62·0	60·6	58·6	40
50	42·6	59·3	62·4	62·7	62·0	60·6	58·5	50
00	1 44·0	6 59·5	11 62·4	16 62·7	21 62·0	26 60·5	31 58·5	00
10	45·2	59·7	62·4	62·7	61·9	60·4	58·4	10
20	46·3	59·9	62·5	62·7	61·9	60·4	58·3	20
30	47·3	60·0	62·5	62·7	61·9	60·3	58·2	30
40	48·3	60·2	62·5	62·7	61·8	60·3	58·2	40
50	49·2	60·3	62·6	62·7	61·8	60·2	58·1	50
00	2 50·0	7 60·5	12 62·6	17 62·7	22 61·7	27 60·1	32 58·0	00
10	50·8	60·6	62·6	62·6	61·7	60·1	57·9	10
20	51·4	60·7	62·6	62·6	61·6	60·0	57·8	20
30	52·1	60·9	62·7	62·6	61·6	59·9	57·8	30
40	52·7	61·0	62·7	62·6	61·5	59·9	57·7	40
50	53·3	61·1	62·7	62·6	61·5	59·8	57·6	50
00	3 53·8	8 61·2	13 62·7	18 62·5	23 61·5	28 59·7	33 57·5	00
10	54·3	61·3	62·7	62·5	61·4	59·7	57·4	10
20	54·8	61·4	62·7	62·5	61·4	59·6	57·4	20
30	55·2	61·5	62·8	62·5	61·3	59·6	57·3	30
40	55·6	61·6	62·8	62·4	61·3	59·5	57·2	40
50	56·0	61·6	62·8	62·4	61·2	59·4	57·1	50
00	4 56·4	9 61·7	14 62·8	19 62·4	24 61·2	29 59·3	34 57·0	00
10	56·7	61·8	62·8	62·3	61·1	59·3	56·9	10
20	57·1	61·9	62·8	62·3	61·1	59·2	56·9	20
30	57·4	61·9	62·8	62·3	61·0	59·1	56·8	30
40	57·7	62·0	62·8	62·2	60·9	59·1	56·7	40
50	57·9	62·1	62·8	62·2	60·9	59·0	56·6	50

H.P.	L	U	L	U	L	U	L	U	L	U	L	U	L	U	H.P.
′	′	′	′	′	′	′	′	′	′	′	′	′	′	′	′
54·0	0·3	0·9	0·3	0·9	0·4	1·0	0·5	1·1	0·6	1·2	0·7	1·3	0·9	1·5	54·0
54·3	0·7	1·1	0·7	1·2	0·7	1·2	0·8	1·3	0·9	1·4	1·1	1·5	1·2	1·7	54·3
54·6	1·1	1·4	1·1	1·4	1·1	1·4	1·2	1·5	1·3	1·6	1·4	1·7	1·5	1·8	54·6
54·9	1·4	1·6	1·5	1·6	1·5	1·6	1·6	1·7	1·6	1·8	1·8	1·9	1·9	2·0	54·9
55·2	1·8	1·8	1·8	1·8	1·9	1·9	1·9	1·9	2·0	2·0	2·1	2·1	2·2	2·2	55·2
55·5	2·2	2·0	2·2	2·0	2·3	2·1	2·3	2·1	2·4	2·2	2·4	2·3	2·5	2·4	55·5
55·8	2·6	2·2	2·6	2·2	2·6	2·3	2·7	2·3	2·7	2·4	2·8	2·4	2·9	2·5	55·8
56·1	3·0	2·4	3·0	2·5	3·0	2·5	3·0	2·5	3·1	2·6	3·1	2·6	3·2	2·7	56·1
56·4	3·4	2·7	3·4	2·7	3·4	2·7	3·4	2·7	3·4	2·8	3·5	2·8	3·5	2·9	56·4
56·7	3·7	2·9	3·7	2·9	3·8	2·9	3·8	2·9	3·8	3·0	3·8	3·0	3·9	3·0	56·7
57·0	4·1	3·1	4·1	3·1	4·1	3·1	4·1	3·1	4·2	3·1	4·2	3·2	4·2	3·2	57·0
57·3	4·5	3·3	4·5	3·3	4·5	3·3	4·5	3·3	4·5	3·3	4·5	3·4	4·6	3·4	57·3
57·6	4·9	3·5	4·9	3·5	4·9	3·5	4·9	3·5	4·9	3·5	4·9	3·5	4·9	3·6	57·6
57·9	5·3	3·8	5·3	3·8	5·2	3·8	5·2	3·7	5·2	3·7	5·2	3·7	5·2	3·7	57·9
58·2	5·6	4·0	5·6	4·0	5·6	4·0	5·6	4·0	5·6	3·9	5·6	3·9	5·6	3·9	58·2
58·5	6·0	4·2	6·0	4·2	6·0	4·2	6·0	4·2	6·0	4·1	5·9	4·1	5·9	4·1	58·5
58·8	6·4	4·4	6·4	4·4	6·4	4·4	6·3	4·4	6·3	4·3	6·3	4·3	6·2	4·2	58·8
59·1	6·8	4·6	6·8	4·6	6·7	4·6	6·7	4·6	6·7	4·5	6·6	4·5	6·6	4·4	59·1
59·4	7·2	4·8	7·1	4·8	7·1	4·8	7·1	4·8	7·0	4·7	7·0	4·7	6·9	4·6	59·4
59·7	7·5	5·1	7·5	5·0	7·5	5·0	7·5	5·0	7·4	4·9	7·3	4·8	7·2	4·7	59·7
60·0	7·9	5·3	7·9	5·3	7·9	5·2	7·8	5·2	7·8	5·1	7·7	5·0	7·6	4·9	60·0
60·3	8·3	5·5	8·3	5·5	8·2	5·4	8·2	5·4	8·1	5·3	8·0	5·2	7·9	5·1	60·3
60·6	8·7	5·7	8·7	5·7	8·6	5·7	8·6	5·6	8·5	5·5	8·4	5·4	8·2	5·3	60·6
60·9	9·1	5·9	9·0	5·9	9·0	5·9	8·9	5·8	8·8	5·7	8·7	5·6	8·6	5·4	60·9
61·2	9·5	6·2	9·4	6·1	9·4	6·1	9·3	6·0	9·2	5·9	9·1	5·8	8·9	5·6	61·2
61·5	9·8	6·4	9·8	6·3	9·7	6·3	9·7	6·2	9·5	6·1	9·4	5·9	9·2	5·8	61·5

DIP

Ht. of Eye	Corrⁿ	Ht. of Eye	Ht. of Eye	Corrⁿ	Ht. of Eye
m		ft.	m		ft.
2·4	′	8·0	9·5	′	31·5
	−2·8			−5·5	
2·6		8·6	9·9		32·7
	−2·9			−5·6	
2·8		9·2	10·3		33·9
	−3·0			−5·7	
3·0		9·8	10·6		35·1
	−3·1			−5·8	
3·2		10·5	11·0		36·3
	−3·2			−5·9	
3·4		11·2	11·4		37·6
	−3·3			−6·0	
3·6		11·9	11·8		38·9
	−3·4			−6·1	
3·8		12·6	12·2		40·1
	−3·5			−6·2	
4·0		13·3	12·6		41·5
	−3·6			−6·3	
4·3		14·1	13·0		42·8
	−3·7			−6·4	
4·5		14·9	13·4		44·2
	−3·8			−6·5	
4·7		15·7	13·8		45·5
	−3·9			−6·6	
5·0		16·5	14·2		46·9
	−4·0			−6·7	
5·2		17·4	14·7		48·4
	−4·1			−6·8	
5·5		18·3	15·1		49·8
	−4·2			−6·9	
5·8		19·1	15·5		51·3
	−4·3			−7·0	
6·1		20·1	16·0		52·8
	−4·4			−7·1	
6·3		21·0	16·5		54·3
	−4·5			−7·2	
6·6		22·0	16·9		55·8
	−4·6			−7·3	
6·9		22·9	17·4		57·4
	−4·7			−7·4	
7·2		23·9	17·9		58·9
	−4·8			−7·5	
7·5		24·9	18·4		60·5
	−4·9			−7·6	
7·9		26·0	18·8		62·1
	−5·0			−7·7	
8·2		27·1	19·3		63·8
	−5·1			−7·8	
8·5		28·1	19·8		65·4
	−5·2			−7·9	
8·8		29·2	20·4		67·1
	−5·3			−8·0	
9·2		30·4	20·9		68·8
	−5·4			−8·1	
9·5		31·5	21·4		70·5

MOON CORRECTION TABLE

The correction is in two parts; the first correction is taken from the upper part of the table with argument apparent altitude, and the second from the lower part, with argument H.P., in the same column as that from which the first correction was taken. Separate corrections are given in the lower part for lower (L) and upper (U) limbs. All corrections are to be **added** to apparent altitude, *but 30′ is to be subtracted from the altitude of the upper limb.*

For corrections for pressure and temperature see page A4.

For bubble sextant observations ignore dip, take the mean of upper and lower limb corrections and subtract 15′ from the altitude.

App. Alt. = Apparent altitude = Sextant altitude corrected for index error and dip.

App. Alt.	35°–39° Corrⁿ	40°–44° Corrⁿ	45°–49° Corrⁿ	50°–54° Corrⁿ	55°–59° Corrⁿ	60°–64° Corrⁿ	65°–69° Corrⁿ	70°–74° Corrⁿ	75°–79° Corrⁿ	80°–84° Corrⁿ	85°–89° Corrⁿ	App. Alt.
′	° ′	° ′	° ′	° ′	° ′	° ′	° ′	° ′	° ′	° ′	° ′	′
00	35 56·5	40 53·7	45 50·5	50 46·9	55 43·1	60 38·9	65 34·6	70 30·1	75 25·3	80 20·5	85 15·6	00
10	56·4	53·6	50·4	46·8	42·9	38·8	34·4	29·9	25·2	20·4	15·5	10
20	56·3	53·5	50·2	46·7	42·8	38·7	34·3	29·7	25·0	20·2	15·3	20
30	56·2	53·4	50·1	46·5	42·7	38·5	34·1	29·6	24·9	20·0	15·1	30
40	56·2	53·3	50·0	46·4	42·5	38·4	34·0	29·4	24·7	19·9	15·0	40
50	56·1	53·2	49·9	46·3	42·4	38·2	33·8	29·3	24·5	19·7	14·8	50
00	36 56·0	41 53·1	46 49·8	51 46·2	56 42·3	61 38·1	66 33·7	71 29·1	76 24·4	81 19·6	86 14·6	00
10	55·9	53·0	49·7	46·0	42·1	37·9	33·5	29·0	24·2	19·4	14·5	10
20	55·8	52·8	49·5	45·9	42·0	37·8	33·4	28·8	24·1	19·2	14·3	20
30	55·7	52·7	49·4	45·8	41·8	37·7	33·2	28·7	23·9	19·1	14·1	30
40	55·6	52·6	49·3	45·7	41·7	37·5	33·1	28·5	23·8	18·9	14·0	40
50	55·5	52·5	49·2	45·5	41·6	37·4	32·9	28·3	23·6	18·7	13·8	50
00	37 55·4	42 52·4	47 49·1	52 45·4	57 41·4	62 37·2	67 32·8	72 28·2	77 23·4	82 18·6	87 13·7	00
10	55·3	52·3	49·0	45·3	41·3	37·1	32·6	28·0	23·3	18·4	13·5	10
20	55·2	52·2	48·8	45·2	41·2	36·9	32·5	27·9	23·1	18·2	13·3	20
30	55·1	52·1	48·7	45·0	41·0	36·8	32·3	27·7	22·9	18·1	13·2	30
40	55·0	52·0	48·6	44·9	40·9	36·6	32·2	27·6	22·8	17·9	13·0	40
50	55·0	51·9	48·5	44·8	40·8	36·5	32·0	27·4	22·6	17·8	12·8	50
00	38 54·9	43 51·8	48 48·4	53 44·6	58 40·6	63 36·4	68 31·9	73 27·2	78 22·5	83 17·6	88 12·7	00
10	54·8	51·7	48·2	44·5	40·5	36·2	31·7	27·1	22·3	17·4	12·5	10
20	54·7	51·6	48·1	44·4	40·3	36·1	31·6	26·9	22·1	17·3	12·3	20
30	54·6	51·5	48·0	44·2	40·2	35·9	31·4	26·8	22·0	17·1	12·2	30
40	54·5	51·4	47·9	44·1	40·1	35·8	31·3	26·6	21·8	16·9	12·0	40
50	54·4	51·2	47·8	44·0	39·9	35·6	31·1	26·5	21·7	16·8	11·8	50
00	39 54·3	44 51·1	49 47·6	54 43·9	59 39·8	64 35·5	69 31·0	74 26·3	79 21·5	84 16·6	89 11·7	00
10	54·2	51·0	47·5	43·7	39·6	35·3	30·8	26·1	21·3	16·5	11·5	10
20	54·1	50·9	47·4	43·6	39·5	35·2	30·7	26·0	21·2	16·3	11·4	20
30	54·0	50·8	47·3	43·5	39·4	35·0	30·5	25·8	21·0	16·1	11·2	30
40	53·9	50·7	47·2	43·3	39·2	34·9	30·4	25·7	20·9	16·0	11·0	40
50	53·8	50·6	47·0	43·2	39·1	34·7	30·2	25·5	20·7	15·8	10·9	50

H.P.	L U	L U	L U	L U	L U	L U	L U	L U	L U	L U	L U	H.P.
′	′ ′	′ ′	′ ′	′ ′	′ ′	′ ′	′ ′	′ ′	′ ′	′ ′	′ ′	′
54·0	1·1 1·7	1·3 1·9	1·5 2·1	1·7 2·4	2·0 2·6	2·3 2·9	2·6 3·2	2·9 3·5	3·2 3·8	3·5 4·1	3·8 4·5	54·0
54·3	1·4 1·8	1·6 2·0	1·8 2·2	2·0 2·5	2·3 2·7	2·5 3·0	2·8 3·2	3·0 3·5	3·3 3·8	3·6 4·1	3·9 4·4	54·3
54·6	1·7 2·0	1·9 2·2	2·1 2·4	2·3 2·6	2·5 2·8	2·7 3·0	3·0 3·3	3·2 3·5	3·5 3·8	3·7 4·1	4·0 4·3	54·6
54·9	2·0 2·2	2·2 2·3	2·3 2·5	2·5 2·7	2·7 2·9	2·9 3·1	3·2 3·3	3·4 3·5	3·6 3·8	3·9 4·0	4·1 4·3	54·9
55·2	2·3 2·3	2·5 2·4	2·6 2·6	2·8 2·8	3·0 2·9	3·2 3·1	3·4 3·3	3·6 3·5	3·8 3·7	4·0 4·0	4·2 4·2	55·2
55·5	2·7 2·5	2·8 2·6	2·9 2·7	3·1 2·9	3·2 3·0	3·4 3·2	3·6 3·4	3·7 3·5	3·9 3·7	4·1 3·9	4·3 4·1	55·5
55·8	3·0 2·6	3·1 2·7	3·2 2·8	3·3 3·0	3·5 3·1	3·6 3·3	3·8 3·4	3·9 3·6	4·1 3·7	4·2 3·9	4·4 4·0	55·8
56·1	3·3 2·8	3·4 2·9	3·5 3·0	3·6 3·1	3·7 3·2	3·8 3·3	4·0 3·4	4·1 3·6	4·2 3·7	4·4 3·8	4·5 4·0	56·1
56·4	3·6 2·9	3·7 3·0	3·8 3·1	3·9 3·2	3·9 3·3	4·0 3·4	4·1 3·5	4·3 3·6	4·4 3·7	4·5 3·8	4·6 3·9	56·4
56·7	3·9 3·1	4·0 3·1	4·1 3·2	4·1 3·3	4·2 3·3	4·3 3·4	4·3 3·5	4·4 3·6	4·5 3·7	4·6 3·8	4·7 3·8	56·7
57·0	4·3 3·2	4·3 3·3	4·3 3·3	4·4 3·4	4·4 3·4	4·5 3·5	4·5 3·5	4·6 3·6	4·7 3·6	4·7 3·7	4·8 3·8	57·0
57·3	4·6 3·4	4·6 3·4	4·6 3·4	4·6 3·5	4·7 3·5	4·7 3·5	4·7 3·6	4·8 3·6	4·8 3·6	4·8 3·7	4·9 3·7	57·3
57·6	4·9 3·6	4·9 3·6	4·9 3·6	4·9 3·6	4·9 3·6	4·9 3·6	4·9 3·6	4·9 3·6	5·0 3·6	5·0 3·6	5·0 3·6	57·6
57·9	5·2 3·7	5·2 3·7	5·2 3·7	5·2 3·7	5·2 3·7	5·1 3·6	5·1 3·6	5·1 3·6	5·1 3·6	5·1 3·6	5·1 3·6	57·9
58·2	5·5 3·9	5·5 3·8	5·5 3·8	5·4 3·8	5·4 3·7	5·4 3·7	5·3 3·7	5·3 3·6	5·2 3·6	5·2 3·5	5·2 3·5	58·2
58·5	5·9 4·0	5·8 4·0	5·8 3·9	5·7 3·9	5·6 3·8	5·6 3·8	5·5 3·7	5·5 3·6	5·4 3·6	5·3 3·5	5·3 3·4	58·5
58·8	6·2 4·2	6·1 4·1	6·0 4·1	6·0 4·0	5·9 3·9	5·8 3·8	5·7 3·7	5·6 3·6	5·5 3·5	5·4 3·5	5·3 3·4	58·8
59·1	6·5 4·3	6·4 4·3	6·3 4·2	6·2 4·1	6·1 4·0	6·0 3·9	5·9 3·8	5·8 3·6	5·7 3·5	5·6 3·4	5·4 3·3	59·1
59·4	6·8 4·5	6·7 4·4	6·6 4·3	6·5 4·2	6·4 4·1	6·2 3·9	6·1 3·8	6·0 3·7	5·8 3·5	5·7 3·4	5·5 3·2	59·4
59·7	7·1 4·6	7·0 4·5	6·9 4·4	6·8 4·3	6·6 4·1	6·5 4·0	6·3 3·8	6·2 3·7	6·0 3·5	5·8 3·3	5·6 3·2	59·7
60·0	7·5 4·8	7·3 4·7	7·2 4·5	7·0 4·4	6·9 4·2	6·7 4·0	6·5 3·9	6·3 3·7	6·1 3·5	5·9 3·3	5·7 3·1	60·0
60·3	7·8 5·0	7·6 4·8	7·5 4·7	7·3 4·5	7·1 4·3	6·9 4·1	6·7 3·9	6·5 3·7	6·3 3·5	6·0 3·2	5·8 3·0	60·3
60·6	8·1 5·1	7·9 5·0	7·7 4·8	7·6 4·6	7·3 4·4	7·1 4·2	6·9 3·9	6·7 3·7	6·4 3·4	6·2 3·2	5·9 2·9	60·6
60·9	8·4 5·3	8·2 5·1	8·0 4·9	7·8 4·7	7·6 4·5	7·3 4·2	7·1 4·0	6·8 3·7	6·6 3·4	6·3 3·2	6·0 2·9	60·9
61·2	8·7 5·4	8·5 5·2	8·3 5·0	8·1 4·8	7·9 4·5	7·6 4·3	7·3 4·0	7·0 3·7	6·7 3·4	6·4 3·1	6·1 2·8	61·2
61·5	9·1 5·6	8·8 5·4	8·6 5·1	8·3 4·9	8·1 4·6	7·8 4·3	7·5 4·0	7·2 3·7	6·9 3·4	6·5 3·1	6·2 2·7	61·5

Appendix D

H.O. 249 Vol. 1 Excerpts

LHA ♈	Hc	Zn	Hc	Zn	Hc	Zn	Hc	Zn	Hc	Zn	Hc	Zn	Hc	Zn
	•CAPELLA		ALDEBARAN		•Diphda		FOMALHAUT		ALTAIR		•VEGA		Kochab	
0	34 20	056	26 36	090	31 06	168	18 49	194	26 41	259	29 50	297	27 43	348
1	34 58	056	27 22	091	31 15	169	18 37	195	25 56	259	29 09	298	27 34	348
2	35 36	056	28 07	092	31 23	170	18 24	196	25 11	260	28 29	298	27 24	348
3	36 14	057	28 53	092	31 30	171	18 11	197	24 26	261	27 49	299	27 15	349
4	36 53	057	29 39	093	31 37	172	17 57	198	23 40	261	27 08	299	27 06	349
5	37 32	057	30 25	094	31 43	174	17 43	199	22 55	262	26 28	300	26 58	349
6	38 10	058	31 11	094	31 47	175	17 28	200	22 09	263	25 49	300	26 49	349
7	38 49	058	31 57	095	31 51	176	17 12	201	21 24	263	25 09	301	26 41	350
8	39 28	058	32 43	096	31 54	177	16 55	201	20 38	264	24 29	301	26 33	350
9	40 07	059	33 28	096	31 56	178	16 38	202	19 52	265	23 50	302	26 25	350
10	40 47	059	34 14	097	31 57	179	16 21	203	19 06	266	23 11	302	26 17	351
11	41 26	059	35 00	098	31 57	180	16 02	204	18 21	266	22 32	302	26 10	351
12	42 06	060	35 45	098	31 57	181	15 43	205	17 35	267	21 53	303	26 03	351
13	42 45	060	36 30	099	31 55	182	15 24	206	16 49	267	21 15	303	25 56	351
14	43 25	060	37 16	100	31 53	184	15 03	206	16 03	268	20 37	304	25 49	352
	•CAPELLA		BETELGEUSE		RIGEL		•Diphda		Enif		•DENEB		Kochab	
15	44 05	060	17 15	095	14 16	114	31 49	185	37 10	249	43 22	299	25 42	352
16	44 45	061	18 01	096	14 57	115	31 45	186	36 27	250	42 42	299	25 36	352
17	45 25	061	18 47	096	15 39	115	31 40	187	35 44	251	42 02	299	25 30	352
18	46 05	061	19 32	097	16 21	116	31 34	188	35 00	252	41 22	300	25 24	353
19	46 46	061	20 18	098	17 02	117	31 27	189	34 17	252	40 42	300	25 18	353
20	47 26	062	21 04	098	17 42	118	31 19	190	33 33	253	40 02	300	25 13	353
21	48 07	062	21 49	099	18 23	118	31 11	191	32 49	254	39 23	301	25 08	354
22	48 47	062	22 34	100	19 03	119	31 01	192	32 04	255	38 43	301	25 03	354
23	49 28	062	23 20	100	19 43	120	30 51	194	31 20	256	38 04	301	24 58	354
24	50 09	063	24 05	101	20 23	121	30 40	195	30 35	256	37 25	302	24 53	354
25	50 50	063	24 50	102	21 02	122	30 28	196	29 51	257	36 46	302	24 49	355
26	51 31	063	25 35	102	21 41	122	30 15	197	29 06	258	36 07	302	24 45	355
27	52 12	063	26 20	103	22 20	123	30 01	198	28 21	258	35 28	303	24 41	355
28	52 53	063	27 04	104	22 58	124	29 47	199	27 36	259	34 50	303	24 38	356
29	53 34	064	27 49	105	23 36	125	29 31	200	26 50	260	34 11	303	24 34	356
	•CAPELLA		BETELGEUSE		RIGEL		•Diphda		Alpheratz		•DENEB		Kochab	
30	54 15	064	28 33	105	24 14	126	29 15	201	64 35	253	33 33	304	24 31	356
31	54 56	064	29 17	106	24 51	126	28 58	202	63 51	254	32 55	304	24 28	357
32	55 38	064	30 02	107	25 28	127	28 41	203	63 06	255	32 17	304	24 25	357
33	56 19	064	30 45	108	26 04	128	28 22	204	62 22	256	31 39	305	24 23	357
34	57 00	064	31 29	108	26 40	129	28 03	205	61 37	257	31 01	305	24 21	357
35	57 42	065	32 13	109	27 16	130	27 43	206	60 52	258	30 24	306	24 19	358
36	58 23	065	32 56	110	27 51	131	27 23	207	60 07	259	29 47	306	24 17	358
37	59 05	065	33 39	111	28 25	132	27 01	208	59 22	260	29 10	306	24 16	358
38	59 47	065	34 22	112	28 59	133	26 39	209	58 36	261	28 33	307	24 14	359
39	60 28	065	35 04	113	29 33	134	26 17	210	57 51	262	27 56	307	24 13	359
40	61 10	065	35 47	113	30 06	135	25 53	211	57 05	263	27 19	308	24 13	359
41	61 52	065	36 29	114	30 38	136	25 29	212	56 20	263	26 43	308	24 12	359
42	62 33	065	37 10	115	31 10	137	25 04	213	55 34	264	26 07	308	24 12	000
43	63 15	065	37 52	116	31 41	137	24 39	214	54 48	265	25 31	309	24 12	000
44	63 57	065	38 33	117	32 12	138	24 13	215	54 03	266	24 55	309	24 12	000
	•Dubhe		POLLUX		SIRIUS		•RIGEL		Diphda		•Alpheratz		DENEB	
45	22 25	026	31 22	078	12 56	125	32 42	140	23 47	216	53 17	266	24 20	310
46	22 45	026	32 07	079	13 33	126	33 12	141	23 19	217	52 31	267	23 45	310
47	23 06	027	32 52	079	14 10	127	33 41	142	22 52	218	51 45	268	23 09	310
48	23 27	027	33 37	080	14 46	128	34 09	143	22 23	218	50 59	268	22 35	311
49	23 48	027	34 22	080	15 23	128	34 37	144	21 54	219	50 13	269	22 00	311
50	24 09	028	35 07	081	15 59	129	35 03	145	21 25	220	49 27	270	21 26	312
51	24 30	028	35 53	081	16 34	130	35 30	146	20 55	221	48 41	270	20 52	312
52	24 52	028	36 38	082	17 09	131	35 55	147	20 25	222	47 55	271	20 18	313
53	25 14	029	37 24	082	17 44	131	36 20	148	19 54	223	47 09	272	19 44	313
54	25 37	029	38 09	083	18 18	132	36 44	149	19 22	224	46 23	272	19 11	314
55	25 59	029	38 55	084	18 52	133	37 07	150	18 50	224	45 37	273	18 37	314
56	26 22	030	39 41	084	19 25	134	37 29	151	18 18	225	44 51	273	18 04	315
57	26 45	030	40 27	085	19 58	135	37 51	153	17 45	226	44 06	274	17 32	315
58	27 08	030	41 12	085	20 30	136	38 11	154	17 11	227	43 20	275	16 59	315
59	27 31	031	41 58	086	21 02	136	38 31	155	16 38	228	42 34	275	16 27	316
	•Dubhe		POLLUX		PROCYON		•SIRIUS		RIGEL		•Hamal		Schedar	
60	27 55	031	42 44	087	29 59	110	21 34	137	38 50	156	60 56	244	53 52	314
61	28 18	031	43 30	087	30 42	111	22 05	138	39 08	157	60 15	245	53 19	314
62	28 42	032	44 16	088	31 25	112	22 35	139	39 26	159	59 33	246	52 46	314
63	29 06	032	45 02	088	32 07	113	23 05	140	39 42	160	58 51	247	52 13	314
64	29 31	032	45 48	089	32 49	114	23 34	141	39 57	161	58 08	248	51 40	314
65	29 55	032	46 34	090	33 31	114	24 03	142	40 12	162	57 25	250	51 07	314
66	30 20	033	47 20	090	34 13	115	24 31	143	40 25	164	56 42	251	50 34	314
67	30 45	033	48 06	091	34 55	116	24 59	144	40 38	165	55 59	252	50 01	314
68	31 10	033	48 52	092	35 36	117	25 26	145	40 49	166	55 15	253	49 28	314
69	31 35	033	49 38	092	36 16	118	25 52	145	40 59	167	54 31	254	48 55	314
70	32 01	034	50 23	093	36 57	119	26 18	146	41 09	169	53 47	254	48 21	314
71	32 26	034	51 09	094	37 37	120	26 43	147	41 17	170	53 02	255	47 48	314
72	32 52	034	51 55	094	38 17	121	27 07	148	41 25	171	52 18	256	47 15	314
73	33 18	034	52 41	095	38 56	122	27 31	149	41 31	173	51 33	257	46 42	314
74	33 44	035	53 27	096	39 35	123	27 54	150	41 36	174	50 48	258	46 09	314
	•Dubhe		POLLUX		PROCYON		•SIRIUS		RIGEL		•Hamal		Schedar	
75	34 10	035	54 12	097	40 14	124	28 17	151	41 41	175	50 03	259	45 36	314
76	34 36	035	54 58	097	40 52	125	28 38	152	41 44	177	49 18	260	45 04	314
77	35 03	035	55 44	098	41 29	126	28 59	153	41 46	178	48 33	260	44 31	315
78	35 29	035	56 29	099	42 06	127	29 20	154	41 47	179	47 47	261	43 58	315
79	35 56	036	57 14	100	42 43	128	29 39	155	41 47	181	47 02	262	43 25	315
80	36 23	036	58 00	101	43 19	129	29 58	156	41 46	182	46 16	263	42 53	315
81	36 50	036	58 45	102	43 55	130	30 16	158	41 44	183	45 31	263	42 20	315
82	37 17	036	59 30	102	44 29	131	30 33	159	41 41	185	44 45	264	41 48	315
83	37 44	036	60 15	103	45 04	132	30 49	160	41 37	186	43 59	265	41 16	315
84	38 11	037	60 59	104	45 38	133	31 05	161	41 31	187	43 13	266	40 43	316
85	38 39	037	61 44	105	46 11	135	31 20	162	41 25	189	42 28	266	40 11	316
86	39 06	037	62 28	106	46 43	136	31 34	163	41 18	190	41 42	267	39 39	316
87	39 34	037	63 12	107	47 15	137	31 47	164	41 09	191	40 56	268	39 07	316
88	40 02	037	63 55	109	47 46	138	31 59	165	41 00	192	40 10	268	38 35	316
89	40 29	037	64 39	110	48 16	140	32 10	166	40 49	194	39 24	269	38 04	316

LAT 40°N

LHA ♈	Hc	Zn	Hc	Zn	Hc	Zn	Hc	Zn	Hc	Zn	Hc	Zn	Hc	Zn
	•Dubhe		REGULUS		PROCYON		•SIRIUS		RIGEL		ALDEBARAN		•Mirfak	
90	40 57	037	29 05	099	48 45	141	32 21	167	40 38	195	60 10	224	61 00	303
91	41 25	037	29 50	100	49 14	142	32 31	168	40 25	196	59 37	226	60 22	303
92	41 53	038	30 35	100	49 42	144	32 39	170	40 12	198	59 04	227	59 43	303
93	42 21	038	31 21	101	50 08	145	32 47	171	39 58	199	58 30	229	59 04	303
94	42 49	038	32 06	102	50 34	146	32 54	172	39 42	200	57 55	230	58 26	303
95	43 17	038	32 51	103	50 59	148	33 00	173	39 26	201	57 19	232	57 47	303
96	43 45	038	33 35	103	51 23	149	33 06	174	39 09	203	56 43	233	57 08	303
97	44 14	038	34 20	104	51 46	151	33 10	175	38 51	204	56 06	234	56 30	303
98	44 42	038	35 05	105	52 08	152	33 13	176	38 32	205	55 28	236	55 51	303
99	45 10	038	35 49	106	52 29	154	33 16	178	38 12	206	54 50	237	55 12	303
100	45 39	038	36 33	106	52 49	155	33 17	179	37 51	207	54 12	238	54 34	303
101	46 07	038	37 17	107	53 08	157	33 18	180	37 30	208	53 32	239	53 55	303
102	46 35	038	38 01	108	53 25	158	33 18	181	37 08	210	52 53	240	53 16	303
103	47 04	038	38 44	109	53 41	160	33 16	182	36 44	211	52 12	241	52 38	303
104	47 32	038	39 28	110	53 56	162	33 14	183	36 20	212	51 32	243	51 59	303
	•Kochab		Denebola		•REGULUS		SIRIUS		RIGEL		•ALDEBARAN		Mirfak	
105	31 26	016	22 56	090	40 11	111	33 11	184	35 56	213	50 51	244	51 21	303
106	31 39	017	23 42	090	40 54	112	33 07	186	35 30	214	50 09	245	50 42	303
107	31 53	017	24 28	091	41 36	112	33 02	187	35 04	215	49 28	246	50 04	303
108	32 06	017	25 14	092	42 19	113	32 57	188	34 37	216	48 46	247	49 26	304
109	32 19	017	26 00	092	43 01	114	32 50	189	34 10	217	48 03	248	48 47	304
110	32 33	017	26 46	093	43 42	115	32 42	190	33 42	218	47 21	249	48 09	304
111	32 47	018	27 32	094	44 24	116	32 34	191	33 13	219	46 38	250	47 31	304
112	33 01	018	28 18	094	45 05	117	32 24	192	32 43	220	45 54	251	46 53	304
113	33 15	018	29 03	095	45 46	118	32 14	193	32 13	221	45 11	251	46 15	304
114	33 29	018	29 49	096	46 26	119	32 03	195	31 42	222	44 27	252	45 37	305
115	33 43	018	30 35	096	47 06	120	31 51	196	31 11	223	43 43	253	45 00	305
116	33 58	018	31 20	097	47 45	121	31 38	197	30 39	224	42 59	254	44 22	305
117	34 12	018	32 06	098	48 24	122	31 25	198	30 07	225	42 15	255	43 44	305
118	34 27	019	32 51	099	49 03	124	31 10	199	29 34	226	41 31	256	43 07	306
119	34 41	019	33 37	099	49 41	125	30 55	200	29 00	227	40 46	256	42 29	306
	•Kochab		Denebola		•REGULUS		SIRIUS		RIGEL		•ALDEBARAN		CAPELLA	
120	34 56	019	34 22	100	50 19	126	30 39	201	28 26	228	40 01	257	59 46	295
121	35 11	019	35 07	101	50 56	127	30 22	202	27 52	229	39 16	258	59 04	295
122	35 26	019	35 52	102	51 32	128	30 04	203	27 17	230	38 31	259	58 22	295
123	35 41	019	36 37	102	52 08	130	29 46	204	26 41	231	37 46	260	57 41	295
124	35 57	019	37 22	103	52 43	131	29 27	205	26 05	232	37 01	260	56 59	296
125	36 12	020	38 07	104	53 17	132	29 07	206	25 29	233	36 16	261	56 18	296
126	36 27	020	38 52	105	53 51	133	28 46	207	24 52	234	35 30	262	55 37	296
127	36 43	020	39 36	105	54 24	135	28 24	208	24 15	234	34 45	262	54 55	296
128	36 58	020	40 20	106	54 56	136	28 02	209	23 37	235	33 59	263	54 14	296
129	37 14	020	41 04	107	55 28	138	27 39	210	22 59	236	33 13	264	53 33	296
130	37 30	020	41 48	108	55 58	139	27 16	211	22 21	237	32 28	265	52 52	297
131	37 46	020	42 32	109	56 28	141	26 52	212	21 43	238	31 42	265	52 10	297
132	38 02	020	43 15	110	56 56	142	26 27	213	21 03	238	30 56	266	51 29	297
133	38 17	020	43 58	111	57 24	144	26 01	214	20 24	239	30 10	267	50 49	297
134	38 33	020	44 41	111	57 50	145	25 35	215	19 44	240	29 24	267	50 08	297
	Kochab		•ARCTURUS		SPICA		REGULUS		•SIRIUS		BETELGEUSE		•CAPELLA	
135	38 49	020	20 37	082	10 22	114	58 16	147	25 08	216	37 24	245	49 27	298
136	39 06	021	21 23	082	11 04	115	58 40	149	24 41	217	36 42	245	48 46	298
137	39 22	021	22 09	083	11 45	116	59 03	151	24 13	218	36 00	246	48 06	298
138	39 38	021	22 54	084	12 27	116	59 25	152	23 44	219	35 18	247	47 25	298
139	39 54	021	23 40	084	13 08	117	59 46	154	23 15	220	34 35	248	46 45	299
140	40 10	021	24 26	085	13 48	118	60 05	156	22 46	221	33 52	249	46 04	299
141	40 26	021	25 11	085	14 29	118	60 23	158	22 16	222	33 10	250	45 24	299
142	40 43	021	25 57	086	15 09	119	60 40	160	21 45	222	32 26	250	44 44	299
143	40 59	021	26 43	087	15 49	120	60 55	162	21 14	223	31 43	251	44 04	300
144	41 15	021	27 29	087	16 29	121	61 09	164	20 42	224	30 59	252	43 24	300
145	41 32	021	28 15	088	17 08	121	61 21	166	20 10	225	30 15	253	42 44	300
146	41 48	021	29 01	088	17 47	122	61 32	168	19 37	226	29 31	254	42 05	300
147	42 04	021	29 47	089	18 26	123	61 41	170	19 04	227	28 47	254	41 25	301
148	42 21	021	30 33	090	19 04	124	61 48	172	18 30	227	28 03	255	40 46	301
149	42 37	021	31 19	090	19 42	125	61 54	174	17 56	228	27 18	256	40 06	301
	•Kochab		ARCTURUS		•SPICA		REGULUS		PROCYON		•POLLUX		CAPELLA	
150	42 53	021	32 05	091	20 20	125	61 58	176	42 57	232	59 46	257	39 27	302
151	43 10	021	32 51	092	20 57	126	62 00	178	42 21	233	59 01	258	38 48	302
152	43 26	021	33 37	092	21 34	127	62 01	180	41 44	234	58 16	259	38 09	302
153	43 42	021	34 22	093	22 11	128	62 00	182	41 06	235	57 30	260	37 31	303
154	43 58	021	35 08	094	22 47	129	61 58	184	40 28	236	56 45	261	36 52	303
155	44 15	021	35 54	094	23 22	130	61 53	186	39 50	237	56 00	262	36 13	303
156	44 31	021	36 40	095	23 58	130	61 47	188	39 11	238	55 14	262	35 35	304
157	44 47	020	37 26	096	24 32	131	61 40	190	38 32	239	54 29	263	34 57	304
158	45 03	020	38 11	096	25 07	132	61 31	192	37 53	240	53 43	264	34 19	304
159	45 19	020	38 57	097	25 41	133	61 20	194	37 13	241	52 57	265	33 41	305
160	45 35	020	39 43	098	26 14	134	61 08	196	36 32	242	52 11	265	33 03	305
161	45 51	020	40 28	099	26 47	135	60 54	198	35 52	243	51 25	266	32 26	305
162	46 06	020	41 14	099	27 19	136	60 39	200	35 11	244	50 40	267	31 48	306
163	46 22	020	41 59	100	27 51	137	60 22	202	34 29	244	49 54	268	31 11	306
164	46 38	020	42 44	101	28 22	138	60 04	204	33 48	245	49 08	268	30 34	306
	•Kochab		ARCTURUS		•SPICA		REGULUS		PROCYON		•POLLUX		CAPELLA	
165	46 53	020	43 29	102	28 53	139	59 44	206	33 06	246	48 22	269	29 57	307
166	47 09	020	44 14	103	29 23	140	59 24	208	32 24	247	47 36	270	29 21	307
167	47 24	019	44 59	103	29 52	141	59 02	210	31 41	248	46 50	270	28 44	308
168	47 39	019	45 43	104	30 21	142	58 38	211	30 58	249	46 04	271	28 08	308
169	47 54	019	46 28	105	30 50	143	58 14	213	30 16	249	45 18	271	27 32	308
170	48 09	019	47 12	106	31 17	144	57 48	215	29 32	250	44 32	272	26 56	309
171	48 24	019	47 56	107	31 44	145	57 22	216	28 49	251	43 46	273	26 20	309
172	48 39	019	48 40	108	32 11	146	56 54	218	28 05	252	43 00	273	25 45	310
173	48 54	018	49 24	109	32 36	147	56 25	219	27 22	253	42 14	274	25 09	310
174	49 08	018	50 07	109	33 01	148	55 56	221	26 38	253	41 28	274	24 34	310
175	49 22	018	50 51	110	33 25	149	55 25	222	25 54	254	40 43	275	23 59	311
176	49 36	018	51 34	111	33 49	150	54 54	224	25 09	255	39 57	276	23 25	311
177	49 50	018	52 16	112	34 11	151	54 21	225	24 25	256	39 11	276	22 50	312
178	50 04	017	52 59	113	34 33	152	53 48	227	23 40	256	38 26	277	22 16	312
179	50 18	017	53 40	115	34 55	153	53 15	228	22 56	257	37 40	277	21 42	313

LHA ♈	Hc	Zn	Hc	Zn	Hc	Zn	Hc	Zn	Hc	Zn	Hc	Zn	Hc	Zn
	•VEGA		Alphecca		ARCTURUS		•SPICA		REGULUS		•POLLUX		CAPELLA	
180	17 55	054	44 04	090	54 22	116	35 15	154	52 40	229	36 54	278	21 08	313
181	18 32	054	44 50	090	55 03	117	35 34	155	52 05	231	36 09	278	20 35	313
182	19 10	055	45 36	091	55 44	118	35 53	157	51 29	232	35 23	279	20 02	314
183	19 47	055	46 22	092	56 25	119	36 11	158	50 53	233	34 38	280	19 29	314
184	20 25	056	47 08	092	57 05	120	36 28	159	50 16	234	33 53	280	18 56	315
185	21 04	056	47 54	093	57 44	122	36 44	160	49 38	235	33 08	281	18 23	315
186	21 42	057	48 40	094	58 23	123	36 59	161	49 00	237	32 22	281	17 51	316
187	22 21	057	49 26	094	59 01	124	37 14	162	48 21	238	31 37	282	17 19	316
188	22 59	058	50 12	095	59 39	126	37 27	164	47 42	239	30 52	282	16 48	317
189	23 38	058	50 57	096	60 16	127	37 40	165	47 03	240	30 08	283	16 16	317
190	24 18	059	51 43	097	60 52	128	37 51	166	46 23	241	29 23	283	15 45	318
191	24 57	059	52 29	097	61 28	130	38 02	167	45 42	242	28 38	284	15 14	318
192	25 37	060	53 14	098	62 03	132	38 11	169	45 02	243	27 54	285	14 44	319
193	26 16	060	54 00	099	62 37	133	38 20	170	44 21	244	27 09	285	14 13	319
194	26 56	061	54 45	100	63 10	135	38 28	171	43 39	245	26 25	286	13 43	320
	DENEB		•VEGA		ARCTURUS		•SPICA		REGULUS		•POLLUX		Dubhe	
195	13 04	041	27 36	061	63 42	137	38 34	172	42 57	246	25 41	286	61 56	331
196	13 35	041	28 17	062	64 13	138	38 40	174	42 15	247	24 57	287	61 33	330
197	14 05	042	28 57	062	64 43	140	38 45	175	41 33	248	24 13	287	61 10	329
198	14 36	042	29 38	062	65 11	142	38 48	176	40 50	249	23 29	288	60 47	329
199	15 07	043	30 19	063	65 39	144	38 51	177	40 07	249	22 45	288	60 23	328
200	15 38	043	31 00	063	66 05	146	38 53	179	39 24	250	22 01	289	59 58	328
201	16 10	044	31 41	064	66 30	148	38 53	180	38 41	251	21 18	289	59 34	327
202	16 42	044	32 22	064	66 53	151	38 53	181	37 57	252	20 35	290	59 09	327
203	17 14	045	33 04	065	67 15	153	38 52	182	37 13	253	19 52	290	58 44	327
204	17 47	045	33 45	065	67 35	155	38 49	184	36 29	254	19 09	291	58 18	326
205	18 19	046	34 27	065	67 54	157	38 46	185	35 45	254	18 26	292	57 53	326
206	18 52	046	35 09	066	68 11	160	38 42	186	35 01	255	17 43	292	57 27	325
207	19 26	047	35 51	066	68 26	162	38 36	187	34 16	256	17 01	293	57 01	325
208	19 59	047	36 33	067	68 39	165	38 30	189	33 32	257	16 18	293	56 34	325
209	20 33	048	37 15	067	68 50	167	38 23	190	32 47	257	15 36	294	56 08	325
	DENEB		•VEGA		Rasalhague		ANTARES		•SPICA		REGULUS		•Dubhe	
210	21 07	048	37 58	068	35 41	105	15 06	146	38 14	191	32 02	258	55 41	324
211	21 41	048	38 40	068	36 26	105	15 32	147	38 05	192	31 17	259	55 14	324
212	22 16	049	39 23	068	37 10	106	15 57	148	37 55	194	30 32	260	54 47	324
213	22 50	049	40 06	069	37 54	107	16 21	148	37 43	195	29 47	260	54 19	324
214	23 25	050	40 48	069	38 38	108	16 45	149	37 31	196	29 01	261	53 52	323
215	24 01	050	41 31	070	39 21	109	17 08	150	37 18	197	28 16	262	53 25	323
216	24 36	051	42 15	070	40 05	110	17 31	151	37 04	198	27 30	263	52 57	323
217	25 12	051	42 58	070	40 48	110	17 53	152	36 49	200	26 45	263	52 29	323
218	25 47	051	43 41	071	41 31	111	18 14	153	36 33	201	25 59	264	52 01	323
219	26 23	052	44 25	071	42 14	112	18 35	153	36 17	202	25 13	265	51 33	323
220	27 00	052	45 08	072	42 56	113	18 55	154	35 59	203	24 27	265	51 05	322
221	27 36	053	45 52	072	43 38	114	19 15	155	35 41	204	23 42	266	50 37	322
222	28 13	053	46 36	072	44 20	115	19 34	156	35 21	205	22 56	267	50 09	322
223	28 49	053	47 20	073	45 01	116	19 52	157	35 01	206	22 10	267	49 41	322
224	29 26	054	48 03	073	45 43	117	20 10	158	34 41	208	21 24	268	49 13	322
	DENEB		•ALTAIR		Rasalhague		ANTARES		•SPICA		Denebola		•Dubhe	
225	30 04	054	19 00	094	46 23	118	20 27	159	34 19	209	41 16	253	48 44	322
226	30 41	055	19 45	095	47 04	119	20 43	160	33 56	210	40 32	254	48 16	322
227	31 18	055	20 31	096	47 44	120	20 59	161	33 33	211	39 48	254	47 48	322
228	31 56	055	21 17	096	48 23	121	21 13	162	33 09	212	39 04	255	47 19	322
229	32 34	056	22 03	097	49 02	122	21 28	163	32 45	213	38 19	256	46 51	322
230	33 12	056	22 48	098	49 41	123	21 41	163	32 19	214	37 35	257	46 23	322
231	33 50	056	23 34	098	50 19	125	21 54	164	31 53	215	36 50	257	45 54	322
232	34 29	057	24 19	099	50 57	126	22 06	165	31 26	216	36 05	258	45 26	322
233	35 07	057	25 04	100	51 34	127	22 17	166	30 59	217	35 20	259	44 58	322
234	35 46	057	25 50	101	52 10	128	22 28	167	30 31	218	34 35	260	44 30	322
235	36 25	058	26 35	101	52 46	129	22 37	168	30 02	219	33 49	260	44 01	322
236	37 04	058	27 20	102	53 21	131	22 47	169	29 33	220	33 04	261	43 33	322
237	37 43	058	28 05	103	53 56	132	22 55	170	29 03	221	32 19	262	43 05	322
238	38 22	059	28 49	103	54 30	133	23 02	171	28 32	222	31 33	263	42 37	322
239	39 01	059	29 34	104	55 03	135	23 09	172	28 01	223	30 47	263	42 09	322
	•DENEB		ALTAIR		Nunki		•ANTARES		ARCTURUS		•Alkaid		Kochab	
240	39 41	059	30 19	105	12 14	141	23 15	173	59 24	235	64 51	303	54 51	352
241	40 20	060	31 03	106	12 43	141	23 20	174	58 46	236	64 12	303	54 44	351
242	41 00	060	31 47	106	13 11	142	23 25	175	58 07	238	63 33	303	54 37	351
243	41 40	060	32 31	107	13 39	143	23 28	176	57 28	239	62 55	302	54 30	351
244	42 20	061	33 15	108	14 06	144	23 31	177	56 49	240	62 16	302	54 22	350
245	43 00	061	33 58	109	14 33	145	23 33	178	56 09	241	61 37	302	54 14	350
246	43 41	061	34 42	110	15 00	145	23 35	179	55 28	243	60 58	302	54 06	349
247	44 21	062	35 25	111	15 25	146	23 35	180	54 47	244	60 19	302	53 58	349
248	45 01	062	36 08	111	15 51	147	23 35	181	54 06	245	59 40	302	53 49	349
249	45 42	062	36 51	112	16 15	148	23 34	182	53 24	246	59 01	302	53 39	348
250	46 23	062	37 33	113	16 39	149	23 32	183	52 42	247	58 22	302	53 30	348
251	47 03	063	38 15	114	17 03	150	23 30	184	51 59	248	57 43	302	53 20	348
252	47 44	063	38 57	115	17 26	150	23 26	185	51 17	249	57 04	302	53 10	347
253	48 25	063	39 39	116	17 48	151	23 22	186	50 33	250	56 25	302	53 00	347
254	49 06	063	40 20	117	18 10	152	23 17	187	49 50	251	55 45	302	52 49	346
	•DENEB		ALTAIR		Nunki		•ANTARES		ARCTURUS		•Alkaid		Kochab	
255	49 48	064	41 01	118	18 31	153	23 12	188	49 07	252	55 06	302	52 38	346
256	50 29	064	41 41	119	18 52	154	23 05	189	48 23	253	54 27	302	52 27	346
257	51 10	064	42 21	120	19 12	155	22 58	190	47 39	254	53 48	302	52 16	345
258	51 52	064	43 01	121	19 31	156	22 50	190	46 55	254	53 10	302	52 04	345
259	52 33	065	43 41	122	19 50	157	22 41	191	46 10	255	52 31	302	51 52	345
260	53 15	065	44 19	123	20 07	157	22 32	192	45 26	256	51 52	302	51 40	345
261	53 56	065	44 58	124	20 25	158	22 21	193	44 41	257	51 13	302	51 28	344
262	54 38	065	45 36	125	20 41	159	22 10	194	43 56	258	50 34	303	51 15	344
263	55 20	065	46 13	126	20 57	160	21 59	195	43 11	259	49 56	303	51 02	344
264	56 01	066	46 50	127	21 12	161	21 46	196	42 26	259	49 17	303	50 49	343
265	56 43	066	47 27	128	21 27	162	21 33	197	41 41	260	48 38	303	50 36	343
266	57 25	066	48 03	129	21 41	163	21 19	198	40 56	261	48 00	303	50 23	343
267	58 07	066	48 38	130	21 54	164	21 05	199	40 10	262	47 22	303	50 09	343
268	58 49	066	49 13	132	22 06	165	20 49	200	39 25	262	46 43	304	49 55	342
269	59 31	066	49 47	133	22 18	166	20 33	201	38 39	263	46 05	304	49 42	342

LHA ♈	Hc	Zn	Hc	Zn	Hc	Zn	Hc	Zn	Hc	Zn	Hc	Zn	Hc	Zn
	•Alpheratz		ALTAIR		Nunki		•ANTARES		ARCTURUS		•Alkaid		Kochab	
270	16 48	066	50 20	134	22 29	167	20 17	202	37 53	264	45 27	304	49 27	342
271	17 30	066	50 53	136	22 39	168	19 59	203	37 08	264	44 49	304	49 13	342
272	18 12	067	51 24	137	22 48	169	19 41	204	36 22	265	44 11	304	48 59	342
273	18 54	068	51 55	138	22 57	170	19 23	204	35 36	266	43 33	305	48 44	341
274	19 37	068	52 26	140	23 05	171	19 03	205	34 50	267	42 55	305	48 30	341
275	20 20	069	52 55	141	23 12	172	18 43	206	34 04	267	42 18	305	48 15	341
276	21 03	069	53 23	142	23 18	173	18 23	207	33 18	268	41 40	305	48 00	341
277	21 46	070	53 51	144	23 24	173	18 02	208	32 32	269	41 03	306	47 45	341
278	22 29	070	54 17	145	23 29	174	17 40	209	31 46	269	40 25	306	47 30	341
279	23 12	071	54 43	147	23 33	175	17 17	210	31 00	270	39 48	306	47 14	341
280	23 56	071	55 08	149	23 36	176	16 54	210	30 14	270	39 11	306	46 59	340
281	24 39	072	55 31	150	23 39	177	16 31	211	29 29	271	38 34	307	46 43	340
282	25 23	072	55 53	152	23 40	178	16 07	212	28 43	272	37 58	307	46 28	340
283	26 07	073	56 14	153	23 41	179	15 42	213	27 57	272	37 21	307	46 12	340
284	26 51	073	56 34	155	23 41	180	15 17	214	27 11	273	36 44	308	45 56	340
	•Alpheratz		Enif		ALTAIR		•Rasalhague		ARCTURUS		Alkaid		•Kochab	
285	27 35	074	42 51	118	56 53	157	56 44	220	26 25	274	36 08	308	45 41	340
286	28 19	074	43 31	119	57 11	159	56 13	222	25 39	274	35 32	308	45 25	340
287	29 03	075	44 11	120	57 27	160	55 42	223	24 53	275	34 56	308	45 09	340
288	29 48	076	44 50	121	57 42	162	55 10	225	24 07	275	34 20	309	44 53	340
289	30 32	076	45 29	122	57 55	164	54 37	226	23 22	276	33 44	309	44 37	339
290	31 17	077	46 08	123	58 07	166	54 04	228	22 36	277	33 09	309	44 21	339
291	32 02	077	46 46	125	58 18	168	53 30	229	21 50	277	32 33	310	44 04	339
292	32 47	078	47 24	126	58 27	169	52 55	230	21 05	278	31 58	310	43 48	339
293	33 32	078	48 01	127	58 35	171	52 19	232	20 19	279	31 23	310	43 32	339
294	34 17	079	48 37	128	58 41	173	51 43	233	19 34	279	30 48	311	43 16	339
295	35 02	079	49 13	129	58 45	175	51 06	234	18 49	280	30 13	311	42 59	339
296	35 47	080	49 49	130	58 49	177	50 28	235	18 03	280	29 39	311	42 43	339
297	36 32	080	50 23	132	58 50	179	49 50	236	17 18	281	29 04	312	42 27	339
298	37 18	081	50 57	133	58 50	181	49 12	238	16 33	281	28 30	312	42 10	339
299	38 03	082	51 31	134	58 49	183	48 33	239	15 48	282	27 56	313	41 54	339
	•Mirfak		Alpheratz		•Enif		ALTAIR		Rasalhague		•Alphecca		Kochab	
300	18 21	039	38 49	082	52 03	136	58 46	185	47 53	240	34 15	278	41 38	339
301	18 50	040	39 34	083	52 35	137	58 42	187	47 13	241	33 29	279	41 21	339
302	19 20	040	40 20	083	53 06	138	58 36	188	46 33	242	32 44	279	41 05	339
303	19 50	041	41 05	084	53 36	140	58 28	190	45 52	243	31 59	280	40 49	339
304	20 20	041	41 51	084	54 05	141	58 19	192	45 11	244	31 13	280	40 32	339
305	20 50	042	42 37	085	54 34	143	58 09	194	44 30	245	30 28	281	40 16	339
306	21 21	042	43 23	085	55 01	144	57 57	196	43 48	246	29 43	281	40 00	339
307	21 52	042	44 09	086	55 28	146	57 44	198	43 06	247	28 58	282	39 44	339
308	22 23	043	44 54	087	55 53	147	57 29	199	42 24	248	28 13	283	39 28	339
309	22 54	043	45 40	087	56 17	149	57 13	201	41 41	248	27 28	283	39 11	339
310	23 26	044	46 26	088	56 40	151	56 56	203	40 58	249	26 44	284	38 55	340
311	23 58	044	47 12	088	57 03	152	56 37	205	40 15	250	25 59	284	38 39	340
312	24 30	044	47 58	089	57 23	154	56 17	206	39 32	251	25 15	285	38 23	340
313	25 02	045	48 44	090	57 43	156	55 57	208	38 48	252	24 30	285	38 07	340
314	25 35	045	49 30	090	58 01	157	55 34	210	38 04	253	23 46	286	37 51	340
	•Mirfak		Hamal		Diphda		•FOMALHAUT		ALTAIR		•VEGA		Kochab	
315	26 08	046	24 41	079	12 10	126	15 13	154	55 11	211	62 27	279	37 36	340
316	26 40	046	25 27	080	12 46	127	15 32	155	54 47	213	61 42	280	37 20	340
317	27 14	046	26 12	080	13 23	128	15 52	156	54 21	214	60 57	280	37 04	340
318	27 47	047	26 57	081	13 59	129	16 10	156	53 55	216	60 11	280	36 49	340
319	28 21	047	27 43	082	14 34	129	16 29	157	53 28	217	59 26	281	36 33	340
320	28 54	047	28 28	082	15 10	130	16 46	158	52 59	219	58 41	281	36 18	340
321	29 28	048	29 14	083	15 45	131	17 03	159	52 30	220	57 56	282	36 02	341
322	30 02	048	30 00	083	16 19	132	17 19	160	52 00	222	57 11	282	35 47	341
323	30 37	049	30 45	084	16 53	133	17 35	161	51 29	223	56 26	282	35 32	341
324	31 11	049	31 31	085	17 27	133	17 49	162	50 57	224	55 41	283	35 17	341
325	31 46	049	32 17	085	18 00	134	18 04	162	50 25	226	54 56	283	35 02	341
326	32 21	050	33 03	086	18 33	135	18 17	163	49 52	227	54 12	284	34 47	341
327	32 56	050	33 48	086	19 05	136	18 30	164	49 18	228	53 27	284	34 32	341
328	33 31	050	34 34	087	19 37	137	18 42	165	48 43	229	52 42	284	34 17	341
329	34 07	050	35 20	088	20 08	138	18 54	166	48 08	231	51 58	285	34 03	342
	•CAPELLA		Hamal		Diphda		•FOMALHAUT		ALTAIR		•VEGA		Kochab	
330	16 48	043	36 06	088	20 39	138	19 04	167	47 32	232	51 13	285	33 48	342
331	17 20	044	36 52	089	21 09	139	19 14	168	46 56	233	50 29	286	33 34	342
332	17 52	044	37 38	090	21 39	140	19 24	169	46 19	234	49 45	286	33 20	342
333	18 24	045	38 24	090	22 08	141	19 33	170	45 42	235	49 01	286	33 06	342
334	18 57	045	39 10	091	22 36	142	19 40	171	45 04	236	48 17	287	32 52	342
335	19 29	046	39 56	091	23 04	143	19 48	171	44 25	237	47 33	287	32 38	343
336	20 02	046	40 42	092	23 32	144	19 54	172	43 46	238	46 49	287	32 24	343
337	20 36	047	41 28	093	23 59	145	20 00	173	43 07	239	46 05	288	32 11	343
338	21 09	047	42 14	093	24 25	146	20 05	174	42 27	240	45 21	288	31 57	343
339	21 43	047	43 00	094	24 51	147	20 09	175	41 47	241	44 38	289	31 44	343
340	22 17	048	43 45	095	25 16	147	20 13	176	41 07	242	43 54	289	31 31	344
341	22 51	048	44 31	096	25 40	148	20 15	177	40 26	243	43 11	289	31 18	344
342	23 25	049	45 17	096	26 04	149	20 18	178	39 45	244	42 28	290	31 05	344
343	24 00	049	46 02	097	26 27	150	20 19	179	39 03	245	41 44	290	30 53	344
344	24 35	050	46 48	098	26 49	151	20 19	180	38 21	246	41 01	291	30 40	344
	•CAPELLA		ALDEBARAN		Diphda		•FOMALHAUT		ALTAIR		•VEGA		Kochab	
345	25 10	050	15 09	081	27 11	152	20 19	181	37 39	247	40 18	291	30 28	345
346	25 45	050	15 54	082	27 32	153	20 18	182	36 57	248	39 36	292	30 15	345
347	26 21	051	16 40	082	27 52	154	20 17	183	36 14	249	38 53	292	30 03	345
348	26 57	051	17 25	083	28 12	155	20 14	183	35 31	249	38 10	292	29 52	345
349	27 33	052	18 11	083	28 31	156	20 11	184	34 48	250	37 28	293	29 40	345
350	28 09	052	18 57	084	28 49	157	20 07	185	34 05	251	36 46	293	29 28	346
351	28 45	052	19 42	085	29 06	158	20 03	186	33 21	252	36 03	294	29 17	346
352	29 21	053	20 28	085	29 23	159	19 57	187	32 38	253	35 21	294	29 06	346
353	29 58	053	21 14	086	29 38	160	19 51	188	31 54	253	34 39	294	28 55	346
354	30 35	054	22 00	087	29 53	162	19 44	189	31 09	254	33 58	295	28 44	347
355	31 12	054	22 46	087	30 07	163	19 37	190	30 25	255	33 16	295	28 33	347
356	31 49	054	23 32	088	30 21	164	19 29	191	29 41	256	32 35	296	28 23	347
357	32 27	055	24 18	088	30 33	165	19 20	192	28 56	256	31 53	296	28 13	347
358	33 04	055	25 04	089	30 45	166	19 10	193	28 11	257	31 12	297	28 03	347
359	33 42	055	25 50	090	30 56	167	19 00	194	27 26	258	30 31	297	27 53	348

LHA ♈	Hc	Zn	Hc	Zn	Hc	Zn	Hc	Zn	Hc	Zn	Hc	Zn	Hc	Zn
	Alpheratz		◆Hamal		RIGEL		◆CANOPUS		Peacock		◆Nunki		Enif	
0	20 56	002	20 03	031	14 03	089	27 36	137	51 47	226	26 33	257	31 02	320
1	20 58	001	20 27	030	14 49	088	28 07	137	51 14	226	25 49	256	30 32	319
2	20 58	000	20 49	029	15 35	088	28 39	136	50 41	226	25 04	256	30 01	318
3	20 57	359	21 11	028	16 21	087	29 11	136	50 09	226	24 20	255	29 30	317
4	20 56	358	21 33	027	17 07	086	29 43	136	49 36	226	23 35	255	28 59	316
5	20 54	357	21 53	026	17 53	086	30 15	135	49 03	226	22 51	254	28 26	315
6	20 52	356	22 13	025	18 38	085	30 47	135	48 30	226	22 07	254	27 54	314
7	20 48	355	22 33	024	19 24	084	31 20	135	47 57	226	21 23	253	27 20	313
8	20 44	354	22 52	024	20 10	084	31 52	134	47 24	226	20 39	252	26 47	312
9	20 39	353	23 10	023	20 56	083	32 25	134	46 51	225	19 55	252	26 12	311
10	20 34	353	23 27	022	21 41	082	32 58	134	46 19	225	19 12	251	25 38	311
11	20 27	352	23 43	021	22 27	082	33 31	134	45 46	225	18 28	251	25 03	310
12	20 20	351	23 59	020	23 12	081	34 05	133	45 13	225	17 45	250	24 27	309
13	20 12	350	24 15	019	23 57	080	34 38	133	44 41	225	17 02	250	23 51	308
14	20 04	349	24 29	018	24 43	080	35 12	133	44 08	225	16 19	249	23 14	307
	◆Hamal		ALDEBARAN		RIGEL		◆CANOPUS		Peacock		◆FOMALHAUT		Alpheratz	
15	24 43	017	14 32	053	25 28	079	35 46	132	43 36	225	62 55	283	19 55	348
16	24 56	016	15 09	052	26 13	078	36 20	132	43 03	225	62 10	282	19 45	347
17	25 08	015	15 45	052	26 58	077	36 54	132	42 31	225	61 25	281	19 34	346
18	25 19	014	16 21	051	27 43	077	37 28	132	41 59	224	60 40	280	19 22	345
19	25 30	013	16 56	050	28 27	076	38 02	131	41 27	224	59 55	279	19 10	344
20	25 40	012	17 31	049	29 12	075	38 37	131	40 55	224	59 09	278	18 58	343
21	25 49	011	18 06	048	29 56	074	39 12	131	40 23	224	58 24	277	18 44	342
22	25 57	010	18 40	048	30 40	074	39 46	131	39 51	224	57 38	277	18 30	342
23	26 04	009	19 14	047	31 24	073	40 21	131	39 19	224	56 53	276	18 15	341
24	26 11	008	19 47	046	32 08	072	40 56	130	38 48	223	56 07	275	18 00	340
25	26 17	007	20 20	045	32 52	071	41 31	130	38 16	223	55 21	274	17 43	339
26	26 22	006	20 52	044	33 35	071	42 07	130	37 45	223	54 35	274	17 26	338
27	26 26	005	21 24	043	34 19	070	42 42	130	37 13	223	53 49	273	17 09	337
28	26 30	004	21 55	043	35 02	069	43 17	129	36 42	223	53 03	272	16 51	336
29	26 32	003	22 26	042	35 44	068	43 53	129	36 11	222	52 17	272	16 32	336
	Hamal		◆ALDEBARAN		RIGEL		SIRIUS		◆CANOPUS		Peacock		◆FOMALHAUT	
30	26 34	002	22 56	041	36 27	067	24 56	091	44 29	129	35 40	222	51 31	271
31	26 35	001	23 26	040	37 09	066	25 42	091	45 04	129	35 09	222	50 46	270
32	26 35	000	23 55	039	37 51	065	26 28	090	45 40	129	34 39	222	50 00	270
33	26 34	359	24 24	038	38 33	065	27 14	089	46 16	129	34 08	221	49 14	269
34	26 33	358	24 52	037	39 14	064	27 59	089	46 52	129	33 38	221	48 28	268
35	26 30	357	25 19	036	39 55	063	28 45	088	47 28	128	33 08	221	47 42	268
36	26 27	356	25 46	035	40 36	062	29 31	087	48 04	128	32 38	221	46 56	267
37	26 23	355	26 12	034	41 16	061	30 17	087	48 40	128	32 08	220	46 10	267
38	26 19	354	26 38	033	41 56	060	31 03	086	49 16	128	31 38	220	45 24	266
39	26 13	352	27 03	032	42 35	059	31 49	085	49 52	128	31 09	220	44 38	265
40	26 06	351	27 27	031	43 14	058	32 35	085	50 29	128	30 40	219	43 52	265
41	25 59	350	27 51	030	43 53	057	33 21	084	51 05	128	30 11	219	43 07	264
42	25 51	349	28 14	029	44 31	056	34 06	083	51 41	128	29 42	219	42 21	264
43	25 42	348	28 36	028	45 09	055	34 52	083	52 17	128	29 13	219	41 35	263
44	25 33	347	28 57	027	45 46	054	35 37	082	52 54	128	28 44	218	40 50	263
	◆ALDEBARAN		RIGEL		SIRIUS		◆CANOPUS		ACHERNAR		◆FOMALHAUT		Hamal	
45	29 18	026	46 23	052	36 23	081	53 30	128	68 09	211	40 04	262	25 22	346
46	29 38	025	46 59	051	37 08	081	54 07	128	67 45	212	39 19	261	25 11	345
47	29 58	024	47 35	050	37 54	080	54 43	128	67 21	213	38 33	261	24 59	344
48	30 16	023	48 10	049	38 39	079	55 19	128	66 56	214	37 48	260	24 47	343
49	30 34	022	48 44	048	39 24	078	55 56	128	66 30	214	37 03	260	24 33	342
50	30 51	021	49 18	046	40 09	077	56 32	128	66 04	215	36 17	259	24 19	342
51	31 07	020	49 50	045	40 53	077	57 08	128	65 37	216	35 32	259	24 04	341
52	31 22	019	50 23	044	41 38	076	57 44	128	65 10	217	34 47	258	23 48	340
53	31 37	018	50 54	043	42 23	075	58 21	128	64 42	217	34 02	258	23 32	339
54	31 51	017	51 25	041	43 07	074	58 57	128	64 14	218	33 17	257	23 15	338
55	32 03	016	51 55	040	43 51	073	59 33	128	63 46	219	32 33	257	22 57	337
56	32 15	015	52 24	038	44 35	073	60 09	129	63 17	219	31 48	256	22 39	336
57	32 27	013	52 52	037	45 19	072	60 45	129	62 48	220	31 03	256	22 19	335
58	32 37	012	53 19	035	46 02	071	61 20	129	62 18	220	30 19	255	22 00	334
59	32 46	011	53 45	034	46 46	070	61 56	129	61 48	221	29 35	255	21 39	333
	ALDEBARAN		◆BETELGEUSE		SIRIUS		◆ACRUX		ACHERNAR		◆FOMALHAUT		Hamal	
60	32 55	010	35 43	036	47 29	069	21 29	157	61 18	221	28 50	254	21 18	332
61	33 02	009	36 09	035	48 11	068	21 48	157	60 48	221	28 06	253	20 56	331
62	33 09	008	36 35	034	48 54	067	22 06	156	60 17	222	27 22	253	20 34	330
63	33 15	007	37 00	033	49 36	066	22 25	156	59 47	222	26 38	252	20 11	329
64	33 20	006	37 25	031	50 18	065	22 44	156	59 16	222	25 55	252	19 47	329
65	33 24	004	37 48	030	50 59	064	23 03	155	58 45	223	25 11	251	19 23	328
66	33 27	003	38 11	029	51 40	063	23 22	155	58 13	223	24 27	251	18 58	327
67	33 29	002	38 33	028	52 21	062	23 42	154	57 42	223	23 44	250	18 32	326
68	33 30	001	38 54	027	53 01	061	24 02	154	57 10	223	23 01	250	18 06	325
69	33 31	000	39 14	026	53 41	059	24 22	154	56 39	224	22 18	249	17 40	324
70	33 30	359	39 34	024	54 20	058	24 42	153	56 07	224	21 35	249	17 13	323
71	33 28	358	39 52	023	54 59	057	25 03	153	55 35	224	20 52	248	16 45	323
72	33 26	356	40 10	022	55 38	056	25 24	153	55 03	224	20 10	248	16 17	322
73	33 23	355	40 26	021	56 15	055	25 45	153	54 31	224	19 27	247	15 48	321
74	33 18	354	40 42	019	56 52	053	26 06	152	53 59	224	18 45	247	15 19	320
	BETELGEUSE		◆SIRIUS		Suhail		◆ACRUX		ACHERNAR		◆Diphda		ALDEBARAN	
75	40 57	018	57 29	052	44 44	116	26 28	152	53 27	225	31 03	272	33 13	353
76	41 10	017	58 05	050	45 25	115	26 50	152	52 54	225	30 17	271	33 07	352
77	41 23	015	58 40	049	46 07	115	27 12	151	52 22	225	29 31	271	33 00	351
78	41 35	014	59 14	047	46 48	115	27 34	151	51 50	225	28 45	270	32 52	350
79	41 46	013	59 47	046	47 30	114	27 56	151	51 17	225	27 59	269	32 43	348
80	41 55	012	60 20	044	48 12	114	28 19	150	50 45	225	27 13	269	32 34	347
81	42 04	010	60 52	043	48 54	114	28 41	150	50 12	225	26 27	268	32 23	346
82	42 12	009	61 22	041	49 36	113	29 04	150	49 40	225	25 41	267	32 12	345
83	42 18	008	61 52	039	50 19	113	29 28	150	49 07	225	24 55	267	31 59	344
84	42 24	006	62 21	038	51 01	113	29 51	149	48 35	225	24 09	266	31 46	343
85	42 28	005	62 48	036	51 43	113	30 15	149	48 03	225	23 24	266	31 32	342
86	42 32	004	63 14	034	52 26	112	30 38	149	47 30	225	22 38	265	31 17	341
87	42 34	002	63 39	032	53 08	112	31 02	149	46 58	225	21 52	264	31 02	340
88	42 35	001	64 03	030	53 51	112	31 26	148	46 26	225	21 06	264	30 45	339
89	42 36	000	64 25	028	54 34	112	31 50	148	45 53	225	20 21	263	30 28	337

LHA ♈	Hc	Zn	Hc	Zn	Hc	Zn	Hc	Zn	Hc	Zn	Hc	Zn	Hc	Zn
	PROCYON		REGULUS		◆Suhail		ACRUX		◆ACHERNAR		RIGEL		◆BETELGEUSE	
90	39 22	033	12 37	062	55 16	111	32 15	148	45 21	224	56 36	339	42 35	358
91	39 46	031	13 17	061	55 59	111	32 39	148	44 49	224	56 19	337	42 33	357
92	40 10	030	13 58	061	56 42	111	33 04	147	44 17	224	56 00	336	42 30	356
93	40 32	029	14 38	060	57 25	111	33 29	147	43 45	224	55 41	334	42 25	354
94	40 54	028	15 17	059	58 08	110	33 54	147	43 13	224	55 20	332	42 20	353
95	41 15	027	15 57	059	58 51	110	34 19	147	42 41	224	54 58	331	42 14	351
96	41 35	025	16 36	058	59 35	110	34 44	147	42 09	224	54 35	329	42 07	350
97	41 54	024	17 14	057	60 18	110	35 10	146	41 37	224	54 11	328	41 58	349
98	42 13	023	17 53	056	61 01	110	35 35	146	41 05	223	53 46	326	41 49	348
99	42 30	021	18 31	055	61 44	109	36 01	146	40 34	223	53 20	325	41 38	346
100	42 46	020	19 08	055	62 28	109	36 27	146	40 02	223	52 53	323	41 27	345
101	43 01	019	19 46	054	63 11	109	36 53	146	39 31	223	52 24	322	41 15	344
102	43 16	017	20 23	053	63 55	109	37 19	145	39 00	223	51 55	320	41 01	342
103	43 29	016	20 59	052	64 38	109	37 45	145	38 29	223	51 26	319	40 47	341
104	43 41	015	21 35	051	65 22	109	38 11	145	37 58	222	50 55	318	40 31	340
	PROCYON		REGULUS		◆Gienah		ACRUX		◆ACHERNAR		RIGEL		◆BETELGEUSE	
105	43 53	013	22 11	051	19 34	097	38 38	145	37 27	222	50 24	316	40 15	339
106	44 03	012	22 46	050	20 19	096	39 04	145	36 56	222	49 51	315	39 58	337
107	44 12	011	23 21	049	21 05	096	39 30	145	36 25	222	49 18	314	39 39	336
108	44 20	009	23 55	048	21 51	095	39 57	145	35 55	222	48 45	312	39 20	335
109	44 27	008	24 29	047	22 37	094	40 24	144	35 24	221	48 11	311	39 00	334
110	44 33	007	25 03	046	23 22	094	40 50	144	34 54	221	47 36	310	38 40	332
111	44 37	005	25 36	045	24 08	093	41 17	144	34 24	221	47 00	309	38 18	331
112	44 41	004	26 08	044	24 54	092	41 44	144	33 54	221	46 24	308	37 55	330
113	44 43	002	26 40	043	25 40	092	42 11	144	33 24	220	45 47	307	37 32	329
114	44 45	001	27 11	043	26 26	091	42 38	144	32 54	220	45 10	305	37 08	328
115	44 45	000	27 42	042	27 12	091	43 05	144	32 25	220	44 32	304	36 43	327
116	44 44	358	28 12	041	27 58	090	43 32	144	31 56	220	43 54	303	36 18	326
117	44 42	357	28 42	040	28 44	089	43 59	144	31 26	219	43 16	302	35 51	324
118	44 39	355	29 11	039	29 30	089	44 26	144	30 57	219	42 36	301	35 24	323
119	44 35	354	29 40	038	30 16	088	44 54	144	30 29	219	41 57	300	34 57	322
	REGULUS		◆SPICA		ACRUX		◆CANOPUS		RIGEL		BETELGEUSE		◆PROCYON	
120	30 07	037	13 51	093	45 21	144	69 16	224	41 17	299	34 28	321	44 29	353
121	30 35	036	14 37	092	45 48	144	68 43	225	40 37	298	33 59	320	44 23	351
122	31 01	035	15 23	092	46 15	144	68 11	226	39 56	297	33 29	319	44 15	350
123	31 27	034	16 08	091	46 42	144	67 38	226	39 15	296	32 59	318	44 06	348
124	31 52	033	16 54	090	47 10	144	67 04	227	38 34	295	32 28	317	43 57	347
125	32 17	032	17 40	090	47 37	144	66 30	228	37 52	295	31 56	316	43 46	346
126	32 40	031	18 26	089	48 04	144	65 56	228	37 10	294	31 24	315	43 34	344
127	33 03	030	19 12	089	48 31	144	65 22	229	36 28	293	30 52	314	43 21	343
128	33 26	028	19 58	088	48 58	144	64 47	229	35 46	292	30 18	313	43 07	342
129	33 47	027	20 44	087	49 25	144	64 12	229	35 03	291	29 45	312	42 52	340
130	34 08	026	21 30	087	49 52	144	63 37	230	34 20	290	29 10	311	42 36	339
131	34 28	025	22 16	086	50 19	144	63 02	230	33 37	289	28 36	310	42 19	338
132	34 47	024	23 02	085	50 46	144	62 27	230	32 53	289	28 01	310	42 02	336
133	35 05	023	23 48	085	51 13	144	61 51	231	32 09	288	27 25	309	41 43	335
134	35 23	022	24 33	084	51 39	145	61 16	231	31 26	287	26 49	308	41 23	334
	◆REGULUS		SPICA		◆ACRUX		CANOPUS		◆RIGEL		BETELGEUSE		PROCYON	
135	35 39	021	25 19	083	52 06	145	60 40	231	30 42	286	26 12	307	41 02	333
136	35 55	019	26 05	082	52 32	145	60 04	231	29 57	286	25 35	306	40 41	332
137	36 10	018	26 50	082	52 59	145	59 28	232	29 13	285	24 58	305	40 19	330
138	36 24	017	27 36	081	53 25	145	58 52	232	28 29	284	24 20	304	39 55	329
139	36 37	016	28 21	080	53 51	145	58 16	232	27 44	283	23 42	304	39 31	328
140	36 49	015	29 06	080	54 17	146	57 40	232	26 59	283	23 04	303	39 07	327
141	37 00	013	29 51	079	54 43	146	57 03	232	26 14	282	22 25	302	38 41	326
142	37 10	012	30 36	078	55 09	146	56 27	232	25 29	281	21 45	301	38 15	324
143	37 19	011	31 21	077	55 34	146	55 51	232	24 44	280	21 06	300	37 47	323
144	37 28	010	32 06	077	55 59	147	55 14	232	23 59	280	20 26	299	37 20	322
145	37 35	009	32 51	076	56 24	147	54 38	232	23 13	279	19 46	299	36 51	321
146	37 41	007	33 35	075	56 49	147	54 02	232	22 28	278	19 05	298	36 22	320
147	37 47	006	34 20	074	57 14	148	53 25	232	21 42	278	18 25	297	35 52	319
148	37 51	005	35 04	074	57 38	148	52 49	232	20 57	277	17 44	296	35 22	318
149	37 55	004	35 48	073	58 02	149	52 13	232	20 11	276	17 02	296	34 50	317
	REGULUS		◆SPICA		ACRUX		◆Miaplacidus		CANOPUS		SIRIUS		◆PROCYON	
150	37 57	002	36 32	072	58 26	149	59 41	188	51 36	232	41 54	284	34 19	316
151	37 59	001	37 15	071	58 50	149	59 35	189	51 00	232	41 09	284	33 46	315
152	37 59	000	37 59	070	59 13	150	59 27	189	50 24	232	40 25	283	33 14	314
153	37 58	359	38 42	070	59 36	150	59 20	190	49 47	232	39 40	282	32 40	313
154	37 57	357	39 25	069	59 58	151	59 11	191	49 11	232	38 55	281	32 06	312
155	37 54	356	40 07	068	60 21	151	59 03	191	48 35	232	38 10	280	31 32	311
156	37 51	355	40 50	067	60 42	152	58 54	192	47 59	232	37 24	280	30 57	310
157	37 46	354	41 32	066	61 04	153	58 44	192	47 23	232	36 39	279	30 21	309
158	37 41	353	42 14	065	61 25	153	58 34	193	46 47	231	35 53	278	29 45	308
159	37 34	351	42 55	064	61 45	154	58 23	194	46 11	231	35 08	278	29 09	307
160	37 27	350	43 37	063	62 05	154	58 12	194	45 35	231	34 22	277	28 32	306
161	37 19	349	44 17	062	62 25	155	58 01	195	44 59	231	33 37	276	27 55	305
162	37 09	348	44 58	061	62 44	156	57 49	195	44 24	231	32 51	275	27 17	305
163	36 59	346	45 38	060	63 03	156	57 37	196	43 48	231	32 05	275	26 39	304
164	36 48	345	46 17	059	63 21	157	57 24	196	43 13	230	31 19	274	26 01	303
	ARCTURUS		◆ANTARES		ACRUX		◆CANOPUS		SIRIUS		PROCYON		◆REGULUS	
165	15 21	047	22 16	106	63 38	158	42 37	230	30 33	273	25 22	302	36 36	344
166	15 54	047	23 01	106	63 55	159	42 02	230	29 48	273	24 43	301	36 22	343
167	16 28	046	23 45	105	64 12	160	41 27	230	29 02	272	24 03	300	36 08	342
168	17 00	045	24 29	105	64 27	161	40 52	230	28 16	271	23 23	300	35 54	341
169	17 33	044	25 14	104	64 42	161	40 17	229	27 30	271	22 43	299	35 38	339
170	18 05	043	25 58	104	64 57	162	39 42	229	26 44	270	22 03	298	35 21	338
171	18 36	043	26 43	103	65 10	163	39 07	229	25 58	270	21 22	297	35 04	337
172	19 07	042	27 28	103	65 23	164	38 32	229	25 12	269	20 41	296	34 45	336
173	19 37	041	28 13	102	65 35	165	37 58	229	24 26	268	20 00	296	34 26	335
174	20 07	040	28 58	101	65 47	166	37 23	228	23 40	268	19 18	295	34 06	334
175	20 36	039	29 43	101	65 57	167	36 49	228	22 54	267	18 36	294	33 45	333
176	21 05	038	30 28	100	66 07	168	36 15	228	22 08	266	17 54	293	33 24	331
177	21 33	037	31 13	100	66 16	169	35 41	227	21 22	266	17 12	293	33 02	330
178	22 01	037	31 59	099	66 24	170	35 07	227	20 37	265	16 29	292	32 38	329
179	22 28	036	32 44	099	66 31	171	34 34	227	19 51	264	15 47	291	32 15	328

LAT 40°S

LHA ♈	Hc	Zn	Hc	Zn	Hc	Zn	Hc	Zn	Hc	Zn	Hc	Zn	Hc	Zn
	ARCTURUS		•ANTARES		RIGIL KENT.		ACRUX		•CANOPUS		Alphard		•REGULUS	
180	22 55	035	33 29	098	58 03	144	66 38	173	34 00	227	43 44	302	31 50	327
181	23 20	034	34 15	098	58 30	144	66 43	174	33 27	226	43 05	301	31 25	326
182	23 46	033	35 01	097	58 57	145	66 48	175	32 54	226	42 25	300	30 59	325
183	24 10	032	35 46	096	59 23	145	66 52	176	32 21	226	41 45	299	30 32	324
184	24 34	031	36 32	096	59 50	145	66 54	177	31 48	225	41 05	298	30 05	323
185	24 58	030	37 18	095	60 15	146	66 56	178	31 15	225	40 24	297	29 37	322
186	25 21	029	38 03	095	60 41	146	66 57	179	30 43	225	39 43	296	29 09	321
187	25 43	028	38 49	094	61 06	147	66 57	181	30 10	225	39 02	295	28 40	320
188	26 04	027	39 35	093	61 31	147	66 56	182	29 38	224	38 20	294	28 10	319
189	26 25	026	40 21	093	61 56	148	66 55	183	29 06	224	37 38	294	27 40	318
190	26 45	025	41 07	092	62 20	149	66 52	184	28 35	224	36 56	293	27 09	317
191	27 04	024	41 53	092	62 44	149	66 48	185	28 03	223	36 13	292	26 37	316
192	27 22	023	42 39	091	63 07	150	66 43	186	27 32	223	35 31	291	26 05	316
193	27 40	022	43 25	090	63 30	151	66 38	187	27 01	222	34 48	290	25 33	315
194	27 57	021	44 11	090	63 52	151	66 32	189	26 30	222	34 04	289	25 00	314
	ARCTURUS		•ANTARES		RIGIL KENT.		•ACRUX		Suhail		•REGULUS		Denebola	
195	28 14	020	44 57	089	64 14	152	66 24	190	47 23	246	24 27	313	32 54	339
196	28 29	019	45 43	088	64 35	153	66 16	191	46 41	245	23 53	312	32 37	338
197	28 44	018	46 29	088	64 56	154	66 07	192	45 59	245	23 18	311	32 20	337
198	28 58	017	47 15	087	65 16	154	65 57	193	45 17	245	22 43	310	32 01	336
199	29 11	016	48 00	086	65 36	155	65 47	194	44 36	244	22 08	309	31 42	335
200	29 23	015	48 46	086	65 55	156	65 36	195	43 54	244	21 32	309	31 23	334
201	29 34	014	49 32	085	66 13	157	65 23	196	43 13	244	20 56	308	31 02	333
202	29 45	013	50 18	084	66 30	158	65 10	197	42 32	243	20 20	307	30 41	332
203	29 55	012	51 04	083	66 47	159	64 57	198	41 51	243	19 43	306	30 18	331
204	30 04	011	51 49	083	67 03	160	64 43	199	41 10	243	19 05	305	29 56	330
205	30 12	010	52 35	082	67 18	161	64 28	199	40 29	242	18 28	305	29 32	329
206	30 19	009	53 20	081	67 33	162	64 12	200	39 49	242	17 50	304	29 08	328
207	30 26	007	54 06	080	67 46	163	63 56	201	39 08	242	17 11	303	28 43	327
208	30 31	006	54 51	079	67 59	165	63 39	202	38 28	241	16 33	302	28 17	326
209	30 36	005	55 36	079	68 11	166	63 21	203	37 47	241	15 53	301	27 51	325
	ARCTURUS		•ANTARES		Peacock		RIGIL KENT.		•ACRUX		Suhail		•SPICA	
210	30 40	004	56 21	078	29 29	141	68 21	167	63 03	203	37 07	241	60 04	342
211	30 42	003	57 06	077	29 58	141	68 31	168	62 45	204	36 27	240	59 50	341
212	30 45	002	57 50	076	30 27	141	68 40	170	62 25	205	35 47	240	59 34	339
213	30 46	001	58 35	075	30 56	140	68 48	171	62 06	206	35 08	240	59 16	337
214	30 46	000	59 19	074	31 26	140	68 54	172	61 46	206	34 28	239	58 57	335
215	30 45	359	60 03	073	31 55	140	69 00	174	61 25	207	33 49	239	58 37	333
216	30 44	358	60 47	072	32 25	140	69 05	175	61 04	207	33 09	238	58 16	331
217	30 41	356	61 31	071	32 55	139	69 08	176	60 43	208	32 30	238	57 53	330
218	30 38	355	62 14	070	33 25	139	69 11	178	60 21	209	31 51	238	57 30	328
219	30 34	354	62 57	069	33 55	139	69 12	179	59 59	209	31 13	237	57 05	326
220	30 29	353	63 39	067	34 26	138	69 12	180	59 36	210	30 34	237	56 39	325
221	30 23	352	64 22	066	34 56	138	69 11	182	59 13	210	29 56	237	56 12	323
222	30 16	351	65 03	065	35 27	138	69 09	183	58 50	211	29 17	236	55 44	322
223	30 09	350	65 45	063	35 58	138	69 06	184	58 27	211	28 39	236	55 15	320
224	30 00	349	66 25	062	36 29	138	69 02	186	58 03	211	28 01	235	54 45	319
	Alphecca		Rasalhague		•Nunki		Peacock		•RIGIL KENT.		•SPICA		ARCTURUS	
225	22 47	008	26 23	043	39 57	093	37 00	137	68 57	187	54 14	317	29 51	348
226	22 53	007	26 54	042	40 43	092	37 31	137	68 51	188	53 43	316	29 41	347
227	22 58	006	27 24	041	41 29	092	38 02	137	68 44	190	53 10	314	29 30	346
228	23 03	005	27 54	040	42 14	091	38 34	137	68 35	191	52 37	313	29 18	345
229	23 07	004	28 23	039	43 00	090	39 06	136	68 26	192	52 03	312	29 06	344
230	23 10	003	28 52	038	43 46	090	39 37	136	68 16	194	51 29	311	28 52	342
231	23 13	002	29 20	037	44 32	089	40 09	136	68 04	195	50 53	309	28 38	341
232	23 14	002	29 48	036	45 18	089	40 41	136	67 52	196	50 17	308	28 23	340
233	23 15	001	30 14	035	46 04	088	41 13	136	67 39	197	49 41	307	28 07	339
234	23 15	000	30 40	034	46 50	087	41 45	136	67 25	198	49 04	306	27 51	338
235	23 14	359	31 06	033	47 36	087	42 17	135	67 10	199	48 26	305	27 33	337
236	23 13	358	31 31	032	48 22	086	42 50	135	66 55	200	47 48	303	27 15	336
237	23 11	357	31 55	031	49 08	085	43 22	135	66 38	201	47 10	302	26 56	335
238	23 07	356	32 18	030	49 53	084	43 54	135	66 21	202	46 31	301	26 37	334
239	23 04	355	32 41	029	50 39	084	44 27	135	66 03	203	45 51	300	26 17	333
	Rasalhague		ALTAIR		•Nunki		Peacock		•RIGIL KENT.		SPICA		•ARCTURUS	
240	33 02	028	17 53	061	51 25	083	44 59	135	65 45	204	45 11	299	25 56	332
241	33 23	027	18 33	060	52 10	082	45 32	135	65 25	205	44 31	298	25 34	331
242	33 44	026	19 13	060	52 56	081	46 05	135	65 05	206	43 50	297	25 12	330
243	34 03	025	19 52	059	53 41	081	46 37	135	64 45	207	43 09	296	24 49	330
244	34 22	023	20 32	058	54 27	080	47 10	135	64 24	208	42 28	295	24 25	329
245	34 40	022	21 11	057	55 12	079	47 43	134	64 02	208	41 46	294	24 01	328
246	34 57	021	21 49	056	55 57	078	48 16	134	63 40	209	41 04	293	23 36	327
247	35 13	020	22 27	056	56 42	077	48 49	134	63 18	210	40 22	293	23 10	326
248	35 28	019	23 05	055	57 26	076	49 21	134	62 54	210	39 39	292	22 44	325
249	35 42	018	23 42	054	58 11	075	49 54	134	62 31	211	38 56	291	22 17	324
250	35 56	016	24 19	053	58 55	074	50 27	134	62 07	212	38 13	290	21 50	323
251	36 08	015	24 56	052	59 39	073	51 00	134	61 43	212	37 30	289	21 22	322
252	36 20	014	25 32	051	60 23	072	51 33	134	61 18	213	36 46	288	20 54	321
253	36 31	013	26 08	051	61 07	071	52 06	134	60 53	213	36 02	287	20 25	320
254	36 41	012	26 43	050	61 50	070	52 39	134	60 27	214	35 18	287	19 55	320
	Rasalhague		•ALTAIR		FOMALHAUT		•ACHERNAR		RIGIL KENT.		•SPICA		ARCTURUS	
255	36 50	011	27 18	049	19 04	113	16 10	154	60 02	214	34 34	286	19 25	319
256	36 58	009	27 52	048	19 46	113	16 30	154	59 35	215	33 50	285	18 55	318
257	37 04	008	28 26	047	20 29	112	16 51	153	59 09	215	33 06	284	18 23	317
258	37 10	007	28 59	046	21 11	112	17 11	153	58 42	216	32 21	284	17 52	316
259	37 15	006	29 32	045	21 54	111	17 32	152	58 16	216	31 36	283	17 20	315
260	37 20	004	30 04	044	22 37	111	17 54	152	57 49	216	30 51	282	16 47	315
261	37 23	003	30 36	043	23 20	110	18 16	152	57 21	217	30 06	281	16 14	314
262	37 25	002	31 07	042	24 04	109	18 38	151	56 54	217	29 21	281	15 41	313
263	37 26	001	31 38	041	24 47	109	19 00	151	56 26	217	28 36	280	15 07	312
264	37 26	000	32 08	040	25 31	108	19 22	150	55 58	217	27 51	279	14 33	311
265	37 25	358	32 37	039	26 14	108	19 45	150	55 30	218	27 05	278	13 58	311
266	37 23	357	33 06	038	26 58	107	20 08	150	55 02	218	26 20	278	13 23	310
267	37 20	356	33 34	037	27 42	107	20 32	149	54 33	218	25 34	277	12 48	309
268	37 17	355	34 01	036	28 26	106	20 56	149	54 05	218	24 48	276	12 12	308
269	37 12	353	34 28	035	29 10	106	21 20	148	53 36	219	24 03	276	11 36	308

LHA ♈	Hc	Zn	Hc	Zn	Hc	Zn	Hc	Zn	Hc	Zn	Hc	Zn	Hc	Zn
	ALTAIR		Enif		•FOMALHAUT		ACHERNAR		•RIGIL KENT.		ANTARES		•Rasalhague	
270	34 54	034	18 15	059	29 55	105	21 44	148	53 08	219	66 41	299	37 06	352
271	35 19	033	18 54	058	30 39	105	22 08	148	52 39	219	66 01	297	36 59	351
272	35 44	032	19 33	058	31 23	104	22 33	147	52 10	219	65 20	296	36 52	350
273	36 08	031	20 12	057	32 08	104	22 58	147	51 41	219	64 38	294	36 43	349
274	36 31	029	20 50	056	32 53	103	23 23	146	51 12	219	63 56	293	36 33	347
275	36 53	028	21 28	055	33 38	103	23 49	146	50 43	219	63 14	292	36 23	346
276	37 14	027	22 06	054	34 22	102	24 15	146	50 14	219	62 31	291	36 11	345
277	37 35	026	22 43	054	35 07	102	24 41	145	49 45	219	61 48	290	35 59	344
278	37 55	025	23 20	053	35 53	101	25 07	145	49 15	219	61 04	289	35 46	343
279	38 13	024	23 56	052	36 38	100	25 33	145	48 46	220	60 21	287	35 31	341
280	38 31	022	24 32	051	37 23	100	26 00	144	48 17	220	59 37	286	35 16	340
281	38 48	021	25 08	050	38 08	099	26 27	144	47 48	220	58 52	285	35 00	339
282	39 05	020	25 43	049	38 54	099	26 54	144	47 18	220	58 08	284	34 44	338
283	39 20	019	26 18	048	39 39	098	27 21	143	46 49	220	57 23	284	34 26	337
284	39 34	018	26 52	048	40 25	098	27 49	143	46 20	220	56 39	283	34 08	336
	ALTAIR		Enif		•FOMALHAUT		ACHERNAR		•RIGIL KENT.		ANTARES		•Rasalhague	
285	39 47	016	27 25	047	41 10	097	28 16	143	45 51	220	55 54	282	33 48	335
286	40 00	015	27 59	046	41 56	097	28 44	142	45 21	219	55 09	281	33 28	334
287	40 11	014	28 31	045	42 41	096	29 12	142	44 52	219	54 23	280	33 07	332
288	40 22	012	29 03	044	43 27	095	29 41	142	44 23	219	53 38	279	32 46	331
289	40 31	011	29 35	043	44 13	095	30 09	142	43 54	219	52 53	278	32 23	330
290	40 39	010	30 06	042	44 59	094	30 38	141	43 25	219	52 07	278	32 00	329
291	40 47	009	30 36	041	45 45	094	31 07	141	42 56	219	51 22	277	31 36	328
292	40 53	007	31 06	040	46 31	093	31 36	141	42 27	219	50 36	276	31 12	327
293	40 59	006	31 35	039	47 16	093	32 05	140	41 58	219	49 50	275	30 46	326
294	41 03	005	32 04	038	48 02	092	32 35	140	41 29	219	49 04	275	30 20	325
295	41 06	003	32 32	037	48 48	091	33 04	140	41 00	219	48 19	274	29 54	324
296	41 08	002	32 59	036	49 34	091	33 34	140	40 32	219	47 33	273	29 27	323
297	41 09	001	33 26	035	50 20	090	34 04	139	40 03	218	46 47	273	28 59	322
298	41 09	359	33 52	034	51 06	089	34 34	139	39 34	218	46 01	272	28 30	321
299	41 08	358	34 17	033	51 52	089	35 04	139	39 06	218	45 15	271	28 01	320
	Enif		•FOMALHAUT		ACHERNAR		•RIGIL KENT.		ANTARES		•Rasalhague		ALTAIR	
300	34 41	032	52 38	088	35 34	139	38 38	218	44 29	271	27 31	319	41 06	357
301	35 05	030	53 24	087	36 05	138	38 09	218	43 43	270	27 01	318	41 03	356
302	35 28	029	54 10	087	36 35	138	37 41	218	42 57	269	26 30	317	40 59	354
303	35 50	028	54 56	086	37 06	138	37 13	217	42 11	269	25 59	316	40 54	353
304	36 11	027	55 42	085	37 37	138	36 45	217	41 25	268	25 27	316	40 48	352
305	36 32	026	56 27	084	38 08	138	36 17	217	40 39	267	24 54	315	40 41	350
306	36 52	025	57 13	084	38 39	137	35 50	217	39 53	267	24 21	314	40 32	349
307	37 11	024	57 59	083	39 10	137	35 22	217	39 07	266	23 48	313	40 23	348
308	37 29	022	58 44	082	39 41	137	34 55	216	38 22	266	23 14	312	40 13	346
309	37 46	021	59 30	081	40 13	137	34 28	216	37 36	265	22 40	311	40 02	345
310	38 02	020	60 15	080	40 44	137	34 00	216	36 50	264	22 05	310	39 49	344
311	38 17	019	61 00	080	41 16	136	33 33	216	36 04	264	21 30	309	39 36	343
312	38 32	018	61 46	079	41 48	136	33 07	216	35 19	263	20 54	309	39 22	341
313	38 45	016	62 31	078	42 19	136	32 40	215	34 33	263	20 18	308	39 07	340
314	38 58	015	63 15	077	42 51	136	32 13	215	33 47	262	19 41	307	38 51	339
	Enif		•Diphda		ACHERNAR		•RIGIL KENT.		ANTARES		Nunki		•ALTAIR	
315	39 09	014	37 30	082	43 23	136	31 47	215	33 02	262	60 37	288	38 34	338
316	39 20	013	38 16	082	43 55	136	31 21	215	32 17	261	59 54	287	38 16	337
317	39 29	011	39 01	081	44 27	136	30 55	214	31 31	260	59 09	286	37 57	335
318	39 38	010	39 46	080	44 59	136	30 29	214	30 46	260	58 25	285	37 38	334
319	39 46	009	40 32	079	45 32	135	30 03	214	30 01	259	57 41	284	37 17	333
320	39 52	008	41 17	079	46 04	135	29 38	214	29 16	259	56 56	283	36 56	332
321	39 58	006	42 02	078	46 36	135	29 13	213	28 31	258	56 11	282	36 34	331
322	40 02	005	42 46	077	47 09	135	28 48	213	27 46	258	55 26	281	36 11	330
323	40 06	004	43 31	076	47 41	135	28 23	213	27 01	257	54 41	281	35 47	328
324	40 08	002	44 16	075	48 13	135	27 58	212	26 16	257	53 56	280	35 23	327
325	40 10	001	45 00	074	48 46	135	27 33	212	25 31	256	53 11	279	34 58	326
326	40 10	000	45 44	074	49 18	135	27 09	212	24 47	255	52 25	278	34 32	325
327	40 10	359	46 28	073	49 51	135	26 45	211	24 02	255	51 40	277	34 05	324
328	40 08	357	47 12	072	50 23	135	26 21	211	23 18	254	50 54	277	33 38	323
329	40 05	356	47 56	071	50 55	135	25 57	211	22 34	254	50 08	276	33 10	322
	•Alpheratz		Diphda		•ACHERNAR		RIGIL KENT.		•Nunki		ALTAIR		Enif	
330	14 51	029	48 39	070	51 28	135	25 34	210	49 23	275	32 42	321	40 02	355
331	15 12	028	49 22	069	52 00	135	25 11	210	48 37	274	32 12	320	39 57	353
332	15 34	027	50 05	068	52 33	135	24 48	210	47 51	274	31 42	319	39 51	352
333	15 54	026	50 47	067	53 05	135	24 25	209	47 05	273	31 12	318	39 44	351
334	16 14	025	51 30	066	53 37	135	24 03	209	46 19	272	30 41	317	39 37	350
335	16 33	024	52 11	065	54 09	136	23 40	209	45 33	272	30 09	316	39 28	348
336	16 52	024	52 53	064	54 41	136	23 18	208	44 47	271	29 37	315	39 18	347
337	17 10	023	53 34	063	55 13	136	22 57	208	44 01	270	29 04	314	39 07	346
338	17 28	022	54 15	062	55 45	136	22 35	208	43 15	270	28 31	313	38 56	345
339	17 44	021	54 55	061	56 17	136	22 14	207	42 29	269	27 57	312	38 43	343
340	18 01	020	55 35	059	56 49	136	21 53	207	41 43	268	27 23	311	38 30	342
341	18 16	019	56 14	058	57 21	137	21 32	207	40 57	268	26 48	310	38 15	341
342	18 31	018	56 53	057	57 52	137	21 12	206	40 11	267	26 13	310	38 00	340
343	18 45	017	57 31	056	58 24	137	20 51	206	39 26	267	25 37	309	37 43	339
344	18 58	017	58 09	054	58 55	137	20 31	206	38 40	266	25 01	308	37 26	337
	•Alpheratz		Diphda		•CANOPUS		RIGIL KENT.		•Nunki		ALTAIR		Enif	
345	19 11	016	58 46	053	20 12	143	20 12	205	37 54	265	24 25	307	37 08	336
346	19 23	015	59 22	051	20 40	142	19 52	205	37 08	265	23 48	306	36 49	335
347	19 35	014	59 57	050	21 08	142	19 33	204	36 22	264	23 10	305	36 29	334
348	19 45	013	60 32	048	21 37	142	19 15	204	35 37	264	22 33	304	36 08	333
349	19 55	012	61 06	047	22 05	141	18 56	204	34 51	263	21 55	304	35 47	332
350	20 04	011	61 39	045	22 34	141	18 38	203	34 05	262	21 16	303	35 25	330
351	20 13	010	62 12	044	23 03	140	18 20	203	33 20	262	20 37	302	35 01	329
352	20 21	009	62 43	042	23 33	140	18 02	202	32 34	261	19 58	301	34 38	328
353	20 28	008	63 13	040	24 02	140	17 45	202	31 49	261	19 19	300	34 13	327
354	20 34	007	63 42	038	24 32	139	17 28	201	31 04	260	18 39	300	33 48	326
355	20 40	007	64 10	036	25 02	139	17 11	201	30 18	260	17 59	299	33 22	325
356	20 45	006	64 36	034	25 32	139	16 55	201	29 33	259	17 19	298	32 55	324
357	20 49	005	65 02	032	26 03	138	16 39	200	28 48	259	16 38	297	32 28	323
358	20 52	004	65 26	030	26 34	138	16 23	200	28 03	258	15 57	297	32 00	322
359	20 55	003	65 48	028	27 05	137	16 08	199	27 18	257	15 16	296	31 31	321

TABLE 5—CORRECTION FOR PRECESSION AND NUTATION

LHA ♈	North latitudes							0°	South latitudes							LHA ♈
	N 89°	N 80°	N 70°	N 60°	N 50°	N 40°	N 20°		S 20°	S 40°	S 50°	S 60°	S 70°	S 80°	S 89°	
								1991								
°	′ °	′ °	′ °	′ °	′ °	′ °	′ °	′ °	′ °	′ °	′ °	′ °	′ °	′ °	′ °	°
0	1 000	1 020	1 040	1 050	1 050	1 060	1 060	2 070	1 070	1 060	1 060	1 050	1 040	1 030	1 010	**0**
30	1 030	1 040	1 050	1 060	1 060	1 070	2 070	2 070	1 070	1 060	1 050	1 040	1 020	1 000	1 340	**30**
60	1 060	1 070	1 070	1 070	1 080	1 080	2 080	1 080	1 070	1 070	1 060	0 —	0 —	0 —	1 310	**60**
90	1 090	1 090	1 090	1 090	1 090	1 090	2 090	1 090	1 090	1 090	0 —	0 —	0 —	0 —	1 270	**90**
120	1 120	1 110	1 100	1 100	1 100	1 100	2 100	1 100	1 100	1 110	1 120	0 —	0 —	0 —	1 240	**120**
150	1 140	1 130	1 120	1 120	1 110	1 110	2 110	1 110	1 110	1 120	1 130	1 140	1 160	1 190	1 210	**150**
180	1 170	1 150	1 140	1 130	1 120	1 120	1 110	2 110	1 120	1 120	1 130	1 130	1 140	1 160	1 180	**180**
210	1 200	1 180	1 160	1 140	1 130	1 120	1 110	2 110	2 110	1 110	1 120	1 120	1 130	1 140	1 150	**210**
240	1 230	0 —	0 —	0 —	1 120	1 110	1 110	1 100	2 100	1 100	1 100	1 110	1 110	1 110	1 120	**240**
270	1 270	0 —	0 —	0 —	0 —	1 090	1 090	1 090	2 090	1 090	1 090	1 090	1 090	1 090	1 090	**270**
300	1 300	0 —	0 —	0 —	1 060	1 070	1 080	1 080	2 080	1 080	1 080	1 080	1 080	1 070	1 060	**300**
330	1 330	1 350	1 020	1 040	1 050	1 060	1 070	1 070	2 070	1 070	1 070	1 060	1 060	1 050	1 040	**330**
360	1 000	1 020	1 040	1 050	1 050	1 060	1 060	2 070	1 070	1 060	1 060	1 050	1 040	1 030	1 010	**360**
								1992								
0	1 000	1 020	1 040	1 050	2 060	2 060	2 070	2 070	2 070	2 060	2 060	1 050	1 040	1 020	1 000	**0**
30	1 030	1 050	1 060	2 060	2 070	2 070	2 070	2 070	2 070	2 060	1 050	1 040	1 020	1 350	1 330	**30**
60	1 060	1 070	2 070	2 080	2 080	2 080	2 080	2 080	2 080	1 070	1 060	1 040	0 —	1 320	1 300	**60**
90	1 090	1 090	2 090	2 090	2 090	2 090	2 090	2 090	2 090	1 090	1 090	0 —	0 —	1 270	1 270	**90**
120	1 120	1 110	2 110	2 100	2 100	2 100	2 100	2 100	2 110	1 110	1 120	1 140	0 —	1 220	1 240	**120**
150	1 150	1 130	1 120	2 120	2 120	2 110	2 110	2 110	2 110	2 120	1 130	1 140	1 160	1 190	1 210	**150**
180	1 180	1 160	1 140	1 130	2 120	2 120	2 110	2 110	2 110	2 120	2 120	1 130	1 140	1 160	1 180	**180**
210	1 210	1 190	1 160	1 140	1 130	2 120	2 110	2 110	2 110	2 110	2 110	2 120	1 120	1 130	1 150	**210**
240	1 240	1 220	0 —	1 140	1 120	1 110	2 100	2 100	2 100	2 100	2 100	2 100	2 110	1 110	1 120	**240**
270	1 270	1 270	0 —	0 —	1 090	1 090	2 090	2 090	2 090	2 090	2 090	2 090	2 090	1 090	1 090	**270**
300	1 300	1 320	0 —	1 040	1 060	1 070	2 070	2 080	2 080	2 080	2 080	2 080	2 070	1 070	1 060	**300**
330	1 330	1 350	1 020	1 040	1 050	2 060	2 070	2 070	2 070	2 070	2 060	2 060	1 060	1 050	1 030	**330**
360	1 000	1 020	1 040	1 050	2 060	2 060	2 070	2 070	2 070	2 060	2 060	1 050	1 040	1 020	1 000	**360**
								1993								
0	1 000	1 020	2 040	2 050	2 060	3 060	3 070	3 070	3 070	3 060	2 060	2 050	2 040	1 020	1 000	**0**
30	1 030	2 050	2 060	2 060	3 070	3 070	3 070	3 070	3 070	2 060	2 050	1 040	1 020	1 350	1 330	**30**
60	1 060	2 070	2 070	3 080	3 080	3 080	3 080	3 080	2 080	2 070	1 060	1 040	1 350	1 310	1 300	**60**
90	1 090	2 090	2 090	3 090	3 090	3 090	3 090	3 090	2 090	1 090	1 090	0 —	0 —	1 270	1 270	**90**
120	1 120	2 110	2 110	3 110	3 100	3 100	3 100	3 100	2 110	2 110	1 120	1 140	1 180	1 220	1 240	**120**
150	1 150	2 140	2 130	2 120	3 120	3 110	3 110	3 110	3 110	2 120	2 130	1 140	1 160	1 180	1 210	**150**
180	1 180	1 160	2 140	2 130	2 120	3 120	3 110	3 110	3 110	3 120	2 120	2 130	2 140	1 160	1 180	**180**
210	1 210	1 190	1 160	1 140	2 130	2 120	3 110	3 110	3 110	3 110	3 110	2 120	2 120	2 130	1 150	**210**
240	1 240	1 230	1 190	1 140	1 120	2 110	2 100	3 100	3 100	3 100	3 100	3 100	2 110	2 110	1 120	**240**
270	1 270	1 270	0 —	0 —	1 090	1 090	2 090	3 090	3 090	3 090	3 090	3 090	2 090	2 090	1 090	**270**
300	1 300	1 320	1 000	1 040	1 060	2 070	2 070	3 080	3 080	3 080	3 080	3 070	2 070	2 070	1 060	**300**
330	1 330	1 000	1 020	1 040	2 050	2 060	3 070	3 070	3 070	3 070	3 060	2 060	2 050	2 040	1 030	**330**
360	1 000	1 020	2 040	2 050	2 060	3 060	3 070	3 070	3 070	3 060	2 060	2 050	2 040	1 020	1 000	**360**
								1994								
0	2 010	2 020	2 040	3 050	3 060	3 060	4 070	4 070	4 060	3 060	3 060	2 050	2 040	2 020	2 000	**0**
30	2 040	2 050	2 060	3 060	3 070	4 070	4 070	4 070	3 070	3 060	2 050	2 040	1 020	1 350	2 330	**30**
60	2 070	2 070	3 070	3 080	4 080	4 080	4 080	4 080	3 080	2 070	1 060	1 040	1 350	1 310	2 300	**60**
90	2 090	2 090	3 090	3 090	4 090	4 090	4 090	4 090	3 090	2 090	1 100	0 —	0 —	1 260	2 270	**90**
120	2 120	2 110	3 110	3 110	3 100	4 100	4 100	4 100	3 110	2 110	2 120	1 140	1 180	1 220	2 240	**120**
150	2 150	2 140	2 130	3 120	3 120	4 110	4 110	4 110	4 110	3 120	2 130	2 140	2 160	1 180	2 200	**150**
180	2 180	2 160	2 140	2 130	3 120	3 120	4 120	4 110	4 110	3 120	3 120	3 130	2 140	2 160	2 170	**180**
210	2 210	1 190	1 160	2 140	2 130	3 120	3 110	4 110	4 110	4 110	3 110	3 120	2 120	2 130	2 140	**210**
240	2 240	1 230	1 190	1 140	1 120	2 110	3 100	4 100	4 100	4 100	4 100	3 100	3 110	2 110	2 110	**240**
270	2 270	1 280	0 —	0 —	1 080	2 090	3 090	4 090	4 090	4 090	4 090	3 090	3 090	2 090	2 090	**270**
300	2 300	1 320	1 000	1 040	2 060	2 070	3 070	4 080	4 080	4 080	3 080	3 070	3 070	2 070	2 060	**300**
330	2 340	1 000	2 020	2 040	2 050	3 060	4 070	4 070	4 070	4 070	3 060	3 060	2 050	2 040	2 030	**330**
360	2 010	2 020	2 040	3 050	3 060	3 060	4 070	4 070	4 060	3 060	3 060	2 050	2 040	2 020	2 000	**360**

Example. In 1993 a position line is obtained in latitude S 52° when LHA♈ is 327°. Entering the table with the year 1993, latitude S 50°, and LHA♈ 330° gives 3 060 which indicates that the position line is to be transferred 3 miles in true bearing 060°.

TABLE 6—CORRECTION (*Q*) FOR *POLARIS*

LHA ♈	Q	LHA ♈	Q	LHA ♈	Q	LHA ♈	Q	LHA ♈	Q	LHA ♈	Q	LHA ♈	Q	LHA ♈	Q
° ′	′	° ′	′	° ′	′	° ′	′	° ′	′	° ′	′	° ′	′	° ′	′
359 24		84 07		119 28		152 37		215 24		279 35		312 36		348 21	
	−38		−30		−4		+22		+46		+20		−6		−32
1 31		85 44		120 42		154 01		227 23		280 58		313 50		350 02	
	−39		−29		−3		+23		+45		+19		−7		−33
3 46		87 18		121 56		155 26		232 23		282 19		315 05		351 46	
	−40		−28		−2		+24		+44		+18		−8		−34
6 11		88 50		123 10		156 53		236 14		283 39		316 20		353 34	
	−41		−27		−1		+25		+43		+17		−9		−35
8 47		90 20		124 24		158 21		239 30		284 59		317 35		355 26	
	−42		−26		0		+26		+42		+16		−10		−36
11 38		91 49		125 39		159 50		242 23		286 18		318 51		357 22	
	−43		−25		+1		+27		+41		+15		−11		−37
14 52		93 17		126 53		161 22		245 01		287 36		320 07		359 24	
	−44		−24		+2		+28		+40		+14		−12		−38
18 40		94 43		128 07		162 54		247 27		288 54		321 23		1 31	
	−45		−23		+3		+29		+39		+13		−13		−39
23 36		96 07		129 21		164 29		249 44		290 11		322 40		3 46	
	−46		−22		+4		+30		+38		+12		−14		−40
35 25		97 31		130 35		166 07		251 53		291 27		323 57		6 11	
	−47		−21		+5		+31		+37		+11		−15		−41
35 24		98 54		131 49		167 46		253 56		292 44		325 15		8 47	
	−46		−20		+6		+32		+36		+10		−16		−42
47 13		100 15		133 04		169 29		255 53		293 59		326 34		11 38	
	−45		−19		+7		+33		+35		+9		−17		−43
52 09		101 36		134 19		171 14		257 46		295 15		327 53		14 52	
	−44		−18		+8		+34		+34		+8		−18		−44
55 57		102 56		135 34		173 03		259 35		296 30		329 13		18 40	
	−43		−17		+9		+35		+33		+7		−19		−45
59 11		104 15		136 50		174 56		261 20		297 45		330 34		23 36	
	−42		−16		+10		+36		+32		+6		−20		−46
62 02		105 34		138 05		176 53		263 03		299 00		331 55		35 25	
	−41		−15		+11		+37		+31		+5		−21		−47
64 38		106 52		139 22		178 56		264 42		300 14		333 18		35 24	
	−40		−14		+12		+38		+30		+4		−22		−46
67 03		108 09		140 38		181 05		266 20		301 28		334 42		47 13	
	−39		−13		+13		+39		+29		+3		−23		−45
69 18		109 26		141 55		183 22		267 55		302 42		336 06		52 09	
	−38		−12		+14		+40		+28		+2		−24		−44
71 25		110 42		143 13		185 48		269 27		303 56		337 32		55 57	
	−37		−11		+15		+41		+27		+1		−25		−43
73 27		111 58		144 31		188 26		270 59		305 10		339 00		59 11	
	−36		−10		+16		+42		+26		0		−26		−42
75 23		113 14		145 50		191 19		272 28		306 25		340 29		62 02	
	−35		−9		+17		+43		+25		−1		−27		−41
77 15		114 29		147 10		194 35		273 56		307 39		341 59		64 38	
	−34		−8		+18		+44		+24		−2		−28		−40
79 03		115 44		148 30		198 26		275 23		308 53		343 31		67 03	
	−33		−7		+19		+45		+23		−3		−29		−39
80 47		116 59		149 51		203 26		276 48		310 07		345 05		69 18	
	−32		−6		+20		+46		+22		−4		−30		−38
82 28		118 13		151 14		215 25		278 12		311 21		346 42		71 25	
	−31		−5		+21		+47		+21		−5		−31		−37
84 07		119 28		152 37		215 24		279 35		312 36		348 21		73 27	

The above table, which does *not* include refraction, gives the quantity *Q* to be applied to the corrected sextant altitude of *Polaris* to give the latitude of the observer. In critical cases ascend.

Polaris: Mag. 2·1, SHA 324° 35′, Dec. N 89° 13′·5

TABLE 7—AZIMUTH OF *POLARIS*

LHA ♈	Latitude 0°	30°	50°	55°	60°	65°	70°	LHA ♈	Latitude 0°	30°	50°	55°	60°	65°	70°
°	°	°	°	°	°	°	°	°	°	°	°	°	°	°	°
0	0·4	0·5	0·7	0·8	0·9	1·1	1·4	180	359·6	359·5	359·3	359·2	359·1	359·0	358·7
10	0·3	0·4	0·5	0·6	0·7	0·8	1·0	190	359·7	359·6	359·5	359·4	359·3	359·2	359·1
20	0·2	0·2	0·3	0·4	0·4	0·5	0·6	200	359·8	259·8	359·7	359·6	359·6	359·5	359·4
30	0·1	0·1	0·1	0·1	0·1	0·2	0·2	210	359·9	359·9	359·9	359·9	359·9	359·8	359·8
40	359·9	359·9	359·9	359·9	359·9	359·8	359·8	220	0·1	0·1	0·1	0·1	0·1	0·1	0·2
50	359·8	359·8	359·7	359·7	359·6	359·5	359·4	230	0·2	0·2	0·3	0·3	0·4	0·4	0·6
60	359·7	359·6	359·5	359·4	359·3	359·2	359·0	240	0·3	0·4	0·5	0·6	0·6	0·7	0·9
70	359·6	359·5	359·3	359·2	359·1	358·9	358·7	250	0·4	0·5	0·7	0·8	0·9	1·0	1·2
80	359·5	359·4	359·1	359·0	358·9	358·7	358·4	260	0·5	0·6	0·8	0·9	1·1	1·3	1·5
90	359·4	359·3	359·0	358·9	358·7	358·5	358·1	270	0·6	0·7	1·0	1·1	1·2	1·5	1·8
100	359·3	359·2	358·9	358·8	358·6	358·3	357·9	280	0·7	0·8	1·1	1·2	1·4	1·6	2·0
110	359·3	359·1	358·8	358·7	358·5	358·2	357·8	290	0·7	0·9	1·2	1·3	1·5	1·8	2·2
120	359·2	359·1	358·8	358·7	358·5	358·2	357·7	300	0·8	0·9	1·2	1·3	1·5	1·8	2·2
130	359·2	359·1	358·8	358·7	358·5	358·2	357·7	310	0·8	0·9	1·2	1·3	1·5	1·8	2·3
140	359·2	359·1	358·8	358·7	358·5	358·2	357·8	320	0·8	0·9	1·2	1·3	1·5	1·8	2·2
150	359·3	359·2	358·9	358·8	358·6	358·4	358·0	330	0·7	0·8	1·1	1·2	1·4	1·7	2·1
160	359·4	359·3	359·0	358·9	358·7	358·5	358·2	340	0·6	0·7	1·0	1·1	1·3	1·5	1·9
170	359·4	359·4	359·2	359·1	358·9	358·7	358·4	350	0·6	0·6	0·9	1·0	1·1	1·3	1·7
180	359·6	359·5	359·3	359·2	359·1	359·0	358·7	360	0·4	0·5	0·7	0·8	0·9	1·1	1·4

When Cassiopeia is left (**right**), *Polaris* is west (east).

Appendix E

H.O. 249 Vol. 2 Excerpts

N. Lat. {LHA greater than 180°.......Zn=Z
{LHA less than 180°..........Zn=360−Z

DECLINATION (0°-14°) SAME NAME AS LATITUDE

LHA	0° Hc	0° d	0° Z	1° Hc	1° d	1° Z	2° Hc	2° d	2° Z	3° Hc	3° d	3° Z	4° Hc	4° d	4° Z
	° ′	′	°	° ′	′	°	° ′	′	°	° ′	′	°	° ′	′	°
0	50 00	+60	180	51 00	+60	180	52 00	+60	180	53 00	+60	180	54 00	+60	180
1	49 59	60	178	50 59	60	178	51 59	60	178	52 59	60	178	53 59	60	178
2	49 58	60	177	50 58	59	177	51 57	60	177	52 57	60	177	53 57	60	177
3	49 54	60	175	50 54	60	175	51 54	60	175	52 54	60	175	53 54	60	175
4	49 50	60	174	50 50	60	174	51 50	59	174	52 49	60	173	53 49	60	173
5	49 45	+59	172	50 44	+60	172	51 44	+59	172	52 43	+60	172	53 43	+60	172
6	49 38	59	171	50 37	60	171	51 37	59	170	52 36	60	170	53 36	59	170
7	49 30	59	169	50 29	59	169	51 28	60	169	52 28	59	169	53 27	59	168
8	49 20	60	168	50 20	59	167	51 19	59	167	52 18	59	167	53 17	59	167
9	49 10	59	166	50 09	59	166	51 08	59	166	52 07	59	165	53 06	58	165
10	48 58	+59	165	49 57	+59	164	50 56	+58	164	51 54	+59	164	52 53	+59	163
11	48 46	58	163	49 44	59	163	50 43	58	163	51 41	58	162	52 39	58	162
12	48 32	58	162	49 30	58	161	50 28	58	161	51 26	58	161	52 24	58	160
13	48 17	58	160	49 15	58	160	50 13	57	159	51 10	58	159	52 08	58	159
14	48 01	57	159	48 58	58	158	49 56	57	158	50 53	58	158	51 51	57	157
15	47 44	+57	157	48 41	+57	157	49 38	+57	157	50 35	+57	156	51 32	+57	156
16	47 25	57	156	48 22	57	156	49 19	57	155	50 16	56	155	51 12	57	154
17	47 06	57	155	48 03	56	154	48 59	57	154	49 56	56	153	50 52	56	153
18	46 46	56	153	47 42	56	153	48 38	56	152	49 34	56	152	50 30	56	151
19	46 25	56	152	47 21	55	151	48 16	56	151	49 12	55	150	50 07	55	150
20	46 03	+55	151	46 58	+55	150	47 53	+55	149	48 48	+55	149	49 43	+55	148
21	45 39	56	149	46 35	54	149	47 29	55	148	48 24	55	147	49 19	54	147
22	45 15	55	148	46 10	55	147	47 05	54	147	47 59	54	146	48 53	54	145
23	44 51	54	147	45 45	54	146	46 39	54	145	47 33	53	145	48 26	54	144
24	44 25	54	145	45 19	53	145	46 12	54	144	47 06	53	143	47 59	53	143
25	43 58	+54	144	44 52	+53	143	45 45	+53	143	46 38	+53	142	47 31	+52	141
26	43 31	53	143	44 24	53	142	45 17	52	142	46 09	53	141	47 02	52	140
27	43 03	52	142	43 55	53	141	44 48	52	140	45 40	52	140	46 32	51	139
28	42 34	52	140	43 26	52	140	44 18	52	139	45 10	51	138	46 01	51	138
29	42 04	52	139	42 56	51	139	43 47	52	138	44 39	51	137	45 30	51	136
30	41 34	+51	138	42 25	+51	137	43 16	+51	137	44 07	+51	136	44 58	+50	135
31	41 03	51	137	41 54	50	136	42 44	51	136	43 35	50	135	44 25	50	134
32	40 31	51	136	41 22	50	135	42 12	50	134	43 02	50	134	43 52	49	133
33	39 59	50	134	40 49	50	133	41 39	49	133	42 28	50	132	43 18	49	132
34	39 26	49	134	40 15	50	133	41 05	49	132	41 54	49	131	42 43	49	131
35	38 52	+49	133	39 41	+50	132	40 31	+48	131	41 19	+49	130	42 08	+48	130
36	38 18	49	132	39 07	49	131	39 56	48	130	40 44	48	129	41 32	48	128
37	37 43	49	131	38 32	48	130	39 20	48	129	40 08	48	128	40 56	47	127
38	37 08	48	129	37 56	48	129	38 44	48	128	39 32	47	127	40 19	47	126
39	36 32	48	128	37 20	48	128	38 08	47	127	38 55	47	126	39 42	46	125
40	35 56	+47	128	36 43	+48	127	37 31	+47	126	38 18	+46	125	39 04	+46	124
41	35 19	47	127	36 06	47	126	36 53	47	125	37 40	46	124	38 26	46	123
42	34 42	47	126	35 29	46	125	36 15	46	124	37 01	46	123	37 47	46	122
43	34 04	47	125	34 51	46	124	35 37	46	123	36 23	45	122	37 08	45	121
44	33 26	46	124	34 12	46	123	34 58	46	122	35 44	45	121	36 29	45	121
45	32 48	+46	123	33 34	+45	122	34 19	+45	121	35 04	+45	120	35 49	+45	120
46	32 09	45	122	32 54	46	121	33 40	44	120	34 24	45	120	35 09	44	119
47	31 30	45	121	32 15	45	120	33 00	44	119	33 44	44	119	34 28	44	118
48	30 50	45	120	31 35	44	119	32 19	45	119	33 04	44	118	33 48	43	117
49	30 10	45	119	30 55	44	118	31 39	44	118	32 23	43	117	33 06	44	116
50	29 30	+44	118	30 14	+44	118	30 58	+44	117	31 42	+43	116	32 25	+43	115
51	28 49	44	118	29 33	44	117	30 17	43	116	31 00	43	115	31 43	43	114
52	28 08	44	117	28 52	43	116	29 35	43	115	30 18	43	114	31 01	43	114
53	27 27	44	116	28 11	43	115	28 54	42	114	29 36	43	114	30 19	42	113
54	26 46	43	115	27 29	43	114	28 12	42	114	28 54	42	113	29 36	42	112
55	26 04	+43	114	26 47	+42	113	27 29	+43	113	28 12	+42	112	28 54	+41	111
56	25 22	42	113	26 04	43	113	26 47	42	112	27 29	42	111	28 11	41	110
57	24 40	42	113	25 22	42	112	26 04	42	111	26 46	41	110	27 27	42	110
58	23 57	42	112	24 39	42	111	25 21	42	110	26 03	41	110	26 44	41	109
59	23 14	42	111	23 56	42	110	24 38	41	110	25 19	41	109	26 00	41	108
60	22 31	+42	110	23 13	+41	110	23 54	+42	109	24 36	+40	108	25 16	+41	107
61	21 48	42	110	22 30	41	109	23 11	41	108	23 52	40	107	24 32	41	106
62	21 05	41	109	21 46	41	108	22 27	41	107	23 08	40	107	23 48	41	106
63	20 21	41	108	21 02	41	107	21 43	41	107	22 23	40	106	23 04	40	105
64	19 37	41	107	20 18	41	107	20 59	40	106	21 39	40	105	22 19	40	104
65	18 53	+41	107	19 34	+41	106	20 15	+40	105	20 55	+40	104	21 35	+40	104
66	18 09	41	106	18 50	40	105	19 30	40	104	20 10	40	103	20 50	40	103
67	17 25	40	105	18 05	41	105	18 46	39	104	19 25	40	103	20 05	40	102
68	16 41	40	104	17 21	40	104	18 01	40	103	18 41	39	102	19 20	40	101
69	15 56	40	104	16 36	40	103	17 16	40	102	17 56	39	102	18 35	39	101

LHA	5° Hc	5° d	5° Z	6° Hc	6° d	6° Z	7° Hc	7° d	7° Z	8° Hc	8° d	8° Z	9° Hc	9° d	9° Z
	° ′	′	°	° ′	′	°	° ′	′	°	° ′	′	°	° ′	′	°
0	55 00	+60	180	56 00	+60	180	57 00	+60	180	58 00	+60	180	59 00	+60	180
1	54 59	60	178	55 59	60	178	56 59	60	178	57 59	60	178	58 59	60	178
2	54 57	60	177	55 57	60	176	56 57	60	176	57 57	60	176	58 57	60	176
3	54 54	60	175	55 54	59	175	56 53	60	175	57 53	60	174	58 53	60	174
4	54 49	60	173	55 49	59	173	56 48	60	173	57 48	60	173	58 48	59	172
5	54 43	+59	171	55 42	+60	171	56 42	+59	171	57 41	+60	171	58 41	+59	171
6	54 35	60	170	55 35	59	169	56 34	59	169	57 33	60	169	58 33	59	169
7	54 26	59	168	55 25	60	168	56 25	59	167	57 24	59	167	58 23	59	167
8	54 16	59	166	55 15	59	166	56 14	59	166	57 13	58	165	58 11	59	165
9	54 04	59	165	55 03	59	164	56 02	58	164	57 00	59	164	57 59	58	163
10	53 52	+58	163	54 50	+58	163	55 48	+59	162	56 47	+58	162	57 45	+58	161
11	53 37	59	161	54 36	58	161	55 34	57	160	56 31	58	160	57 29	58	160
12	53 22	58	160	54 20	58	159	55 18	57	159	56 15	57	158	57 12	58	158
13	53 06	57	158	54 03	57	158	55 00	57	157	55 57	57	157	56 54	57	156
14	52 48	57	157	53 45	57	156	54 42	56	156	55 38	57	155	56 35	56	154
15	52 29	+57	155	53 26	+56	154	54 22	+56	154	55 18	+57	153	56 15	+55	153
16	52 09	56	153	53 05	56	153	54 01	56	152	54 57	56	152	55 53	55	151
17	51 48	56	152	52 44	55	151	53 39	56	151	54 35	55	150	55 30	55	149
18	51 26	55	150	52 21	56	150	53 16	55	149	54 11	55	149	55 06	54	148
19	51 02	55	149	51 57	55	148	52 52	55	148	53 47	54	147	54 41	54	146
20	50 38	+55	148	51 33	+54	147	52 27	+54	146	53 21	+54	145	54 15	+53	145
21	50 13	54	146	51 07	54	145	52 01	54	145	52 55	53	144	53 48	53	143
22	49 47	54	145	50 41	53	144	51 34	53	143	52 27	53	143	53 20	52	142
23	49 20	53	143	50 13	53	143	51 06	53	142	51 59	52	141	52 51	52	140
24	48 52	53	142	49 45	52	141	50 37	52	141	51 29	52	140	52 21	51	139
25	48 23	+53	141	49 16	+51	140	50 07	+52	139	50 59	+51	138	51 50	+51	138
26	47 54	52	139	48 46	51	139	49 37	51	138	50 28	51	137	51 19	50	136
27	47 23	52	138	48 15	51	137	49 06	50	137	49 56	51	136	50 47	50	135
28	46 52	51	137	47 43	51	136	48 34	50	135	49 24	50	134	50 14	49	134
29	46 21	50	136	47 11	50	135	48 01	50	134	48 51	49	133	49 40	49	132
30	45 48	+50	134	46 38	+50	134	47 28	+49	133	48 17	+49	132	49 06	+48	131
31	45 15	49	133	46 04	50	132	46 54	48	132	47 42	49	131	48 31	48	130
32	44 41	49	132	45 30	49	131	46 19	48	130	47 07	48	130	47 55	48	129
33	44 07	48	131	44 55	49	130	45 44	48	129	46 32	47	128	47 19	48	128
34	43 32	48	130	44 20	48	129	45 08	47	128	45 55	47	127	46 42	47	126
35	42 56	+48	129	43 44	+47	128	44 31	+47	127	45 18	+47	126	46 05	+46	125
36	42 20	47	128	43 07	47	127	43 54	47	126	44 41	46	125	45 27	46	124
37	41 43	47	127	42 30	47	126	43 17	46	125	44 03	46	124	44 49	45	123
38	41 06	47	126	41 53	46	125	42 39	46	124	43 25	45	123	44 10	45	122
39	40 28	47	125	41 15	45	124	42 00	46	123	42 46	45	122	43 31	45	121
40	39 50	+46	124	40 36	+46	123	41 22	+45	122	42 07	+44	121	42 51	+45	120
41	39 12	45	123	39 57	45	122	40 42	45	121	41 27	44	120	42 11	44	119
42	38 33	45	122	39 18	45	121	40 03	44	120	40 47	44	119	41 31	44	118
43	37 53	45	121	38 38	45	120	39 23	44	119	40 07	43	118	40 50	43	117
44	37 14	44	120	37 58	44	119	38 42	44	118	39 26	43	117	40 09	43	116
45	36 34	+44	119	37 18	+43	118	38 01	+44	117	38 45	+43	116	39 28	+42	115
46	35 53	44	118	36 37	43	117	37 20	43	116	38 03	43	115	38 46	42	114
47	35 12	44	117	35 56	43	116	36 39	43	115	37 22	42	114	38 04	42	113
48	34 31	43	116	35 14	43	115	35 57	43	114	36 40	42	114	37 22	41	113
49	33 50	43	115	34 33	42	114	35 15	42	113	35 57	42	113	36 39	41	112
50	33 08	+43	114	33 51	+42	114	34 33	+42	113	35 15	+41	112	35 56	+41	111
51	32 26	42	114	33 08	42	113	33 50	42	112	34 32	41	111	35 13	41	110
52	31 44	42	113	32 26	41	112	33 07	42	111	33 49	41	110	34 30	41	109
53	31 01	42	112	31 43	41	111	32 24	42	110	33 06	40	109	33 46	41	108
54	30 18	42	111	31 00	41	110	31 41	41	109	32 22	41	109	33 03	40	108
55	29 35	+42	110	30 17	+41	109	30 58	+40	109	31 38	+41	108	32 19	+40	107
56	28 52	41	109	29 33	41	109	30 14	41	108	30 55	40	107	31 35	39	106
57	28 09	40	109	28 49	41	108	29 30	40	107	30 10	40	106	30 50	40	105
58	27 25	41	108	28 06	40	107	28 46	40	106	29 26	40	105	30 06	39	105
59	26 41	41	107	27 22	40	106	28 02	40	106	28 42	39	105	29 21	40	104
60	25 57	+40	106	26 37	+40	106	27 17	+40	105	27 57	+40	104	28 37	+39	103
61	25 13	40	106	25 53	40	105	26 33	40	104	27 13	39	103	27 52	39	102
62	24 29	40	105	25 09	39	104	25 48	40	103	26 28	39	102	27 07	39	102
63	23 44	40	104	24 24	39	103	25 03	40	103	25 43	39	102	26 22	39	101
64	22 59	40	103	23 39	40	103	24 19	39	102	24 58	39	101	25 37	38	100
65	22 15	+39	103	22 54	+39	102	23 33	+40	101	24 13	+38	100	24 51	+39	99
66	21 30	39	102	22 09	39	101	22 48	39	100	23 27	39	99	24 06	38	99
67	20 45	39	101	21 24	39	101	22 03	39	100	22 42	38	99	23 20	39	98
68	20 00	39	101	20 39	39	100	21 18	38	99	21 56	39	98	22 35	38	98
69	19 14	39	100	19 53	39	99	20 32	39	98	21 11	38	98	21 49	38	97

LHA	10° Hc	10° d	10° Z	11° Hc	11° d	11° Z	12° Hc	12° d	12° Z	13° Hc	13° d	13° Z	14° Hc	14° d	14° Z	LHA
	° ′	′	°	° ′	′	°	° ′	′	°	° ′	′	°	° ′	′	°	
0	60 00	+60	180	61 00	+60	180	62 00	+60	180	63 00	+60	180	64 00	+60	180	360
1	59 59	60	178	60 59	60	178	61 59	60	178	62 59	60	178	63 59	60	178	359
2	59 57	60	176	60 57	60	176	61 57	60	176	62 57	60	176	63 57	59	176	358
3	59 53	60	174	60 53	60	174	61 53	59	174	62 52	60	174	63 52	60	173	357
4	59 47	60	172	60 47	60	172	61 47	59	172	62 46	60	172	63 46	59	171	356
5	59 40	+60	170	60 40	+59	170	61 39	+60	170	62 39	+59	169	63 38	+59	169	355
6	59 32	59	168	60 31	59	168	61 30	59	168	62 29	59	167	63 28	59	167	354
7	59 22	59	166	60 21	59	166	61 20	58	166	62 18	59	165	63 17	59	165	353
8	59 10	59	165	60 09	58	164	61 07	59	164	62 06	58	163	63 04	59	163	352
9	58 57	58	163	59 55	59	162	60 54	58	162	61 52	58	161	62 50	58	161	351
10	58 43	+58	161	59 41	+58	160	60 39	+57	160	61 36	+58	159	62 34	+57	159	350
11	58 27	57	159	59 24	58	158	60 22	57	158	61 19	57	157	62 16	57	157	349
12	58 10	57	157	59 07	57	157	60 04	57	156	61 01	56	155	61 57	56	155	348
13	57 51	57	155	58 48	56	155	59 44	57	154	60 41	56	153	61 37	56	153	347
14	57 31	57	154	58 28	56	153	59 24	55	152	60 19	56	152	61 15	55	151	346
15	57 10	+56	152	58 06	+56	151	59 02	+55	151	59 57	+55	150	60 52	+55	149	345
16	56 48	56	150	57 44	54	150	58 38	55	149	59 33	55	148	60 28	54	147	344
17	56 25	55	149	57 20	54	148	58 14	54	147	59 08	54	146	60 02	53	145	343
18	56 00	55	147	56 55	54	146	57 49	53	145	58 42	53	145	59 35	53	144	342
19	55 35	54	145	56 29	53	145	57 22	53	144	58 15	53	143	59 08	52	142	341
20	55 08	+53	144	56 01	+53	143	56 54	+53	142	57 47	+52	141	58 39	+51	140	340
21	54 41	52	142	55 33	53	142	56 26	52	141	57 18	51	140	58 09	51	139	339
22	54 12	52	141	55 04	52	140	55 56	51	139	56 47	51	138	57 38	51	137	338
23	53 43	51	139	54 34	52	139	55 26	50	138	56 16	51	137	57 07	49	136	337
24	53 12	52	138	54 04	50	137	54 54	50	136	55 44	50	135	56 34	49	134	336
25	52 41	+51	137	53 32	+50	136	54 22	+50	135	55 12	+49	134	56 01	+48	133	335
26	52 09	50	135	52 59	50	134	53 49	49	133	54 38	49	132	55 27	48	131	334
27	51 37	49	134	52 26	49	133	53 15	49	132	54 04	48	131	54 52	47	130	333
28	51 03	49	133	51 52	49	132	52 41	48	131	53 29	47	130	54 16	47	129	332
29	50 29	49	131	51 18	48	130	52 06	47	130	52 53	47	129	53 40	47	127	331
30	49 54	+48	130	50 42	+48	129	51 30	+47	128	52 17	+46	127	53 03	+46	126	330
31	49 19	47	129	50 06	47	128	50 53	47	127	51 40	46	126	52 26	45	125	329
32	48 43	47	128	49 30	46	127	50 16	46	126	51 02	46	125	51 48	45	124	328
33	48 06	47	127	48 53	46	126	49 39	45	125	50 24	46	124	51 10	44	123	327
34	47 29	46	125	48 15	46	125	49 01	45	124	49 46	45	123	50 31	44	121	326
35	46 51	+46	124	47 37	+45	123	48 22	+45	122	49 07	+44	121	49 51	+44	120	325
36	46 13	45	123	46 58	45	122	47 43	44	121	48 27	44	120	49 11	43	119	324
37	45 34	45	122	46 19	45	121	47 04	44	120	47 48	43	119	48 31	43	118	323
38	44 55	45	121	45 40	44	120	46 24	43	119	47 07	43	118	47 50	43	117	322
39	44 16	44	120	45 00	43	119	45 43	44	118	46 27	42	117	47 09	42	116	321
40	43 36	+43	119	44 19	+44	118	45 03	+42	117	45 45	+43	116	46 28	+41	115	320
41	42 55	43	118	43 39	43	117	44 22	42	116	45 04	42	115	45 46	41	114	319
42	42 15	43	117	42 58	42	116	43 40	42	115	44 22	42	114	45 04	41	113	318
43	41 33	43	116	42 16	42	115	42 53	42	114	43 40	41	113	44 21	41	112	317
44	40 52	42	115	41 34	42	114	42 16	42	113	42 58	41	112	43 39	40	111	316
45	40 10	+42	114	40 52	+42	113	41 34	+41	112	42 15	+41	111	42 56	+40	110	315
46	39 28	42	113	40 10	41	113	40 51	41	112	41 32	41	111	42 13	39	110	314
47	38 46	41	113	39 27	41	112	40 08	41	111	40 49	40	110	41 29	40	109	313
48	38 03	42	112	38 45	40	111	39 25	41	110	40 06	40	109	40 46	39	108	312
49	37 20	41	111	38 01	41	110	38 42	40	109	39 22	40	108	40 02	39	107	311
50	36 37	+41	110	37 18	+40	109	37 58	+40	108	38 38	+40	107	39 18	+38	106	310
51	35 54	41	109	36 35	40	108	37 15	39	107	37 54	39	106	38 33	39	105	309
52	35 11	40	108	35 51	40	107	36 31	39	107	37 10	39	106	37 49	38	105	308
53	34 27	40	108	35 07	39	107	35 46	40	106	36 26	38	105	37 04	39	104	307
54	33 43	40	107	34 23	39	106	35 02	39	105	35 41	39	104	36 20	38	103	306
55	32 59	+39	106	33 38	+40	105	34 18	+38	104	34 56	+39	103	35 35	+38	102	305
56	32 14	40	105	32 54	39	104	33 33	39	103	34 12	38	102	34 50	38	102	304
57	31 30	39	104	32 09	39	104	32 48	39	103	33 27	38	102	34 05	37	101	303
58	30 45	40	104	31 25	38	103	32 03	39	102	32 42	38	101	33 20	37	100	302
59	30 01	39	103	30 40	38	102	31 18	38	101	31 56	38	100	32 34	38	99	301
60	29 16	+39	102	29 55	+38	101	30 33	+38	100	31 11	+38	100	31 49	+37	99	300
61	28 31	38	101	29 09	39	101	29 48	38	100	30 26	37	99	31 03	38	98	299
62	27 46	38	101	28 24	38	100	29 02	38	99	29 40	38	98	30 18	37	97	298
63	27 01	38	100	27 39	38	99	28 17	38	98	28 55	37	97	29 32	37	97	297
64	26 15	38	99	26 53	38	98	27 31	38	98	28 09	37	97	28 46	37	96	296
65	25 30	+38	99	26 08	+38	98	26 46	+37	97	27 23	+38	96	28 01	+37	95	295
66	24 44	38	98	25 22	38	97	26 00	38	96	26 38	37	95	27 15	37	94	294
67	23 59	38	97	24 37	37	96	25 14	38	96	25 52	37	95	26 29	37	94	293
68	23 13	38	97	23 51	38	96	24 29	37	95	25 06	37	94	25 43	37	93	292
69	22 27	38	96	23 05	38	95	23 43	37	94	24 20	37	93	24 57	37	92	291

S. Lat. {LHA greater than 180°.......Zn=180−Z
{LHA less than 180°..........Zn=180+Z

LAT 40°

N. Lat. {LHA greater than 180°....... Zn=Z / LHA less than 180°........... Zn=360−Z}

DECLINATION (0°-14°) SAME NAME AS LATITUDE

LHA	0°			1°			2°			3°			4°			5°			6°			7°			8°			9°			10°			11°			12°			13°			14°			LHA
	Hc	d	Z	Hc	d	Z	Hc	d	Z	Hc	d	Z	Hc	d	Z	Hc	d	Z	Hc	d	Z	Hc	d	Z	Hc	d	Z	Hc	d	Z	Hc	d	Z	Hc	d	Z	Hc	d	Z	Hc	d	Z	Hc	d	Z	
	° ′	′	°	° ′	′	°	° ′	′	°	° ′	′	°	° ′	′	°	° ′	′	°	° ′	′	°	° ′	′	°	° ′	′	°	° ′	′	°	° ′	′	°	° ′	′	°	° ′	′	°	° ′	′	°	° ′	′	°	
70	15 11	+40	103	15 51	+40	102	16 31	+40	102	17 11	+39	101	17 50	+39	100	18 29	+39	99	19 08	+39	98	19 47	+38	98	20 25	+39	97	21 04	+38	96	21 42	+37	95	22 19	+38	94	22 57	+37	94	23 34	+37	93	24 11	+37	92	290
71	14 27	39	103	15 06	40	102	15 46	39	101	16 25	40	100	17 05	39	99	17 44	39	99	18 23	38	98	19 01	39	97	19 40	38	96	20 18	38	95	20 56	38	95	21 34	37	94	22 11	37	93	22 48	37	92	23 25	37	91	289
72	13 42	39	102	14 21	40	101	15 01	39	100	15 40	39	100	16 19	39	99	16 58	39	98	17 37	39	97	18 16	38	96	18 54	38	96	19 32	38	95	20 10	38	94	20 48	37	93	21 25	37	92	22 02	37	91	22 39	37	91	288
73	12 57	39	101	13 36	39	100	14 15	40	100	14 55	39	99	15 34	39	98	16 13	38	97	16 51	39	96	17 30	38	96	18 08	38	95	18 46	38	94	19 24	38	93	20 02	37	92	20 39	37	92	21 16	37	91	21 53	37	90	287
74	12 11	40	100	12 51	39	100	13 30	39	99	14 09	39	98	14 48	39	97	15 27	39	97	16 06	38	96	16 44	38	95	17 22	38	94	18 00	38	93	18 38	38	93	19 16	37	92	19 53	38	91	20 31	36	90	21 07	37	89	286
75	11 26	+39	100	12 05	+40	99	12 45	+39	98	13 24	+39	97	14 03	+38	97	14 41	+39	96	15 20	+38	95	15 58	+38	94	16 36	+38	94	17 14	+38	93	17 52	+38	92	18 30	+37	91	19 07	+38	90	19 45	+37	89	20 22	+36	89	285
76	10 41	39	99	11 20	39	98	11 59	39	98	12 38	39	97	13 17	39	96	13 56	38	95	14 34	38	94	15 12	39	94	15 51	38	93	16 29	37	92	17 06	38	91	17 44	37	90	18 21	38	90	18 59	37	89	19 36	36	88	284
77	09 55	40	98	10 35	39	98	11 14	38	97	11 52	39	96	12 31	39	95	13 10	38	95	13 48	38	94	14 26	39	93	15 05	38	92	15 43	37	91	16 20	38	91	16 58	37	90	17 35	38	89	18 13	37	88	18 50	36	87	283
78	09 10	39	98	09 49	39	97	10 28	39	96	11 07	38	96	11 45	39	95	12 24	38	94	13 02	39	93	13 41	38	92	14 19	38	92	14 57	37	91	15 34	38	90	16 12	37	89	16 49	38	88	17 27	37	88	18 04	37	87	282
79	08 24	39	97	09 03	39	96	09 42	39	96	10 21	38	95	10 59	39	94	11 38	38	93	12 16	39	93	12 55	38	92	13 33	38	91	14 11	37	90	14 48	38	89	15 26	38	89	16 04	37	88	16 41	37	87	17 18	37	86	281
80	07 39	+39	97	08 18	+38	96	08 56	+39	95	09 35	+39	94	10 14	+38	93	10 52	+38	93	11 30	+39	92	12 09	+38	91	12 47	+38	90	13 25	+38	89	14 03	+37	89	14 40	+38	88	15 18	+37	87	15 55	+37	86	16 32	+37	85	280
81	06 53	39	96	07 32	39	95	08 11	38	94	08 49	39	94	09 28	38	93	10 06	39	92	10 45	38	91	11 23	38	90	12 01	38	90	12 39	38	89	13 17	37	88	13 54	38	87	14 32	37	86	15 09	37	86	15 46	37	85	279
82	06 07	39	95	06 46	39	94	07 25	38	94	08 03	39	93	08 42	38	92	09 20	39	91	09 59	38	91	10 37	38	90	11 15	38	89	11 53	38	88	12 31	37	87	13 08	38	87	13 46	37	86	14 23	38	85	15 01	37	84	278
83	05 21	39	95	06 00	39	94	06 39	38	93	07 17	39	92	07 56	38	91	08 34	39	91	09 13	38	90	09 51	38	89	10 29	38	88	11 07	38	88	11 45	37	87	12 22	38	86	13 00	38	85	13 38	37	84	14 15	37	84	277
84	04 36	38	94	05 14	39	93	05 53	38	92	06 31	39	92	07 10	38	91	07 48	39	90	08 27	38	89	09 05	38	89	09 43	38	88	10 21	38	87	10 59	38	86	11 37	37	85	12 14	38	85	12 52	37	84	13 29	37	83	276
85	03 50	+38	93	04 28	+39	92	05 07	+38	92	05 45	+39	91	06 24	+38	90	07 02	+39	89	07 41	+38	89	08 19	+38	88	08 57	+38	87	09 35	+38	86	10 13	+38	86	10 51	+38	85	11 29	+37	84	12 06	+38	83	12 44	+37	82	275
86	03 04	38	93	03 42	39	92	04 21	38	91	04 59	39	90	05 38	38	90	06 16	39	89	06 55	38	88	07 33	38	87	08 11	38	86	08 49	38	86	09 27	38	85	10 05	38	84	10 43	38	83	11 21	37	83	11 58	37	82	274
87	02 18	38	92	02 56	39	91	03 35	39	90	04 14	38	90	04 52	38	89	05 30	39	88	06 09	38	87	06 47	38	87	07 25	38	86	08 03	38	85	08 41	38	84	09 19	38	83	09 57	38	83	10 35	38	82	11 13	37	81	273
88	01 32	38	91	02 10	39	91	02 49	39	90	03 28	38	89	04 06	38	88	04 44	39	87	05 23	38	87	06 01	38	86	06 39	39	85	07 18	38	84	07 56	38	84	08 34	38	83	09 12	38	82	09 50	37	81	10 27	38	80	272
89	00 46	39	91	01 25	38	90	02 03	39	89	02 42	38	88	03 20	39	88	03 59	38	87	04 37	38	86	05 15	39	85	05 54	38	85	06 32	38	84	07 10	38	83	07 48	38	82	08 26	38	81	09 04	38	81	09 42	38	80	271
90	00 00	+39	90	00 39	+38	89	01 17	+39	88	01 56	+38	88	02 34	+39	87	03 13	+38	86	03 51	+39	85	04 30	+38	85	05 08	+38	84	05 46	+39	83	06 25	+38	82	07 03	+38	82	07 41	+38	81	08 19	+38	80	08 57	+38	79	270
91	−0 46	39	89	−0 07	38	89	00 31	39	88	01 10	38	87	01 48	39	86	02 27	38	86	03 05	39	85	03 44	38	84	04 22	39	83	05 01	38	82	05 39	38	82	06 17	39	81	06 56	38	80	07 34	38	79	08 12	38	79	269
92	−1 32	39	89	−0 53	38	88	−0 15	39	87	00 24	38	86	01 02	39	86	01 41	39	85	02 20	38	84	02 58	39	83	03 37	38	83	04 15	39	82	04 54	38	81	05 32	38	80	06 10	39	80	06 49	38	79	07 27	38	78	268
93	−2 18	39	88	−1 39	38	87	−1 01	39	87	−0 22	39	86	00 17	38	85	00 55	39	84	01 34	39	83	02 13	38	83	02 51	39	82	03 30	38	81	04 08	39	80	04 47	38	80	05 25	39	79	06 04	38	78	06 42	38	77	267
94	−3 04	39	87	−2 25	39	87	−1 46	38	86	−1 08	39	85	−0 29	39	84	00 10	38	84	00 48	39	83	01 27	39	82	02 06	38	81	02 44	39	81	03 23	39	80	04 02	38	79	04 40	39	78	05 19	38	78	05 57	38	77	266
95	−3 50	+39	87	−3 11	+39	86	−2 32	+38	85	−1 54	+39	84	−1 15	+39	84	−0 36	+39	83	00 03	+38	82	00 41	+39	81	01 20	+39	81	01 59	+39	80	02 38	+38	79	03 16	+39	78	03 55	+39	78	04 34	+38	77	05 12	+39	76	265
96	−4 36	39	86	−3 57	39	85	−3 18	39	85	−2 39	39	84	−2 00	38	83	−1 22	39	82	−0 43	39	82	−0 04	39	81	00 35	39	80	01 14	39	79	01 53	38	79	02 31	39	78	03 10	39	77	03 49	39	76	04 28	38	75	264
97	−5 21	38	85	−4 43	39	85	−4 04	39	84	−3 25	39	83	−2 46	39	82	−2 07	39	82	−1 28	39	81	−0 49	39	80	−0 10	39	79	00 29	39	79	01 08	39	78	01 47	39	77	02 26	38	76	03 04	39	76	03 43	39	75	263
98	−6 07	39	85	−5 28	38	84	−4 50	39	83	−4 11	39	83	−3 32	39	82	−2 53	39	81	−2 14	40	80	−1 34	39	80	−0 55	39	79	−0 16	39	78	00 23	39	77	01 02	39	76	01 41	39	76	02 20	39	75	02 59	39	74	262
99				−6 14	39	83	−5 35	39	83	−4 56	39	82	−4 17	39	81	−3 38	39	80	−2 59	39	80	−2 20	40	79	−1 40	39	78	−1 01	39	77	−0 22	39	77	00 17	39	76	00 56	40	75	01 36	39	74	02 15	39	74	261
100							−6 21	+39	82	−5 42	+40	81	−5 02	+39	80	−4 23	+39	80	−3 44	+39	79	−3 05	+40	78	−2 25	+39	77	−1 46	+39	77	−1 07	+40	76	−0 27	+39	75	00 12	+40	74	00 52	+39	74	01 31	+39	73	260
101										−6 27	39	81	−5 48	40	80	−5 08	39	79	−4 29	39	78	−3 50	40	78	−3 10	39	77	−2 31	40	76	−1 51	39	75	−1 12	40	75	−0 32	39	74	00 07	40	73	00 47	40	72	259
102																−5 54	40	78	−5 14	40	78	−4 34	39	77	−3 55	40	76	−3 15	39	75	−2 36	40	75	−1 56	40	74	−1 16	40	73	−0 36	39	72	00 03	40	72	258
103																			−5 59	40	77	−5 19	40	76	−4 39	39	75	−4 00	40	75	−3 20	40	74	−2 40	40	73	−2 00	40	72	−1 20	40	72	−0 40	40	71	257
104																						−6 04	40	76	−5 24	40	75	−4 44	40	74	−4 04	40	73	−3 24	40	73	−2 44	40	72	−2 04	40	71	−1 24	41	70	256
105																									−6 08	+40	74	−5 28	+40	73	−4 48	+40	73	−4 08	+41	72	−3 27	+40	71	−2 47	+40	70	−2 07	+41	70	255
106																												−6 12	40	73	−5 32	41	72	−4 51	40	71	−4 11	41	71	−3 30	40	70	−2 50	41	69	254
107																															−6 15	40	71	−5 35	41	71	−4 54	41	70	−4 13	40	69	−3 33	41	68	253
108																																		−6 18	41	70	−5 37	41	69	−4 56	41	68	−4 15	41	68	252
109																																					−6 20	41	68	−5 39	41	68	−4 58	42	67	251
110																																								−6 21	+41	67	−5 40	+42	66	250
111																																											−6 22	42	66	249

LHA	0°			1°			2°			3°			4°			5°			6°			7°			8°			9°			10°			11°			12°			13°			14°			LHA
	Hc	d	Z	Hc	d	Z	Hc	d	Z	Hc	d	Z	Hc	d	Z	Hc	d	Z	Hc	d	Z	Hc	d	Z	Hc	d	Z	Hc	d	Z	Hc	d	Z	Hc	d	Z	Hc	d	Z	Hc	d	Z	Hc	d	Z	
98	−6 07	39	85																																											262
97	−5 21	39	85	−6 00	39	86																																								263
96	−4 36	38	86	−5 14	39	87	−5 53	38	88	−6 31	39	88																																		264
95	−3 50	−38	87	−4 28	−39	88	−5 07	−38	88	−5 45	−39	89	−6 24	−38	90																															265
94	−3 04	38	87	−3 42	39	88	−4 21	38	89	−4 59	39	90	−5 38	38	90	−6 16	39	91																												266
93	−2 18	38	88	−2 56	39	89	−3 35	39	90	−4 14	38	90	−4 52	38	91	−5 30	39	92	−6 09	38	93																									267
92	−1 32	38	89	−2 10	39	89	−2 49	39	90	−3 28	38	91	−4 06	38	92	−4 44	39	93	−5 23	38	93	−6 01	38	94																						268
91	−0 46	39	89	−1 25	38	90	−2 03	39	91	−2 42	38	92	−3 20	39	92	−3 59	38	93	−4 37	38	94	−5 15	39	95	−5 54	38	95																			269
90	00 00	−39	90	−0 39	−38	91	−1 17	−39	92	−1 56	−38	92	−2 34	−39	93	−3 13	−38	94	−3 51	−39	95	−4 30	−38	95	−5 08	−38	96	−5 46	−39	97	−6 25	−38	98													270
89	00 46	39	91	00 07	38	91	−0 31	39	92	−1 10	38	93	−1 48	39	94	−2 27	38	94	−3 05	39	95	−3 44	38	96	−4 22	39	97	−5 01	38	98	−5 39	38	98	−6 17	39	99										271
88	01 32	39	91	00 53	38	92	00 15	39	93	−0 24	38	94	−1 02	39	94	−1 41	39	95	−2 20	38	96	−2 58	39	97	−3 37	38	97	−4 15	39	98	−4 54	38	99	−5 32	38	100	−6 10	39	100							272
87	02 18	39	92	01 39	38	93	01 01	39	93	00 22	39	94	−0 17	38	95	−0 55	39	96	−1 34	39	97	−2 13	38	97	−2 51	39	98	−3 30	38	99	−4 08	39	100	−4 47	38	100	−5 25	39	101	−6 04	38	102				273
86	03 04	39	93	02 25	39	93	01 46	38	94	01 08	39	95	00 29	39	96	−0 10	38	96	−0 48	39	97	−1 27	39	98	−2 06	38	99	−2 44	39	99	−3 23	39	100	−4 02	38	101	−4 40	39	102	−5 19	38	102	−5 57	38	103	274
85	03 50	−39	93	03 11	−39	94	02 32	−38	95	01 54	−39	96	01 15	−39	96	00 36	−39	97	−0 03	−38	98	−0 41	−39	99	−1 20	−39	99	−1 59	−39	100	−2 38	−38	101	−3 16	−39	102	−3 55	−39	102	−4 34	−38	103	−5 12	−39	104	275
84	04 36	39	94	03 57	39	95	03 18	39	95	02 39	39	96	02 00	38	97	01 22	39	98	00 43	39	98	00 04	39	99	−0 35	39	100	−1 14	39	101	−1 53	38	101	−2 31	39	102	−3 10	39	103	−3 49	39	104	−4 28	38	105	276
83	05 21	38	95	04 43	39	95	04 04	39	96	03 25	39	97	02 46	39	98	02 07	39	98	01 28	39	99	00 49	39	100	00 10	39	101	−0 29	39	101	−1 08	39	102	−1 47	39	103	−2 26	38	104	−3 04	39	104	−3 43	39	105	277
82	06 07	39	95	05 28	38	96	04 50	39	97	04 11	39	97	03 32	39	98	02 53	39	99	02 14	40	100	01 34	39	100	00 55	39	101	00 16	39	102	−0 23	39	103	−1 02	39	104	−1 41	39	104	−2 20	39	105	−2 59	39	106	278
81	06 53	39	96	06 14	39	97	05 35	39	97	04 56	39	98	04 17	39	99	03 38	39	100	02 59	39	100	02 20	40	101	01 40	39	102	01 01	39	103	00 22	39	103	−0 17	39	104	−0 56	40	105	−1 36	39	106	−2 15	39	106	279
80	07 39	−39	97	07 00	−39	97	06 21	−39	98	05 42	−40	99	05 02	−39	100	04 23	−39	100	03 44	−39	101	03 05	−40	102	02 25	−39	103	01 46	−39	103	01 07	−40	104	00 27	−39	105	−0 12	−40	106	−0 52	−39	106	−1 31	−39	107	280
79	08 24	39	97	07 45	39	98	07 06	39	99	06 27	39	99	05 48	40	100	05 08	39	101	04 29	39	102	03 50	40	102	03 10	39	103	02 31	40	104	01 51	39	105	01 12	40	105	00 32	39	106	−0 07	40	107	−0 47	40	108	281
78	09 10	39	98	08 31	39	99	07 52	40	99	07 12	39	100	06 33	39	101	05 54	40	102	05 14	40	102	04 34	39	103	03 55	40	104	03 15	39	105	02 36	40	105	01 56	40	106	01 16	40	107	00 36	39	108	−0 03	40	108	282
77	09 55	39	98	09 16	39	99	08 37	39	100	07 58	40	101	07 18	40	102	06 38	39	102	05 59	40	103	05 19	40	104	04 39	39	105	04 00	40	105	03 20	40	106	02 40	40	107	02 00	40	108	01 20	40	108	00 40	40	109	283
76	10 41	39	99	10 02	40	100	09 22	39	101	08 43	40	101	08 03	40	102	07 23	39	103	06 44	40	104	06 04	40	104	05 24	40	105	04 44	40	106	04 04	40	107	03 24	40	107	02 44	40	108	02 04	40	109	01 24	41	110	284
75	11 26	−39	100	10 47	−40	101	10 07	−39	101	09 28	−40	102	08 48	−40	103	08 08	−40	104	07 28	−40	104	06 48	−40	105	06 08	−40	106	05 28	−40	107	04 48	−40	107	04 08	−41	108	03 27	−40	109	02 47	−40	110	02 07	−41	110	285
74	12 11	39	100	11 32	40	101	10 52	39	102	10 13	40	103	09 33	40	104	08 53	40	104	08 13	40	105	07 33	41	106	06 52	40	107	06 12	40	107	05 32	41	108	04 51	40	109	04 11	41	109	03 30	40	110	02 50	41	111	286
73	12 57	40	101	12 17	40	102	11 37	40	103	10 57	40	103	10 17	40	104	09 37	40	105	08 57	40	106	08 17	41	106	07 36	40	107	06 56	41	108	06 15	40	109	05 35	41	109	04 54	41	110	04 13	40	111	03 33	41	112	287
72	13 42	40	102	13 02	40	103	12 22	40	103	11 42	40	104	11 02	40	105	10 22	41	106	09 41	40	106	09 01	41	107	08 20	40	108	07 40	41	109	06 59	41	109	06 18	41	110	05 37	41	111	04 56	41	112	04 15	41	112	288
71	14 27	40	103	13 47	40	103	13 07	41	104	12 26	40	105	11 46	40	106	11 06	41	106	10 25	40	107	09 45	41	108	09 04	41	109	08 23	41	109	07 42	41	110	07 01	41	111	06 20	41	112	05 39	41	112	04 58	42	113	289
70	15 11	−40	103	14 31	−40	104	13 51	−40	105	13 11	−41	106	12 30	−40	106	11 50	−41	107	11 09	−41	108	10 28	−41	109	09 47	−41	109	09 06	−41	110	08 25	−41	111	07 44	−41	111	07 03	−42	112	06 21	−41	113	05 40	−42	114	290

S. Lat. {LHA greater than 180°....... Zn=180−Z / LHA less than 180°........... Zn=180+Z}

DECLINATION (0°-14°) CONTRARY NAME TO LATITUDE

LAT 40°

N. Lat. {LHA greater than 180°........ Zn=Z; LHA less than 180°........... Zn=360−Z}

DECLINATION (0°-14°) CONTRARY NAME TO LATITUDE

LAT 40°

LHA	0° Hc	d	Z	1° Hc	d	Z	2° Hc	d	Z	3° Hc	d	Z	4° Hc	d	Z	5° Hc	d	Z	6° Hc	d	Z	7° Hc	d	Z
	° ′	′	°	° ′	′	°	° ′	′	°	° ′	′	°	° ′	′	°	° ′	′	°	° ′	′	°	° ′	′	°
69	15 56	40	104	15 16	40	105	14 36	41	105	13 55	41	106	13 14	40	107	12 34	41	108	11 53	41	108	11 12	41	109
68	16 41	41	105	16 00	40	105	15 20	41	106	14 39	41	107	13 58	41	108	13 17	41	108	12 36	41	109	11 55	41	110
67	17 25	40	105	16 45	41	106	16 04	41	107	15 23	41	108	14 42	41	108	14 01	41	109	13 20	42	110	12 38	41	111
66	18 09	40	106	17 29	41	107	16 48	41	108	16 07	41	108	15 26	42	109	14 44	41	110	14 03	42	111	13 21	42	111
65	18 53	−40	107	18 13	−41	108	17 32	−42	108	16 50	−41	109	16 09	−42	110	15 27	−41	111	14 46	−42	111	14 04	−42	112
64	19 37	41	107	18 56	41	108	18 15	41	109	17 34	42	110	16 52	42	111	16 10	42	111	15 28	42	112	14 46	42	113
63	20 21	41	108	19 40	42	109	18 58	41	110	18 17	42	110	17 35	42	111	16 53	42	112	16 11	42	113	15 29	43	113
62	21 05	42	109	20 23	41	110	19 42	42	110	19 00	42	111	18 18	42	112	17 36	43	113	16 53	42	113	16 11	43	114
61	21 48	42	110	21 06	41	110	20 25	42	111	19 43	43	112	19 00	42	113	18 18	43	113	17 35	42	114	16 53	43	115
60	22 31	−42	110	21 49	−42	111	21 07	−42	112	20 25	−42	113	19 43	−43	113	19 00	−43	114	18 17	−43	115	17 34	−43	116
59	23 14	42	111	22 32	42	112	21 50	43	113	21 07	42	113	20 25	43	114	19 42	43	115	18 59	44	116	18 15	43	116
58	23 57	42	112	23 15	43	113	22 32	43	113	21 49	43	114	21 06	43	115	20 23	43	116	19 40	44	116	18 56	43	117
57	24 40	43	113	23 57	43	113	23 14	43	114	22 31	43	115	21 48	43	116	21 05	44	116	20 21	44	117	19 37	44	118
56	25 22	43	113	24 39	43	114	23 56	43	115	23 13	44	116	22 29	43	117	21 46	44	117	21 02	44	118	20 18	45	119
55	26 04	−43	114	25 21	−43	115	24 38	−44	116	23 54	−44	117	23 10	−44	117	22 26	−44	118	21 42	−44	119	20 58	−45	120
54	26 46	44	115	26 02	43	116	25 19	44	117	24 35	44	117	23 51	44	118	23 07	45	119	22 22	44	120	21 38	45	120
53	27 27	43	116	26 44	44	117	26 00	44	117	25 16	45	118	24 31	44	119	23 47	45	120	23 02	45	120	22 17	45	121
52	28 08	43	117	27 25	45	117	26 40	44	118	25 56	44	119	25 12	45	120	24 27	45	120	23 42	46	121	22 56	45	122
51	28 49	44	118	28 05	44	118	27 21	45	119	26 36	45	120	25 51	45	121	25 06	45	121	24 21	46	122	23 35	45	123
50	29 30	−44	118	28 46	−45	119	28 01	−45	120	27 16	−45	121	26 31	−46	121	25 45	−45	122	25 00	−46	123	24 14	−46	124
49	30 10	44	119	29 26	45	120	28 41	46	121	27 55	45	122	27 10	46	122	26 24	46	123	25 38	46	124	24 52	46	124
48	30 50	45	120	30 05	45	121	29 20	46	122	28 34	46	122	27 48	46	123	27 02	46	124	26 16	46	125	25 30	47	125
47	31 30	46	121	30 44	45	122	29 59	46	123	29 13	46	123	28 27	47	124	27 40	46	125	26 54	47	125	26 07	47	126
46	32 09	46	122	31 23	46	123	30 37	46	123	29 51	46	124	29 05	47	125	28 18	47	126	27 31	47	126	26 44	47	127
45	32 48	−46	123	32 02	−46	124	31 16	−47	124	30 29	−47	125	29 42	−47	126	28 55	−47	126	28 08	−47	127	27 21	−48	128
44	33 26	46	124	32 40	47	124	31 53	46	125	31 07	48	126	30 19	47	127	29 32	48	127	28 44	47	128	27 57	48	129
43	34 04	46	125	33 18	47	125	32 31	47	126	31 44	48	127	30 56	48	128	30 08	48	128	29 20	48	129	28 32	48	130
42	34 42	47	126	33 55	47	126	33 08	48	127	32 20	48	128	31 32	48	128	30 44	48	129	29 56	49	130	29 07	48	131
41	35 19	47	127	34 32	48	127	33 44	48	128	32 56	48	129	32 08	48	129	31 20	49	130	30 31	49	131	29 42	49	131
40	35 56	−48	128	35 08	−48	128	34 20	−48	129	33 32	−49	130	32 43	−48	130	31 55	−49	131	31 06	−50	132	30 16	−49	132
39	36 32	48	128	35 44	48	129	34 56	49	130	34 07	49	131	33 18	49	131	32 29	49	132	31 40	50	133	30 50	50	133
38	37 08	49	129	36 19	48	130	35 31	49	131	34 42	50	132	33 52	49	132	33 03	50	133	32 13	50	134	31 23	50	134
37	37 43	49	131	36 54	49	131	36 05	49	132	35 16	50	133	34 26	50	133	33 36	50	134	32 46	50	135	31 56	51	135
36	38 18	49	132	37 29	50	132	36 39	50	133	35 49	50	134	34 59	50	134	34 09	50	135	33 19	51	136	32 28	51	136
35	38 52	−50	133	38 02	−50	133	37 12	−50	134	36 22	−50	135	35 32	−51	135	34 41	−51	136	33 50	−51	137	32 59	−51	137
34	39 26	50	134	38 36	51	134	37 45	50	135	36 55	51	136	36 04	51	136	35 13	51	137	34 22	52	138	33 30	51	138
33	39 59	51	135	39 08	51	135	38 17	51	136	37 26	51	137	36 35	51	137	35 44	52	138	34 52	51	139	34 01	52	139
32	40 31	51	136	39 40	51	137	38 49	51	137	37 58	52	138	37 06	52	139	36 14	52	139	35 22	52	140	34 30	52	140
31	41 03	52	137	40 11	51	138	39 20	52	138	38 28	52	139	37 36	52	140	36 44	52	140	35 52	53	141	34 59	52	141
30	41 34	−52	138	40 42	−52	139	39 50	−52	139	38 58	−52	140	38 06	−53	141	37 13	−52	141	36 21	−53	142	35 28	−53	143
29	42 04	52	139	41 12	52	140	40 20	53	141	39 27	53	141	38 34	52	142	37 42	53	142	36 49	54	143	35 55	53	144
28	42 34	53	140	41 41	52	141	40 49	53	142	39 56	53	142	39 03	54	143	38 09	53	144	37 16	54	144	36 22	54	145
27	43 03	53	142	42 10	53	142	41 17	54	143	40 23	53	144	39 30	54	144	38 36	54	145	37 42	54	145	36 48	54	146
26	43 31	53	143	42 38	54	143	41 44	54	144	40 50	53	145	39 57	55	145	39 02	54	146	38 08	54	146	37 14	54	147
25	43 58	−53	144	43 05	−54	145	42 11	−54	145	41 17	−55	146	40 22	−54	146	39 28	−55	147	38 33	−54	148	37 39	−55	148
24	44 25	54	145	43 31	55	146	42 36	54	147	41 42	55	147	40 47	54	148	39 53	55	148	38 58	55	149	38 03	55	149
23	44 51	55	147	43 56	55	147	43 01	54	148	42 07	55	148	41 12	56	149	40 16	55	149	39 21	55	150	38 26	56	150
22	45 15	54	148	44 21	55	148	43 26	56	149	42 30	55	150	41 35	55	150	40 40	56	151	39 44	56	151	38 48	56	152
21	45 39	55	149	44 44	55	150	43 49	56	150	42 53	55	151	41 58	56	151	41 02	56	152	40 06	56	152	39 10	56	153
20	46 03	−56	151	45 07	−56	151	44 11	−56	152	43 15	−56	152	42 19	−56	153	41 23	−56	153	40 27	−57	154	39 30	−56	154
19	46 25	56	152	45 29	56	152	44 33	57	153	43 36	56	153	42 40	57	154	41 43	56	154	40 47	57	155	39 50	57	155
18	46 46	56	153	45 50	57	154	44 53	57	154	43 56	56	155	43 00	57	155	42 03	57	156	41 06	57	156	40 09	57	156
17	47 06	57	155	46 09	56	155	45 13	57	156	44 16	57	156	43 19	57	156	42 22	58	157	41 24	57	157	40 27	57	158
16	47 25	57	156	46 28	57	156	45 31	57	157	44 34	57	157	43 37	58	158	42 39	57	158	41 42	58	159	40 44	58	159
15	47 44	−58	157	46 46	−57	158	45 49	−58	158	44 51	−57	159	43 54	−58	159	42 56	−58	159	41 58	−58	160	41 00	−58	160
14	48 01	58	159	47 03	58	159	46 05	58	160	45 07	57	160	44 10	58	160	43 12	59	161	42 13	58	161	41 15	58	161
13	48 17	58	160	47 19	58	161	46 21	58	161	45 23	59	161	44 24	58	162	43 26	58	162	42 28	58	162	41 30	59	163
12	48 32	58	162	47 34	59	162	46 35	58	162	45 37	59	163	44 38	58	163	43 40	59	163	42 41	58	164	41 43	59	164
11	48 46	59	163	47 47	58	164	46 49	59	164	45 50	59	164	44 51	58	164	43 53	59	165	42 54	59	165	41 55	59	165
10	48 58	−58	165	48 00	−59	165	47 01	−59	165	46 02	−59	166	45 03	−59	166	44 04	−59	166	43 05	−59	166	42 06	−59	167
9	49 10	59	166	48 11	59	166	47 12	59	167	46 13	59	167	45 14	59	167	44 15	59	167	43 16	60	168	42 16	59	168
8	49 20	59	168	48 21	59	168	47 22	59	168	46 23	60	168	45 23	59	169	44 24	59	169	43 25	60	169	42 25	59	169
7	49 30	60	169	48 30	59	169	47 31	60	170	46 31	59	170	45 32	59	170	44 33	60	170	43 33	59	170	42 34	60	171
6	49 38	60	171	48 38	59	171	47 39	60	171	46 39	60	171	45 39	59	171	44 40	60	172	43 40	59	172	42 41	60	172
5	49 45	−60	172	48 45	−60	172	47 45	−60	173	46 45	−59	173	45 46	−60	173	44 46	−60	173	43 46	−60	173	42 46	−59	173
4	49 50	60	174	48 50	60	174	47 50	59	174	46 51	60	174	45 51	60	174	44 51	60	174	43 51	60	175	42 51	59	175
3	49 54	59	175	48 55	60	175	47 55	60	176	46 55	60	176	45 55	60	176	44 55	60	176	43 55	60	176	42 55	60	176
2	49 58	60	177	48 58	60	177	47 58	60	177	46 58	60	177	45 58	60	177	44 58	60	177	43 58	60	177	42 58	60	177
1	49 59	60	178	48 59	60	179	47 59	60	179	46 59	60	179	45 59	60	179	44 59	60	179	43 59	59	179	43 00	60	179
0	50 00	−60	180	49 00	−60	180	48 00	−60	180	47 00	−60	180	46 00	−60	180	45 00	−60	180	44 00	−60	180	43 00	−60	180

LHA	8° Hc	d	Z	9° Hc	d	Z	10° Hc	d	Z	11° Hc	d	Z	12° Hc	d	Z	13° Hc	d	Z	14° Hc	d	Z	LHA
	° ′	′	°	° ′	′	°	° ′	′	°	° ′	′	°	° ′	′	°	° ′	′	°	° ′	′	°	
69	10 31	42	110	09 49	41	111	09 08	41	111	08 27	42	112	07 45	41	113	07 04	42	114	06 22	42	114	291
68	11 14	42	111	10 32	41	111	09 51	42	112	09 09	42	113	08 27	41	114	07 46	42	114	07 04	42	115	292
67	11 57	42	111	11 15	42	112	10 33	42	113	09 51	42	114	09 09	42	114	08 27	42	115	07 45	42	116	293
66	12 39	41	112	11 58	42	113	11 16	43	114	10 33	42	114	09 51	42	115	09 09	42	116	08 27	43	116	294
65	13 22	−42	113	12 40	−42	113	11 58	−43	114	11 15	−42	115	10 33	−43	116	09 50	−42	116	09 08	−43	117	295
64	14 04	42	113	13 22	43	114	12 39	42	115	11 57	43	116	11 14	43	116	10 31	43	117	09 48	42	118	296
63	14 46	42	114	14 04	43	115	13 21	43	116	12 38	43	116	11 55	43	117	11 12	43	118	10 29	43	119	297
62	15 28	43	115	14 45	43	116	14 02	43	116	13 19	43	117	12 36	43	118	11 53	44	119	11 09	43	119	298
61	16 10	43	116	15 27	44	116	14 43	43	117	14 00	43	118	13 17	44	119	12 33	44	119	11 49	43	120	299
60	16 51	−43	116	16 08	−44	117	15 24	−43	118	14 41	−44	119	13 57	−44	119	13 13	−44	120	12 29	−44	121	300
59	17 32	44	117	16 48	43	118	16 05	44	119	15 21	44	119	14 37	44	120	13 53	45	121	13 08	44	121	301
58	18 13	44	118	17 29	44	119	16 45	44	119	16 01	45	120	15 16	44	121	14 32	44	121	13 48	45	122	302
57	18 53	44	119	18 09	44	119	17 25	45	120	16 40	44	121	15 56	45	121	15 11	45	122	14 26	45	123	303
56	19 33	44	119	18 49	45	120	18 04	44	121	17 20	45	122	16 35	45	122	15 50	45	123	15 05	46	124	304
55	20 13	−44	120	19 29	−45	121	18 44	−45	122	17 59	−45	122	17 14	−46	123	16 28	−45	124	15 43	−46	124	305
54	20 53	45	121	20 08	45	122	19 23	46	122	18 37	45	123	17 52	46	124	17 06	45	124	16 21	46	125	306
53	21 32	45	122	20 47	46	123	20 01	45	123	19 16	46	124	18 30	46	125	17 44	46	125	16 58	46	126	307
52	22 11	46	123	21 25	45	123	20 40	46	124	19 54	46	125	19 08	47	125	18 21	46	126	17 35	46	127	308
51	22 50	46	123	22 04	46	124	21 18	47	125	20 31	46	126	19 45	47	126	18 58	46	127	18 12	47	128	309
50	23 28	−46	124	22 42	−47	125	21 55	−46	126	21 09	−47	126	20 22	−47	127	19 35	−47	128	18 48	−47	128	310
49	24 06	47	125	23 19	47	126	22 32	46	126	21 46	48	127	20 58	47	128	20 11	47	128	19 24	48	129	311
48	24 43	47	126	23 56	47	127	23 09	47	127	22 22	47	128	21 35	48	129	20 47	48	129	19 59	47	130	312
47	25 20	47	127	24 33	47	127	23 46	48	128	22 58	48	129	22 10	48	129	21 22	48	130	20 34	48	131	313
46	25 57	48	128	25 09	47	128	24 22	48	129	23 34	48	130	22 46	49	130	21 57	48	131	21 09	48	132	314
45	26 33	−48	129	25 45	−48	129	24 57	−48	130	24 09	−49	131	23 20	−48	131	22 32	−49	132	21 43	−49	132	315
44	27 09	49	129	26 20	48	130	25 32	48	131	24 44	49	131	23 55	49	132	23 06	49	133	22 17	49	133	316
43	27 44	49	130	26 55	48	131	26 07	49	132	25 18	49	132	24 29	49	133	23 40	50	134	22 50	49	134	317
42	28 19	49	131	27 30	49	132	26 41	49	133	25 52	50	133	25 02	49	134	24 13	50	134	23 23	50	135	318
41	28 53	49	132	28 04	49	133	27 15	50	133	26 25	50	134	25 35	50	135	24 45	50	135	23 55	50	136	319
40	29 27	−50	133	28 37	−49	134	27 48	−50	134	26 58	−50	135	26 08	−51	136	25 17	−50	136	24 27	−51	137	320
39	30 00	50	134	29 10	50	135	28 20	50	135	27 30	50	136	26 40	51	137	25 49	51	137	24 58	51	138	321
38	30 33	50	135	29 43	51	136	28 52	50	136	28 02	51	137	27 11	51	137	26 20	51	138	25 29	51	139	322
37	31 05	50	136	30 15	51	137	29 24	51	137	28 33	51	138	27 42	51	138	26 51	52	139	25 59	52	140	323
36	31 37	51	137	30 46	51	138	29 55	51	138	29 04	52	139	28 12	52	139	27 20	51	140	26 29	52	140	324
35	32 08	−51	138	31 17	−52	139	30 25	−51	139	29 34	−52	140	28 42	−52	140	27 50	−52	141	26 58	−53	141	325
34	32 39	52	139	31 47	52	140	30 55	52	140	30 03	52	141	29 11	52	141	28 19	53	142	27 26	53	142	326
33	33 09	52	140	32 17	53	141	31 24	52	141	30 32	53	142	29 39	52	142	28 47	53	143	27 54	53	143	327
32	33 38	52	141	32 46	53	142	31 53	53	142	31 00	53	143	30 07	53	143	29 14	53	144	28 21	53	144	328
31	34 07	53	142	33 14	53	143	32 21	53	143	31 28	54	144	30 34	53	144	29 41	53	145	28 48	54	145	329
30	34 35	−54	143	33 41	−53	144	32 48	−53	144	31 55	−54	145	31 01	−54	145	30 07	−54	146	29 13	−54	146	330
29	35 02	54	144	34 08	53	145	33 15	54	145	32 21	54	146	31 27	54	146	30 33	54	147	29 39	55	147	331
28	35 28	53	145	34 35	54	146	33 41	55	146	32 46	54	147	31 52	54	147	30 58	55	148	30 03	54	148	332
27	35 54	54	146	35 00	54	147	34 06	55	147	33 11	54	148	32 17	55	148	31 22	55	149	30 27	55	149	333
26	36 20	55	147	35 25	55	148	34 30	55	148	33 35	55	149	32 40	55	149	31 45	55	150	30 50	55	150	334
25	36 44	−55	149	35 49	−55	149	34 54	−55	150	33 59	−56	150	33 03	−55	150	32 08	−55	151	31 13	−56	151	335
24	37 08	56	150	36 12	55	150	35 17	56	151	34 21	55	151	33 26	56	152	32 30	56	152	31 34	56	152	336
23	37 30	55	151	36 35	56	151	35 39	56	152	34 43	56	152	33 47	56	153	32 51	56	153	31 55	56	154	337
22	37 52	56	152	36 56	56	152	36 00	56	153	35 04	56	153	34 08	56	154	33 12	57	154	32 15	56	155	338
21	38 14	57	153	37 17	56	154	36 21	57	154	35 24	56	154	34 28	57	155	33 31	56	155	32 35	57	156	339
20	38 34	−57	154	37 37	−56	155	36 41	−57	155	35 44	−57	156	34 47	−57	156	33 50	−57	156	32 53	−57	157	340
19	38 53	57	156	37 56	57	156	36 59	57	156	36 02	57	157	35 05	57	157	34 08	57	158	33 11	57	158	341
18	39 12	57	157	38 15	57	157	37 18	58	158	36 20	57	158	35 23	58	158	34 25	57	159	33 28	58	159	342
17	39 30	58	158	38 32	57	158	37 35	58	159	36 37	58	159	35 39	57	159	34 42	58	160	33 44	58	160	343
16	39 46	57	159	38 49	58	160	37 51	58	160	36 53	58	160	35 55	58	161	34 57	58	161	33 59	58	161	344
15	40 02	−58	160	39 04	−58	161	38 06	−58	161	37 08	−58	161	36 10	−58	162	35 12	−58	162	34 14	−59	162	345
14	40 17	58	162	39 19	58	162	38 21	59	162	37 22	58	163	36 24	58	163	35 26	59	163	34 27	58	164	346
13	40 31	58	163	39 33	59	163	38 34	58	164	37 36	59	164	36 37	59	164	35 38	58	164	34 40	59	165	347
12	40 44	59	164	39 45	58	165	38 47	59	165	37 48	59	165	36 49	59	165	35 50	58	166	34 52	59	166	348
11	40 56	59	166	39 57	59	166	38 58	59	166	37 59	59	166	37 00	59	167	36 01	59	167	35 02	59	167	349
10	41 07	−59	167	40 08	−59	167	39 09	−59	167	38 10	−59	168	37 11	−60	168	36 11	−59	168	35 12	−59	168	350
9	41 17	59	168	40 18	59	168	39 19	60	169	38 19	59	169	37 20	59	169	36 21	60	169	35 21	59	169	351
8	41 26	59	169	40 27	60	170	39 27	59	170	38 28	60	170	37 28	59	170	36 29	60	170	35 29	59	171	352
7	41 34	60	171	40 34	59	171	39 35	60	171	38 35	59	171	37 36	60	171	36 36	59	172	35 37	60	172	353
6	41 41	60	172	40 41	59	172	39 42	60	172	38 42	60	172	37 42	60	173	36 42	59	173	35 43	60	173	354
5	41 47	−60	173	40 47	−60	174	39 47	−60	174	38 47	−59	174	37 48	−60	174	36 48	−60	174	35 48	−60	174	355
4	41 52	60	175	40 52	60	175	39 52	60	175	38 52	60	175	37 52	60	175	36 52	60	175	35 52	60	175	356
3	41 55	60	176	40 55	60	176	39 55	60	176	38 55	59	176	37 56	60	176	36 56	60	176	35 56	60	176	357
2	41 58	60	177	40 58	60	177	39 58	60	177	38 58	60	178	37 58	60	178	36 58	60	178	35 58	60	178	358
1	42 00	60	179	41 00	60	179	40 00	60	179	39 00	60	179	38 00	60	179	37 00	60	179	36 00	60	179	359
0	42 00	−60	180	41 00	−60	180	40 00	−60	180	39 00	−60	180	38 00	−60	180	37 00	−60	180	36 00	−60	180	360

S. Lat. {LHA greater than 180°........ Zn=180−Z; LHA less than 180°........... Zn=180+Z}

DECLINATION (0°-14°) CONTRARY NAME TO LATITUDE

N. Lat. {LHA greater than 180°....... Zn=Z
{LHA less than 180°.......... Zn=360−Z

DECLINATION (15°-29°) SAME NAME AS LATITUDE

LHA	15° Hc	d	Z	16° Hc	d	Z	17° Hc	d	Z	18° Hc	d	Z	19° Hc	d	Z
	° ′	′	°	° ′	′	°	° ′	′	°	° ′	′	°	° ′	′	°
0	65 00	+60	180	66 00	+60	180	67 00	+60	180	68 00	+60	180	69 00	+60	180
1	64 59	60	178	65 59	60	178	66 59	60	178	67 59	60	178	68 59	60	177
2	64 56	60	175	65 56	60	175	66 56	60	175	67 56	60	175	68 56	60	175
3	64 52	60	173	65 52	59	173	66 51	60	173	67 51	60	172	68 51	59	172
4	64 45	60	171	65 45	59	171	66 44	60	170	67 44	59	170	68 43	60	170
5	64 37	+60	169	65 37	+59	168	66 36	+59	168	67 35	+59	167	68 34	+59	167
6	64 27	59	167	65 26	59	166	66 25	59	166	67 24	59	165	68 23	58	164
7	64 16	58	164	65 14	59	164	66 13	58	163	67 11	58	163	68 09	58	162
8	64 03	58	162	65 01	58	162	65 59	57	161	66 56	58	160	67 54	58	160
9	63 48	57	160	64 45	58	159	65 43	57	159	66 40	57	158	67 37	57	157
10	63 31	+57	158	64 28	+57	157	65 25	+57	157	66 22	+56	156	67 18	+57	155
11	63 13	57	156	64 10	56	155	65 06	56	154	66 02	56	154	66 58	55	153
12	62 53	57	154	63 50	55	153	64 45	56	152	65 41	55	151	66 36	55	150
13	62 33	55	152	63 28	55	151	64 23	55	150	65 18	55	149	66 13	54	148
14	62 10	55	150	63 05	55	149	64 00	54	148	64 54	54	147	65 48	53	146
15	61 47	+54	148	62 41	+54	147	63 35	+53	146	64 28	+53	145	65 21	+53	144
16	61 22	53	146	62 15	54	145	63 09	52	144	64 01	53	143	64 54	51	142
17	60 55	54	145	61 49	52	144	62 41	52	143	63 33	52	141	64 25	51	140
18	60 28	53	143	61 21	52	142	62 13	51	141	63 04	51	140	63 55	50	138
19	60 00	52	141	60 52	51	140	61 43	51	139	62 34	50	138	63 24	49	137
20	59 30	+52	139	60 22	+50	138	61 12	+50	137	62 02	+50	136	62 52	+48	135
21	59 00	51	138	59 51	49	137	60 40	50	136	61 30	49	134	62 19	48	133
22	58 29	50	136	59 19	49	135	60 08	49	134	60 57	48	133	61 45	47	132
23	57 56	50	135	58 46	48	134	59 34	48	133	60 22	48	131	61 10	47	130
24	57 23	49	133	58 12	48	132	59 00	48	131	59 48	46	130	60 34	46	129
25	56 49	+48	132	57 37	+48	131	58 25	+47	130	59 12	+46	128	59 58	+45	127
26	56 15	47	130	57 02	47	129	57 49	46	128	58 35	46	127	59 21	45	126
27	55 39	47	129	56 26	47	128	57 13	45	127	57 58	45	126	58 43	44	124
28	55 03	47	128	55 50	45	127	56 35	46	125	57 21	44	124	58 05	44	123
29	54 27	45	126	55 12	46	125	55 58	44	124	56 42	44	123	57 26	43	122
30	53 49	+46	125	54 35	+44	124	55 19	+44	123	56 03	+44	122	56 47	+42	120
31	53 11	45	124	53 56	44	123	54 40	44	122	55 24	43	120	56 07	42	119
32	52 33	44	123	53 17	44	122	54 01	43	120	54 44	42	119	55 26	42	118
33	51 54	44	122	52 38	43	120	53 21	43	119	54 04	42	118	54 46	41	117
34	51 15	43	120	51 58	43	119	52 41	42	118	53 23	41	117	54 04	41	116
35	50 35	+43	119	51 18	+42	118	52 00	+42	117	52 42	+41	116	53 23	+40	115
36	49 54	43	118	50 37	42	117	51 19	41	116	52 00	41	115	52 41	40	114
37	49 14	42	117	49 56	41	116	50 37	41	115	51 18	40	114	51 58	40	113
38	48 33	41	116	49 14	42	115	49 56	40	114	50 36	40	113	51 16	39	112
39	47 51	42	115	48 33	40	114	49 13	41	113	49 54	39	112	50 33	39	111
40	47 09	+41	114	47 50	+41	113	48 31	+40	112	49 11	+39	111	49 50	+38	110
41	46 27	41	113	47 08	40	112	47 48	40	111	48 28	38	110	49 06	38	109
42	45 45	40	112	46 25	40	111	47 05	39	110	47 44	39	109	48 23	38	108
43	45 02	40	111	45 42	40	110	46 22	39	109	47 01	38	108	47 39	37	107
44	44 19	40	110	44 59	39	109	45 38	39	108	46 17	38	107	46 55	37	106
45	43 36	+39	109	44 15	+39	108	44 54	+39	107	45 33	+37	106	46 10	+37	105
46	42 52	40	109	43 32	38	108	44 10	38	106	44 48	38	105	45 26	37	104
47	42 09	39	108	42 48	38	107	43 26	38	106	44 04	37	105	44 41	37	103
48	41 25	39	107	42 04	38	106	42 42	38	105	43 20	37	104	43 57	36	103
49	40 41	38	106	41 19	38	105	41 57	38	104	42 35	37	103	43 12	36	102
50	39 56	+39	105	40 35	+38	104	41 13	+37	103	41 50	+37	102	42 27	+36	101
51	39 12	38	104	39 50	38	103	40 28	37	102	41 05	36	101	41 41	36	100
52	38 27	38	104	39 05	38	103	39 43	37	102	40 20	36	101	40 56	36	100
53	37 43	37	103	38 20	38	102	38 58	37	101	39 35	36	100	40 11	35	99
54	36 58	37	102	37 35	38	101	38 13	36	100	38 49	36	99	39 25	36	98
55	36 13	+37	101	36 50	+37	100	37 27	+37	99	38 04	+36	98	38 40	+35	97
56	35 28	37	101	36 05	37	100	36 42	36	99	37 18	36	98	37 54	35	97
57	34 42	38	100	35 20	36	99	35 56	37	98	36 33	35	97	37 08	36	96
58	33 57	37	99	34 34	37	98	35 11	36	97	35 47	36	96	36 23	35	95
59	33 12	37	98	33 49	36	97	34 25	36	96	35 01	36	96	35 37	35	95
60	32 26	+37	98	33 03	+36	97	33 39	+36	96	34 15	+36	95	34 51	+35	94
61	31 41	36	97	32 17	37	96	32 54	36	95	33 30	35	94	34 05	35	93
62	30 55	37	96	31 32	36	95	32 08	36	94	32 44	35	93	33 19	35	93
63	30 09	37	96	30 46	36	95	31 22	36	94	31 58	35	93	32 33	35	92
64	29 23	37	95	30 00	36	94	30 36	36	93	31 12	35	92	31 47	35	91
65	28 38	+36	94	29 14	+36	93	29 50	+36	92	30 26	+35	91	31 01	+35	91
66	27 52	36	94	28 28	36	93	29 04	36	92	29 40	35	91	30 15	35	90
67	27 06	36	93	27 42	36	92	28 18	36	91	28 54	35	90	29 29	35	89
68	26 20	36	92	26 56	36	91	27 32	36	90	28 08	36	90	28 44	35	89
69	25 34	36	92	26 10	36	91	26 46	36	90	27 22	36	89	27 58	35	88

LHA	20° Hc	d	Z	21° Hc	d	Z	22° Hc	d	Z	23° Hc	d	Z	24° Hc	d	Z
	° ′	′	°	° ′	′	°	° ′	′	°	° ′	′	°	° ′	′	°
0	70 00	+60	180	71 00	+60	180	72 00	+60	180	73 00	+60	180	74 00	+60	180
1	69 59	60	177	70 59	60	177	71 59	60	177	72 59	60	177	73 59	60	177
2	69 56	59	175	70 55	60	174	71 55	60	174	72 55	60	174	73 55	59	173
3	69 50	60	172	70 50	59	171	71 49	60	171	72 49	59	171	73 48	59	170
4	69 43	59	169	70 42	59	169	71 41	59	168	72 40	59	168	73 39	59	167
5	69 33	+59	166	70 32	+58	166	71 30	+59	165	72 29	+58	165	73 27	+59	164
6	69 21	58	164	70 19	59	163	71 18	58	162	72 16	57	162	73 13	58	161
7	69 07	58	161	70 05	58	161	71 03	57	160	72 00	57	159	72 57	57	158
8	68 52	57	159	69 49	57	158	70 46	56	157	71 42	56	156	72 38	56	155
9	68 34	56	156	69 30	57	155	70 27	55	154	71 22	56	153	72 18	55	152
10	68 15	+55	154	69 10	+56	153	70 06	+55	152	71 01	+54	151	71 55	+54	149
11	67 53	56	152	68 49	54	151	69 43	54	149	70 37	54	148	71 31	53	147
12	67 31	54	149	68 25	54	148	69 19	53	147	70 12	53	146	71 05	52	144
13	67 07	53	147	68 00	53	146	68 53	52	145	69 45	52	143	70 37	51	142
14	66 41	53	145	67 34	52	144	68 26	51	142	69 17	51	141	70 08	50	139
15	66 14	+52	143	67 06	+51	142	67 57	+51	140	68 48	+49	139	69 37	+49	137
16	65 45	52	141	66 37	50	140	67 27	50	138	68 17	49	137	69 06	47	135
17	65 16	50	139	66 06	50	138	66 56	49	136	67 45	48	135	68 33	46	133
18	64 45	50	137	65 35	48	136	66 23	48	134	67 11	47	133	67 58	46	131
19	64 13	49	135	65 02	48	134	65 50	47	133	66 37	46	131	67 23	45	129
20	63 40	+48	134	64 28	+48	132	65 16	+46	131	66 02	+45	129	66 47	+45	128
21	63 07	47	132	63 54	46	131	64 40	46	129	65 26	45	128	66 11	43	126
22	62 32	47	130	63 19	45	129	64 04	45	127	64 49	44	126	65 33	43	124
23	61 57	45	129	62 42	45	127	63 27	44	126	64 11	44	124	64 55	41	123
24	61 20	45	127	62 05	45	126	62 50	43	124	63 33	43	123	64 16	41	121
25	60 43	+45	126	61 28	+44	124	62 12	+42	123	62 54	+42	121	63 36	+40	120
26	60 06	44	124	60 50	43	123	61 33	42	122	62 15	41	120	62 56	40	118
27	59 27	44	123	60 11	42	122	60 53	42	120	61 35	40	119	62 15	39	117
28	58 49	42	122	59 31	42	120	60 13	41	119	60 54	40	117	61 34	39	116
29	58 09	42	120	58 51	42	119	59 33	40	118	60 13	39	116	60 52	38	115
30	57 29	+42	119	58 11	+41	118	58 52	+39	116	59 31	+39	115	60 10	+38	113
31	56 49	41	118	57 30	40	117	58 10	40	115	58 50	38	114	59 28	37	112
32	56 08	41	117	56 49	39	115	57 28	39	114	58 07	38	113	58 45	37	111
33	55 27	40	116	56 07	39	114	56 46	39	113	57 25	37	111	58 02	36	110
34	54 45	40	114	55 25	39	113	56 04	38	112	56 42	37	110	57 19	36	109
35	54 03	+39	113	54 42	+39	112	55 21	+37	111	55 58	+37	109	56 35	+36	108
36	53 21	39	112	54 00	38	111	54 38	37	110	55 15	36	108	55 51	36	107
37	52 38	39	111	53 17	37	110	53 54	37	109	54 31	36	107	55 07	35	106
38	51 55	38	110	52 33	38	109	53 11	36	108	53 47	36	106	54 23	35	105
39	51 12	38	109	51 50	37	108	52 27	36	107	53 03	35	106	53 38	35	104
40	50 28	+38	108	51 06	+37	107	51 43	+36	106	52 19	+35	105	52 54	+34	103
41	49 44	38	108	50 22	36	106	50 58	36	105	51 34	35	104	52 09	34	102
42	49 01	37	107	49 38	36	105	50 14	35	104	50 49	35	103	51 24	34	102
43	48 16	37	106	48 53	36	105	49 29	35	103	50 04	35	102	50 39	33	101
44	47 32	37	105	48 09	35	104	48 44	35	102	49 19	35	101	49 54	33	100
45	46 47	+37	104	47 24	+35	103	47 59	+35	102	48 34	+34	100	49 08	+34	99
46	46 03	36	103	46 39	35	102	47 14	35	101	47 49	34	100	48 23	33	98
47	45 18	36	102	45 54	35	101	46 29	35	100	47 04	33	99	47 37	33	98
48	44 33	36	102	45 09	35	100	45 44	34	99	46 18	34	98	46 52	33	97
49	43 48	35	101	44 23	35	100	44 58	35	99	45 33	33	97	46 06	33	96
50	43 03	+35	100	43 38	+35	99	44 13	+34	98	44 47	+33	97	45 20	+33	95
51	42 17	36	99	42 53	34	98	43 27	34	97	44 01	34	96	44 35	32	95
52	41 32	35	98	42 07	35	97	42 42	34	96	43 16	33	95	43 49	32	94
53	40 46	35	98	41 21	35	97	41 56	34	96	42 30	33	94	43 03	32	93
54	40 01	35	97	40 36	34	96	41 10	34	95	41 44	33	94	42 17	32	93
55	39 15	+35	96	39 50	+34	95	40 24	+34	94	40 58	+33	93	41 31	+32	92
56	38 29	35	96	39 04	34	95	39 38	34	93	40 12	33	92	40 45	33	91
57	37 44	34	95	38 18	35	94	38 53	33	93	39 26	33	92	39 59	33	91
58	36 58	35	94	37 33	34	93	38 07	33	92	38 40	33	91	39 13	33	90
59	36 12	35	94	36 47	34	93	37 21	33	91	37 54	33	90	38 27	33	89
60	35 26	+35	93	36 01	+34	92	36 35	+33	91	37 08	+33	90	37 41	+33	89
61	34 40	35	92	35 15	34	91	35 49	33	90	36 22	33	89	36 55	33	88
62	33 54	35	92	34 29	34	91	35 03	33	90	35 36	33	89	36 09	33	88
63	33 08	35	91	33 43	34	90	34 17	33	89	34 50	34	88	35 24	32	87
64	32 22	35	90	32 57	34	89	33 31	34	88	34 05	33	87	34 38	32	86
65	31 36	+35	90	32 11	+34	89	32 45	+34	88	33 19	+33	87	33 52	+32	86
66	30 50	35	89	31 25	34	88	31 59	34	87	32 33	33	86	33 06	33	85
67	30 04	35	88	30 39	34	87	31 13	34	86	31 47	33	85	32 20	33	84
68	29 19	34	88	29 53	34	87	30 27	34	86	31 01	33	85	31 34	33	84
69	28 33	34	87	29 07	35	86	29 42	33	85	30 15	34	84	30 49	33	83

25° Hc	d	Z	26° Hc	d	Z	27° Hc	d	Z	28° Hc	d	Z	29° Hc	d	Z	LHA
° ′	′	°	° ′	′	°	° ′	′	°	° ′	′	°	° ′	′	°	
75 00	+60	180	76 00	+60	180	77 00	+60	180	78 00	+60	180	79 00	+60	180	360
74 59	60	177	75 59	59	176	76 58	60	176	77 58	60	176	78 58	60	176	359
74 54	60	173	75 54	60	173	76 54	59	172	77 53	60	172	78 53	59	171	358
74 47	60	170	75 47	59	169	76 46	59	168	77 45	59	167	78 44	58	167	357
74 38	59	166	75 37	58	165	76 35	58	165	77 33	58	163	78 31	58	162	356
74 26	+58	163	75 24	+57	162	76 21	+58	161	77 19	+57	160	78 16	+56	158	355
74 11	57	160	75 08	57	159	76 05	56	157	77 01	56	156	77 57	55	154	354
73 54	56	157	74 50	56	155	75 46	55	154	76 41	54	152	77 35	54	150	353
73 34	56	154	74 30	54	152	75 24	54	151	76 18	53	149	77 11	52	147	352
73 13	54	151	74 07	53	149	75 01	52	147	75 53	52	146	76 45	51	143	351
72 49	+53	148	73 42	+53	146	74 35	+51	144	75 26	+51	142	76 17	+49	140	350
72 24	52	145	73 16	51	143	74 07	50	142	74 57	49	140	75 46	48	138	349
71 57	51	143	72 48	50	141	73 38	49	139	74 27	47	137	75 14	46	135	348
71 28	50	140	72 18	49	138	73 07	47	136	73 54	47	134	74 41	44	132	347
70 58	49	138	71 47	47	136	72 34	47	134	73 21	45	132	74 06	43	130	346
70 26	+48	136	71 14	+47	134	72 01	+45	132	72 46	+44	130	73 30	+42	127	345
69 53	47	133	70 40	46	132	71 26	44	130	72 10	43	127	72 53	41	125	344
69 19	46	131	70 05	45	130	70 50	43	128	71 33	42	125	72 15	39	123	343
68 44	45	129	69 29	44	128	70 13	42	126	70 55	41	123	71 36	38	121	342
68 08	44	128	68 52	43	126	69 35	41	124	70 16	40	122	70 56	38	119	341
67 32	+43	126	68 15	+41	124	68 56	+41	122	69 37	+39	120	70 16	+36	118	340
66 54	42	124	67 36	41	122	68 17	40	120	68 57	38	118	69 35	36	116	339
66 16	41	123	66 57	40	121	67 37	39	119	68 16	37	117	68 53	35	115	338
65 36	41	121	66 17	40	119	66 57	38	117	67 35	36	115	68 11	35	113	337
64 57	40	120	65 37	38	118	66 15	38	116	66 53	35	114	67 28	35	112	336
64 16	+40	118	64 56	+38	116	65 34	+36	115	66 10	+36	113	66 46	+33	111	335
63 36	38	117	64 14	38	115	64 52	36	113	65 28	34	111	66 02	33	109	334
62 54	38	115	63 32	37	114	64 09	36	112	64 45	34	110	65 19	32	108	333
62 13	37	114	62 50	37	112	63 27	34	111	64 01	34	109	64 35	32	107	332
61 30	38	113	62 08	35	111	62 43	35	110	63 18	33	108	63 51	31	106	331
60 48	+37	112	61 25	+35	110	62 00	+34	108	62 34	+33	107	63 07	+31	105	330
60 05	36	111	60 41	35	109	61 16	34	107	61 50	32	106	62 22	31	104	329
59 22	36	110	59 58	34	108	60 32	33	106	61 05	32	105	61 37	31	103	328
58 38	36	109	59 14	34	107	59 48	33	105	60 21	31	104	60 52	31	102	327
57 55	35	107	58 30	33	106	59 03	33	104	59 36	31	103	60 07	30	101	326
57 11	+34	106	57 45	+34	105	58 19	+32	103	58 51	+31	102	59 22	+30	100	325
56 27	34	106	57 01	33	104	57 34	32	102	58 06	31	101	58 37	29	99	324
55 42	34	105	56 16	33	103	56 49	32	102	57 21	30	100	57 51	30	98	323
54 58	33	104	55 31	33	102	56 04	32	101	56 36	30	99	57 06	29	98	322
54 13	33	103	54 46	33	101	55 19	31	100	55 50	30	98	56 20	29	97	321
53 28	+33	102	54 01	+32	101	54 33	+32	99	55 05	+30	98	55 35	+28	96	320
52 43	33	101	53 16	32	100	53 48	31	98	54 19	30	97	54 49	29	95	319
51 58	33	100	52 31	31	99	53 02	31	97	53 33	30	96	54 03	29	95	318
51 12	33	99	51 45	32	98	52 17	31	97	52 48	29	95	53 17	29	94	317
50 27	33	99	51 00	31	97	51 31	31	96	52 02	29	95	52 31	29	93	316
49 42	+32	98	50 14	+31	97	50 45	+31	95	51 16	+29	94	51 45	+29	92	315
48 56	32	97	49 28	32	96	50 00	30	95	50 30	30	93	51 00	28	92	314
48 10	32	96	48 42	32	95	49 14	30	94	49 44	30	92	50 14	28	91	313
47 25	32	96	47 57	31	94	48 28	30	93	48 58	30	92	49 28	28	90	312
46 39	32	95	47 11	31	94	47 42	30	92	48 12	30	91	48 42	28	90	311
45 53	+32	94	46 25	+31	93	46 56	+30	92	47 26	+30	90	47 56	+28	89	310
45 07	32	94	45 39	31	92	46 10	30	91	46 40	30	90	47 10	28	89	309
44 21	32	93	44 53	31	92	45 24	30	90	45 54	30	89	46 24	28	88	308
43 35	32	92	44 07	31	91	44 38	30	90	45 08	30	89	45 38	29	87	307
42 49	32	92	43 21	31	90	43 52	30	89	44 22	30	88	44 52	29	87	306
42 03	+32	91	42 35	+31	90	43 06	+31	89	43 37	+29	87	44 06	+29	86	305
41 18	31	90	41 49	31	89	42 20	31	88	42 51	29	87	43 20	29	86	304
40 32	31	90	41 03	31	88	41 34	31	87	42 05	29	86	42 34	29	85	303
39 46	31	89	40 17	31	88	40 48	31	87	41 19	30	86	41 49	29	84	302
39 00	31	88	39 31	32	87	40 03	30	86	40 33	30	85	41 03	29	84	301
38 14	+32	88	38 46	+31	87	39 17	+30	85	39 47	+30	84	40 17	+30	83	300
37 28	32	87	38 00	31	86	38 31	31	85	39 02	30	84	39 32	29	83	299
36 42	32	86	37 14	31	85	37 45	31	84	38 16	30	83	38 46	30	82	298
35 56	32	86	36 28	32	85	37 00	30	84	37 30	31	83	38 01	29	82	297
35 10	32	85	35 42	32	84	36 14	31	83	36 45	30	82	37 15	30	81	296
34 24	+33	85	34 57	+31	84	35 28	+31	83	35 59	+31	82	36 30	+30	80	295
33 39	32	84	34 11	32	83	34 43	31	82	35 14	31	81	35 45	30	80	294
32 53	32	83	33 25	32	82	33 57	32	81	34 29	30	80	34 59	31	79	293
32 07	33	83	32 40	32	82	33 12	31	81	33 43	31	80	34 14	31	79	292
31 22	32	82	31 54	32	81	32 26	32	80	32 58	31	79	33 29	31	78	291

15°	16°	17°	18°	19°	20°	21°	22°	23°	24°	25°	26°	27°	28°	29°

S. Lat. {LHA greater than 180°....... Zn=180−Z
{LHA less than 180°.......... Zn=180+Z

DECLINATION (15°-29°) SAME NAME AS LATITUDE

LAT 40°

N. Lat. {LHA greater than 180°....... Zn=Z; LHA less than 180°............ Zn=360−Z}

DECLINATION (15°-29°) SAME NAME AS LATITUDE

LAT 40°

(Each cell: Hc ° ′, d ′, Z °)

LHA	15°	16°	17°	18°	19°	20°	21°	22°	23°	24°	25°	26°	27°	28°	29°	LHA
70	24 48 +36 91	25 24 +36 90	26 00 +36 89	26 36 +36 88	27 12 +35 87	27 47 +34 86	28 21 +35 86	28 56 +34 85	29 30 +33 84	30 03 +33 83	30 36 +33 82	31 09 +32 81	31 41 +32 80	32 13 +31 79	32 44 +31 78	290
71	24 02 36 90	24 38 37 89	25 15 35 89	25 50 36 88	26 26 35 87	27 01 35 86	27 36 34 85	28 10 34 84	28 44 34 83	29 18 33 82	29 51 33 81	30 24 32 80	30 56 32 79	31 28 31 78	31 59 31 77	289
72	23 16 36 90	23 52 37 89	24 29 35 88	25 04 36 87	25 40 35 86	26 15 35 85	26 50 34 84	27 24 34 83	27 58 34 82	28 32 34 82	29 06 32 81	29 38 33 80	30 11 32 79	30 43 32 78	31 15 31 77	288
73	22 30 37 89	23 07 36 88	23 43 36 87	24 19 35 86	24 54 35 86	25 29 35 85	26 04 35 84	26 39 34 83	27 13 34 82	27 47 33 81	28 20 33 80	28 53 33 79	29 26 32 78	29 58 32 77	30 30 31 76	287
74	21 44 37 88	22 21 36 88	22 57 36 87	23 33 35 86	24 08 36 85	24 44 35 84	25 19 34 83	25 53 34 82	26 27 34 81	27 01 34 80	27 35 33 79	28 08 33 78	28 41 32 78	29 13 32 77	29 45 32 76	286
75	20 58 +37 88	21 35 +36 87	22 11 +36 86	22 47 +36 85	23 23 +35 84	23 58 +35 83	24 33 +35 83	25 08 +34 82	25 42 +34 81	26 16 +34 80	26 50 +33 79	27 23 +33 78	27 56 +33 77	28 29 +32 76	29 01 +32 75	285
76	20 12 37 87	20 49 36 86	21 25 36 85	22 01 36 85	22 37 35 84	23 12 35 83	23 47 35 82	24 22 35 81	24 57 34 80	25 31 34 79	26 05 33 78	26 38 33 77	27 11 33 76	27 44 33 76	28 17 32 75	284
77	19 26 37 86	20 03 36 86	20 39 36 85	21 15 36 84	21 51 36 83	22 27 35 82	23 02 35 81	23 37 35 80	24 12 34 80	24 46 34 79	25 20 34 78	25 54 33 77	26 27 33 76	27 00 32 75	27 32 33 74	283
78	18 41 36 86	19 17 37 85	19 54 36 84	20 30 36 83	21 06 35 82	21 41 36 82	22 17 35 81	22 52 34 80	23 26 35 79	24 01 34 78	24 35 34 77	25 09 33 76	25 42 33 75	26 15 33 74	26 48 33 73	282
79	17 55 36 85	18 31 37 84	19 08 36 84	19 44 36 83	20 20 36 82	20 56 35 81	21 31 35 80	22 06 35 79	22 41 35 78	23 16 34 78	23 50 34 77	24 24 34 76	24 58 33 75	25 31 33 74	26 04 33 73	281
80	17 09 +37 85	17 46 +36 84	18 22 +37 83	18 59 +36 82	19 35 +35 81	20 10 +36 80	20 46 +35 80	21 21 +35 79	21 56 +35 78	22 31 +35 77	23 06 +34 76	23 40 +34 75	24 14 +33 74	24 47 +33 73	25 20 +33 72	280
81	16 23 37 84	17 00 37 83	17 37 36 82	18 13 36 82	18 49 36 81	19 25 36 80	20 01 35 79	20 36 35 78	21 11 35 77	21 46 35 76	22 21 34 75	22 55 34 74	23 29 34 74	24 03 34 73	24 37 33 72	279
82	15 38 36 83	16 14 37 83	16 51 37 82	17 28 36 81	18 04 36 80	18 40 36 79	19 16 35 78	19 51 36 78	20 27 35 77	21 02 35 76	21 37 34 75	22 11 34 74	22 45 34 73	23 19 34 72	23 53 33 71	278
83	14 52 37 83	15 29 37 82	16 06 36 81	16 42 37 80	17 19 36 79	17 55 36 79	18 31 36 78	19 07 35 77	19 42 35 76	20 17 35 75	20 52 35 74	21 27 35 73	22 02 34 73	22 36 34 72	23 10 33 71	277
84	14 06 37 82	14 43 37 81	15 20 37 81	15 57 37 80	16 34 36 79	17 10 36 78	17 46 36 77	18 22 35 76	18 57 36 76	19 33 35 75	20 08 35 74	20 43 35 73	21 18 34 72	21 52 34 71	22 26 34 70	276
85	13 21 +37 82	13 58 +37 81	14 35 +37 80	15 12 +36 79	15 48 +37 78	16 25 +36 77	17 01 +36 77	17 37 +36 76	18 13 +36 75	18 49 +35 74	19 24 +35 73	19 59 +35 72	20 34 +35 71	21 09 +34 71	21 43 +34 70	275
86	12 35 38 81	13 13 37 80	13 50 37 79	14 27 37 78	15 04 36 77	15 40 37 77	16 17 36 76	16 53 36 75	17 29 36 74	18 05 35 74	18 40 35 73	19 15 36 72	19 51 34 71	20 25 35 70	21 00 34 69	274
87	11 50 37 80	12 27 38 79	13 05 37 79	13 42 37 78	14 19 36 77	14 55 37 76	15 32 36 75	16 08 37 75	16 45 36 74	17 21 35 73	17 56 36 72	18 32 35 71	19 07 35 70	19 42 35 70	20 17 35 69	273
88	11 05 37 80	11 42 38 79	12 20 37 78	12 57 37 77	13 34 37 76	14 11 37 76	14 48 36 75	15 24 37 74	16 01 36 73	16 37 36 72	17 13 35 72	17 48 36 71	18 24 35 70	18 59 35 69	19 34 35 68	272
89	10 20 37 79	10 57 38 78	11 35 37 77	12 12 37 77	12 49 37 76	13 26 37 75	14 03 37 74	14 40 37 73	15 17 36 73	15 53 36 72	16 29 36 71	17 05 36 70	17 41 36 69	18 17 35 68	18 52 35 68	271
90	09 35 +37 78	10 12 +38 78	10 50 +37 77	11 27 +38 76	12 05 +37 75	12 42 +37 74	13 19 +37 74	13 56 +37 73	14 33 +36 72	15 09 +37 71	15 46 +36 70	16 22 +36 70	16 58 +36 69	17 34 +35 68	18 09 +36 67	270
91	08 50 38 78	09 28 37 77	10 05 38 76	10 43 37 75	11 20 38 75	11 58 37 74	12 35 37 73	13 12 37 72	13 49 37 71	14 26 37 71	15 03 36 70	15 39 36 69	16 15 36 68	16 51 36 67	17 27 36 66	269
92	08 05 38 77	08 43 38 76	09 21 38 76	09 59 37 75	10 36 38 74	11 14 37 73	11 51 38 72	12 29 37 72	13 06 37 71	13 43 37 70	14 20 36 69	14 56 37 68	15 33 36 68	16 09 36 67	16 45 36 66	268
93	07 20 38 77	07 58 38 76	08 36 38 75	09 14 38 74	09 52 38 74	10 30 37 73	11 07 38 72	11 45 37 71	12 22 38 70	13 00 37 69	13 37 37 69	14 14 36 68	14 50 37 67	15 27 36 66	16 03 37 65	267
94	06 35 39 76	07 14 38 75	07 52 38 74	08 30 38 74	09 08 38 73	09 46 38 72	10 24 38 71	11 02 37 70	11 39 38 70	12 17 37 69	12 54 37 68	13 31 37 67	14 08 37 66	14 45 37 66	15 22 36 65	266
95	05 51 +38 75	06 29 +39 75	07 08 +38 74	07 46 +38 73	08 24 +38 72	09 02 +38 71	09 40 +38 71	10 18 +38 70	10 56 +38 69	11 34 +37 68	12 11 +38 68	12 49 +37 67	13 26 +37 66	14 03 +37 65	14 40 +37 64	265
96	05 06 39 75	05 45 39 74	06 24 38 73	07 02 39 72	07 41 38 72	08 19 38 71	08 57 38 70	09 35 38 69	10 13 38 69	10 51 38 68	11 29 38 67	12 07 37 66	12 44 38 65	13 22 37 65	13 59 37 64	264
97	04 22 39 74	05 01 39 73	05 40 38 73	06 18 39 72	06 57 39 71	07 36 38 70	08 14 38 69	08 52 39 69	09 31 38 68	10 09 38 67	10 47 38 66	11 25 38 66	12 03 37 65	12 40 38 64	13 18 37 63	263
98	03 38 39 73	04 17 39 73	04 56 39 72	05 35 39 71	06 14 38 70	06 52 39 70	07 31 39 69	08 10 38 68	08 48 39 68	09 27 38 67	10 05 38 66	10 43 38 65	11 21 38 64	11 59 38 63	12 37 38 63	262
99	02 54 39 73	03 33 39 72	04 12 39 71	04 51 39 71	05 30 39 70	06 09 39 69	06 48 39 68	07 27 39 68	08 06 39 67	08 45 38 66	09 23 39 65	10 02 38 64	10 40 38 64	11 18 38 63	11 56 38 62	261
100	02 10 +40 72	02 50 +39 71	03 29 +39 71	04 08 +39 70	04 47 +40 69	05 27 +39 68	06 06 +39 68	06 45 +39 67	07 24 +39 66	08 03 +38 65	08 41 +39 65	09 20 +39 64	09 59 +38 63	10 37 +39 62	11 16 +38 61	260
101	01 27 39 72	02 06 40 71	02 46 39 70	03 25 40 69	04 05 39 69	04 44 39 68	05 23 40 67	06 03 39 66	06 42 39 66	07 21 39 65	08 00 39 64	08 39 39 63	09 13 39 62	09 57 38 62	10 35 39 61	259
102	00 43 40 71	01 23 39 70	02 02 40 69	02 42 40 69	03 22 39 68	04 01 40 67	04 41 40 66	05 21 39 66	06 00 40 65	06 40 39 64	07 19 39 63	07 58 39 63	08 37 39 62	09 16 39 61	09 55 39 60	258
103	00 00 40 70	00 40 40 70	01 20 39 69	01 59 40 68	02 39 40 67	03 19 40 67	03 59 40 66	04 39 40 65	05 19 39 64	05 58 40 64	06 38 39 63	07 17 40 62	07 57 39 61	08 36 40 61	09 16 39 60	257
104	−0 43 40 70	−0 03 40 69	00 37 40 68	01 17 40 67	01 57 40 67	02 37 40 66	03 17 40 65	03 57 40 64	04 37 40 64	05 17 40 63	05 57 40 62	06 37 40 61	07 17 39 61	07 56 40 60	08 36 40 59	256
105	−1 26 +40 69	−0 46 +40 68	−0 06 +41 67	00 35 +40 67	01 15 +40 66	01 55 +41 65	02 36 +40 65	03 16 +40 64	03 56 +40 63	04 36 +41 62	05 17 +40 62	05 57 +40 61	06 37 +40 60	07 17 +40 59	07 57 +40 59	255
106	−2 09 40 68	−1 29 41 68	−0 48 40 67	−0 08 41 66	00 33 41 65	01 14 40 65	01 54 41 64	02 35 40 63	03 15 41 62	03 56 40 62	04 36 41 61	05 17 40 60	05 57 40 59	06 37 41 59	07 18 40 58	254
107	−2 52 41 68	−2 11 41 67	−1 30 41 66	−0 49 40 65	−0 09 41 65	00 32 41 64	01 13 41 63	01 54 41 63	02 35 41 62	03 16 40 61	03 56 41 60	04 37 41 60	05 18 40 59	05 58 41 58	06 39 40 57	253
108	−3 34 41 67	−2 53 41 66	−2 12 41 66	−1 31 41 65	−0 50 41 64	−0 09 41 63	00 32 41 63	01 13 41 62	01 54 41 61	02 35 41 60	03 16 41 60	03 57 41 59	04 38 41 58	05 19 41 57	06 00 41 57	252
109	−4 16 41 66	−3 35 41 66	−2 54 41 65	−2 13 42 64	−1 31 41 63	−0 50 41 63	−0 09 42 62	00 33 41 61	01 14 42 61	01 56 41 60	02 37 41 59	03 18 41 58	03 59 42 58	04 41 41 57	05 22 41 56	251
110	−4 58 +41 66	−4 17 +42 65	−3 35 +41 64	−2 54 +42 63	−2 12 +41 63	−1 31 +42 62	−0 49 +42 61	−0 07 +41 61	00 34 +42 60	01 16 +42 59	01 58 +41 58	02 39 +42 58	03 21 +41 57	04 02 +42 56	04 44 +41 56	250
111	−5 40 42 65	−4 58 41 64	−4 17 42 64	−3 35 42 63	−2 53 42 62	−2 11 42 61	−1 29 42 61	−0 47 42 60	−0 05 42 59	00 37 42 59	01 19 41 58	02 00 42 57	02 42 42 56	03 24 42 56	04 06 42 55	249
112	−6 22 42 64	−5 40 42 64	−4 58 42 63	−4 16 43 62	−3 33 42 62	−2 51 42 61	−2 09 42 60	−1 27 42 59	−0 45 43 59	−0 02 42 58	00 40 42 57	01 22 42 56	02 04 42 56	02 46 43 55	03 29 42 54	248
113		−6 21 42 63	−5 39 43 62	−4 56 42 61	−4 14 43 61	−3 31 42 60	−2 49 43 59	−2 06 42 59	−1 24 43 58	−0 41 42 57	00 01 43 57	00 44 42 56	01 26 43 55	02 09 42 54	02 51 43 54	247
114			−6 19 43 61	−5 36 42 61	−4 54 43 60	−4 11 43 59	−3 28 43 59	−2 45 42 58	−2 03 43 57	−1 20 43 57	−0 37 43 56	00 06 43 55	00 49 43 54	01 32 43 54	02 15 42 53	246
115				−6 16 +43 60	−5 33 +43 59	−4 50 +43 59	−4 07 +43 58	−3 24 +43 57	−2 41 +43 57	−1 58 +43 56	−1 15 +43 55	−0 32 +44 55	00 12 +43 54	00 55 +43 53	01 38 +43 53	245
116					−6 13 43 59	−5 30 44 58	−4 46 43 57	−4 03 44 57	−3 19 43 56	−2 36 44 55	−1 52 43 55	−1 09 44 54	−0 25 43 53	00 18 44 53	01 02 43 52	244
117						−6 08 43 57	−5 25 44 57	−4 41 44 56	−3 57 43 55	−3 14 44 55	−2 30 44 54	−1 46 44 53	−1 02 44 53	−0 18 44 52	00 26 44 51	243
118							−6 03 44 56	−5 19 44 55	−4 35 43 55	−3 51 44 54	−3 07 44 53	−2 23 45 53	−1 38 44 52	−0 54 44 51	−0 10 44 51	242
119								−5 57 45 55	−5 12 44 54	−4 28 45 53	−3 43 44 53	−2 59 45 52	−2 14 44 51	−1 30 45 51	−0 45 44 50	241
120									−5 49 +44 53	−5 05 +45 53	−4 20 +45 52	−3 35 +45 51	−2 50 +45 51	−2 05 +45 50	−1 20 +45 49	240
121										−5 41 45 52	−4 56 45 51	−4 11 46 51	−3 25 45 50	−2 40 45 49	−1 55 45 49	239
122											−5 31 45 51	−4 46 46 50	−4 00 45 49	−3 15 46 49	−2 29 45 48	238
123											−6 07 46 50	−5 21 46 49	−4 35 46 49	−3 49 46 48	−3 03 46 47	237
124												−5 56 47 48	−5 09 46 48	−4 23 46 47	−3 37 47 47	236
125													−5 43 +46 47	−4 57 +47 47	−4 10 +47 46	235
126														−5 30 47 46	−4 43 47 45	234
127														−6 03 48 45	−5 15 47 44	233
128															−5 47 47 44	232

DECLINATION (15°-29°) CONTRARY NAME TO LATITUDE

LHA	15°	16°	17°	18°	19°	20°	21°	22°	23°	24°	25°	26°	27°	28°	29°	LHA
85	−5 51 −38 105															275
84	−5 06 39 105	−5 45 39 106	−6 24 38 107													276
83	−4 22 39 106	−5 01 39 107	−5 40 38 107	−6 18 39 108												277
82	−3 38 39 107	−4 17 39 107	−4 56 39 108	−5 35 39 109	−6 14 38 110											278
81	−2 54 39 107	−3 33 39 108	−4 12 39 109	−4 51 39 109	−5 30 39 110	−6 09 39 111										279
80	−2 10 −40 108	−2 50 −39 109	−3 29 −39 109	−4 08 −39 110	−4 47 −40 111	−5 27 −39 112	−6 06 −39 112									280
79	−1 27 39 108	−2 06 40 109	−2 46 39 110	−3 25 40 111	−4 05 39 111	−4 44 39 112	−5 23 40 113	−6 03 39 114								281
78	−0 43 40 109	−1 23 39 110	−2 02 40 111	−2 42 40 111	−3 22 39 112	−4 01 40 113	−4 41 40 114	−5 21 39 114	−6 00 40 115							282
77	00 00 40 110	−0 40 40 110	−1 20 39 111	−1 59 40 112	−2 39 40 113	−3 19 40 113	−3 59 40 114	−4 39 40 115	−5 19 39 116	−5 58 40 116						283
76	00 43 40 110	00 03 40 111	−0 37 40 112	−1 17 40 113	−1 57 40 113	−2 37 40 114	−3 17 40 115	−3 57 40 116	−4 37 40 116	−5 17 40 117	−5 57 40 118					284
75	01 26 −40 111	00 46 −40 112	00 06 −41 113	−0 35 −40 113	−1 15 −40 114	−1 55 −41 115	−2 36 −40 115	−3 16 −40 116	−3 56 −40 117	−4 36 −41 118	−5 17 −40 118	−5 57 −40 119				285
74	02 09 40 112	01 29 41 112	00 48 40 113	00 08 41 114	−0 33 41 115	−1 14 40 115	−1 54 41 116	−2 35 40 117	−3 15 41 118	−3 56 40 118	−4 36 41 119	−5 17 40 120	−5 57 40 121			286
73	02 52 41 112	02 11 41 113	01 30 41 114	00 49 40 115	00 09 41 115	−0 32 41 116	−1 13 41 117	−1 54 41 117	−2 35 41 118	−3 16 40 119	−3 56 41 120	−4 37 41 120	−5 18 40 121	−5 58 41 122		287
72	03 34 41 113	02 53 41 114	02 12 41 115	01 31 41 115	00 50 41 116	00 09 41 117	−0 32 41 117	−1 13 41 118	−1 54 41 119	−2 35 41 120	−3 16 41 120	−3 57 41 121	−4 38 41 122	−5 19 41 122	−6 00 41 123	288
71	04 16 41 114	03 35 41 114	02 54 41 115	02 13 42 116	01 31 41 117	00 50 41 117	00 09 42 118	−0 33 41 119	−1 14 42 119	−1 56 41 120	−2 37 41 121	−3 18 41 122	−3 59 42 122	−4 41 41 123	−5 22 41 124	289
70	04 58 −41 114	04 17 −42 115	03 35 −41 116	02 54 −42 117	02 12 −41 117	01 31 −42 118	00 49 −42 119	00 07 −41 119	−0 34 −42 120	−1 16 −42 121	−1 58 −41 122	−2 39 −42 122	−3 21 −41 123	−4 02 −42 124	−4 44 −41 124	290

S. Lat. {LHA greater than 180°....... Zn=180−Z; LHA less than 180°............ Zn=180+Z}

N. Lat. {LHA greater than 180°....... Zn=Z; LHA less than 180°........... Zn=360−Z}

DECLINATION (15°-29°) CONTRARY NAME TO LATITUDE

LHA	15° Hc	d	Z	16° Hc	d	Z	17° Hc	d	Z	18° Hc	d	Z	19° Hc	d	Z
	° ′	′	°	° ′	′	°	° ′	′	°	° ′	′	°	° ′	′	°
69	05 40	42	115	04 58	41	116	04 17	42	116	03 35	42	117	02 53	42	118
68	06 22	42	116	05 40	42	116	04 58	42	117	04 16	43	118	03 33	42	119
67	07 03	42	116	06 21	42	117	05 39	43	118	04 56	42	119	04 14	43	119
66	07 44	42	117	07 02	43	118	06 19	43	119	05 36	42	119	04 54	43	120
65	08 25	−43	118	07 42	−43	119	06 59	−43	119	06 16	−43	120	05 33	−43	121
64	09 06	44	119	08 22	43	119	07 39	43	120	06 56	43	121	06 13	43	121
63	09 46	44	119	09 02	43	120	08 19	43	121	07 36	44	121	06 52	44	122
62	10 26	44	120	09 42	44	121	08 58	43	121	08 15	44	122	07 31	44	123
61	11 06	44	121	10 22	44	121	09 38	44	122	08 54	45	123	08 09	44	123
60	11 45	−44	121	11 01	−45	122	10 16	−44	123	09 32	−44	123	08 48	−45	124
59	12 24	44	122	11 40	45	123	10 55	45	123	10 10	44	124	09 26	45	125
58	13 03	45	123	12 18	45	124	11 33	45	124	10 48	45	125	10 03	45	126
57	13 41	45	124	12 56	45	124	12 11	45	125	11 26	46	126	10 40	45	126
56	14 19	45	124	13 34	45	125	12 49	46	126	12 03	46	126	11 17	45	127
55	14 57	−45	125	14 12	−46	126	13 26	−46	126	12 40	−46	127	11 54	−46	128
54	15 35	46	126	14 49	46	126	14 03	47	127	13 16	46	128	12 30	46	128
53	16 12	46	127	15 26	47	127	14 39	46	128	13 53	47	129	13 06	47	129
52	16 49	47	127	16 02	47	128	15 15	47	129	14 28	47	129	13 41	47	130
51	17 25	47	128	16 38	47	129	15 51	47	129	15 04	48	130	14 16	47	131
50	18 01	−47	129	17 14	−48	130	16 26	−47	130	15 39	−48	131	14 51	−48	132
49	18 36	47	130	17 49	48	130	17 01	48	131	16 13	48	132	15 25	48	132
48	19 12	48	131	18 24	48	131	17 36	49	132	16 47	48	132	15 59	48	133
47	19 46	48	131	18 58	48	132	18 10	49	133	17 21	48	133	16 33	49	134
46	20 21	49	132	19 32	49	133	18 43	49	133	17 54	49	134	17 05	49	135
45	20 54	−48	133	20 06	−50	134	19 16	−49	134	18 27	−49	135	17 38	−49	135
44	21 28	49	134	20 39	50	135	19 49	49	135	19 00	50	136	18 10	50	136
43	22 01	50	135	21 11	50	135	20 21	50	136	19 31	50	137	18 41	50	137
42	22 33	50	136	21 43	50	136	20 53	50	137	20 03	50	137	19 13	51	138
41	23 05	50	137	22 15	51	137	21 24	50	138	20 34	51	138	19 43	51	139
40	23 36	−50	137	22 46	−51	138	21 55	−51	139	21 04	−51	139	20 13	−51	140
39	24 07	51	138	23 16	51	139	22 25	51	139	21 34	51	140	20 43	52	141
38	24 38	52	139	23 46	51	140	22 55	52	140	22 03	51	141	21 12	52	141
37	25 07	51	140	24 16	52	141	23 24	52	141	22 32	52	142	21 40	52	142
36	25 37	52	141	24 45	52	142	23 53	53	142	23 00	52	143	22 08	53	143
35	26 05	−52	142	25 13	−53	143	24 20	−52	143	23 28	−53	144	22 35	−53	144
34	26 33	52	143	25 41	53	143	24 48	53	144	23 55	53	144	23 02	53	145
33	27 01	53	144	26 08	53	144	25 15	54	145	24 21	53	145	23 28	54	146
32	27 28	54	145	26 34	53	145	25 41	54	146	24 47	54	146	23 53	53	147
31	27 54	54	146	27 00	54	146	26 06	54	147	25 12	54	147	24 18	54	148
30	28 19	−54	147	27 25	−54	147	26 31	−54	148	25 37	−54	148	24 43	−55	149
29	28 44	54	148	27 50	55	148	26 55	54	149	26 01	55	149	25 06	56	150
28	29 09	55	149	28 14	55	149	27 19	55	150	26 24	55	150	25 29	55	151
27	29 32	55	150	28 37	55	150	27 42	55	151	26 47	56	151	25 51	55	152
26	29 55	56	151	28 59	55	151	28 04	55	152	27 09	56	152	26 13	56	153
25	30 17	−56	152	29 21	−55	152	28 26	−56	153	27 30	−56	153	26 34	−56	154
24	30 38	56	153	29 42	56	153	28 46	56	154	27 50	56	154	26 54	56	155
23	30 59	56	154	30 03	57	154	29 06	56	155	28 10	57	155	27 13	56	156
22	31 19	57	155	30 22	56	155	29 26	57	156	28 29	57	156	27 32	57	157
21	31 38	57	156	30 41	57	156	29 44	57	157	28 47	57	157	27 50	57	158
20	31 56	−57	157	30 59	−57	157	30 02	−57	158	29 05	−58	158	28 07	−57	159
19	32 14	58	158	31 16	57	159	30 19	58	159	29 21	57	159	28 24	58	160
18	32 30	57	159	31 33	58	160	30 35	58	160	29 37	57	160	28 40	58	161
17	32 46	58	160	31 48	58	161	30 50	58	161	29 52	58	161	28 54	58	162
16	33 01	58	162	32 03	58	162	31 05	58	162	30 07	58	162	29 09	59	163
15	33 15	−58	163	32 17	−58	163	31 19	−59	163	30 20	−58	163	29 22	−59	164
14	33 29	59	164	32 30	58	164	31 32	59	164	30 33	59	165	29 34	58	165
13	33 41	59	165	32 42	58	165	31 44	59	165	30 45	59	166	29 46	59	166
12	33 53	59	166	32 54	59	166	31 55	59	167	30 56	59	167	29 57	59	167
11	34 03	59	167	33 04	59	167	32 05	59	168	31 06	59	168	30 07	59	168
10	34 13	−59	168	33 14	−59	169	32 15	−60	169	31 15	−59	169	30 16	−59	169
9	34 22	59	170	33 23	60	170	32 23	59	170	31 24	60	170	30 24	59	170
8	34 30	60	171	33 30	59	171	32 31	60	171	31 31	59	171	30 32	60	171
7	34 37	60	172	33 37	59	172	32 38	60	172	31 38	60	172	30 38	59	172
6	34 43	60	173	33 43	59	173	32 44	60	173	31 44	60	173	30 44	60	173
5	34 48	−60	174	33 48	−59	174	32 49	−60	174	31 49	−60	174	30 49	−60	175
4	34 52	59	175	33 53	60	175	32 53	60	175	31 53	60	176	30 53	60	176
3	34 56	60	177	33 56	60	177	32 56	60	177	31 56	60	177	30 56	60	177
2	34 58	60	178	33 58	60	178	32 58	60	178	31 58	60	178	30 58	60	178
1	35 00	60	179	34 00	60	179	33 00	60	179	32 00	60	179	31 00	60	179
0	35 00	−60	180	34 00	−60	180	33 00	−60	180	32 00	−60	180	31 00	−60	180

LHA	20° Hc	d	Z	21° Hc	d	Z	22° Hc	d	Z	23° Hc	d	Z	24° Hc	d	Z
	° ′	′	°	° ′	′	°	° ′	′	°	° ′	′	°	° ′	′	°
69	02 11	42	119	01 29	42	119	00 47	42	120	00 05	42	121	−0 37	42	121
68	02 51	42	119	02 09	42	120	01 27	42	121	00 45	43	121	00 02	42	122
67	03 31	42	120	02 49	43	121	02 06	42	121	01 24	43	122	00 41	42	123
66	04 11	43	121	03 28	43	121	02 45	42	122	02 03	43	123	01 20	43	123
65	04 50	−43	121	04 07	−43	122	03 24	−43	123	02 41	−43	123	01 58	−43	124
64	05 30	44	122	04 46	43	123	04 03	44	123	03 19	43	124	02 36	44	125
63	06 08	43	123	05 25	44	123	04 41	44	124	03 57	43	125	03 14	44	125
62	06 47	44	123	06 03	44	124	05 19	44	125	04 35	44	125	03 51	44	126
61	07 25	44	124	06 41	44	125	05 57	45	125	05 12	44	126	04 28	45	127
60	08 03	−44	125	07 19	−45	125	06 34	−45	126	05 49	−44	127	05 05	−45	127
59	08 41	45	125	07 56	45	126	07 11	45	127	06 26	45	127	05 41	45	128
58	09 18	45	126	08 33	45	127	07 48	46	128	07 02	45	128	06 17	46	129
57	09 55	45	127	09 10	46	128	08 24	46	128	07 38	45	129	06 53	46	130
56	10 32	46	128	09 46	46	128	09 00	46	129	08 14	46	130	07 28	46	130
55	11 08	−46	128	10 22	−47	129	09 35	−46	130	08 49	−46	130	08 03	−47	131
54	11 44	47	129	10 57	46	130	10 11	47	130	09 24	47	131	08 37	46	132
53	12 19	47	130	11 32	46	130	10 46	47	131	09 59	48	132	09 11	47	132
52	12 54	47	131	12 07	47	131	11 20	47	132	10 33	48	133	09 45	47	133
51	13 29	47	131	12 42	48	132	11 54	48	133	11 06	47	133	10 19	48	134
50	14 03	−47	132	13 16	−48	133	12 28	−48	133	11 40	−48	134	10 52	−49	135
49	14 37	48	133	13 49	48	134	13 01	48	134	12 13	49	135	11 24	48	135
48	15 11	49	134	14 22	48	134	13 34	49	135	12 45	49	136	11 56	49	136
47	15 44	49	134	14 55	49	135	14 06	49	136	13 17	49	136	12 28	49	137
46	16 16	49	135	15 27	49	136	14 38	49	136	13 49	50	137	12 59	49	138
45	16 49	−50	136	15 59	−50	137	15 09	−49	137	14 20	−50	138	13 30	−50	138
44	17 20	50	137	16 30	50	137	15 40	50	138	14 50	50	139	14 00	50	139
43	17 51	50	138	17 01	50	138	16 11	51	139	15 20	50	139	14 30	51	140
42	18 22	50	139	17 32	51	139	16 41	51	140	15 50	51	140	14 59	51	141
41	18 52	51	139	18 01	51	140	17 10	51	141	16 19	51	141	15 28	51	142
40	19 22	−51	140	18 31	−52	141	17 39	−51	141	16 48	−51	142	15 57	−52	142
39	19 51	51	141	19 00	52	142	18 08	52	142	17 16	52	143	16 24	52	143
38	20 20	52	142	19 28	52	142	18 36	52	143	17 44	52	144	16 52	53	144
37	20 48	52	143	19 56	53	143	19 03	52	144	18 11	53	144	17 18	52	145
36	21 15	52	144	20 23	53	144	19 30	53	145	18 37	52	145	17 45	53	146
35	21 42	−53	145	20 49	−53	145	19 56	−53	146	19 03	−53	146	18 10	−53	147
34	22 09	54	145	21 15	53	146	20 22	53	146	19 29	54	147	18 35	53	147
33	22 34	53	146	21 41	54	147	20 47	53	147	19 54	54	148	19 00	54	148
32	23 00	54	147	22 06	54	148	21 12	54	148	20 18	54	149	19 24	55	149
31	23 24	54	148	22 30	54	149	21 36	55	149	20 41	54	150	19 47	55	150
30	23 48	−54	149	22 54	−55	150	21 59	−55	150	21 04	−54	150	20 10	−55	151
29	24 11	54	150	23 17	55	151	22 22	56	151	21 27	56	151	20 32	56	152
28	24 34	55	151	23 39	56	151	22 44	56	152	21 48	55	152	20 53	55	153
27	24 56	55	152	24 01	56	152	23 05	56	153	22 09	55	153	21 14	56	154
26	25 17	55	153	24 22	56	153	23 26	56	154	22 30	56	154	21 34	56	155
25	25 38	−56	154	24 42	−56	154	23 46	−56	155	22 50	−57	155	21 53	−56	155
24	25 58	57	155	25 01	56	155	24 05	56	156	23 09	57	156	22 12	56	156
23	26 17	57	156	25 20	56	156	24 24	57	157	23 27	57	157	22 30	57	157
22	26 35	56	157	25 39	57	157	24 42	57	158	23 45	57	158	22 48	57	158
21	26 53	57	158	25 56	57	158	24 59	57	159	24 02	58	159	23 04	57	159
20	27 10	−57	159	26 13	−58	159	25 15	−57	160	24 18	−58	160	23 20	−57	160
19	27 26	57	160	26 29	58	160	25 31	58	161	24 33	57	161	23 36	58	161
18	27 42	58	161	26 44	58	161	25 46	58	161	24 48	58	162	23 50	58	162
17	27 56	58	162	26 58	58	162	26 00	58	162	25 02	58	163	24 04	58	163
16	28 10	58	163	27 12	58	163	26 14	59	163	25 15	58	164	24 17	58	164
15	28 23	−58	164	27 25	−59	164	26 26	−58	165	25 28	−59	165	24 29	−58	165
14	28 36	59	165	27 37	59	165	26 38	58	166	25 40	59	166	24 41	59	166
13	28 47	59	166	27 48	58	166	26 50	59	167	25 51	59	167	24 52	59	167
12	28 58	59	167	27 59	59	167	27 00	59	168	26 01	59	168	25 02	59	168
11	29 08	59	168	28 09	60	168	27 09	59	169	26 10	59	169	25 11	59	169
10	29 17	−60	169	28 17	−59	169	27 18	−59	170	26 19	−59	170	25 20	−60	170
9	29 25	59	170	28 26	60	170	27 26	59	171	26 27	60	171	25 27	59	171
8	29 32	59	171	28 33	60	172	27 33	59	172	26 34	60	172	25 34	60	172
7	29 39	60	172	28 39	60	173	27 39	59	173	26 40	60	173	25 40	60	173
6	29 44	59	174	28 45	60	174	27 45	60	174	26 45	60	174	25 45	59	174
5	29 49	−60	175	28 49	−59	175	27 50	−60	175	26 50	−60	175	25 50	−60	175
4	29 53	60	176	28 53	60	176	27 53	60	176	26 53	59	176	25 54	60	176
3	29 56	60	177	28 56	60	177	27 56	60	177	26 56	60	177	25 56	60	177
2	29 58	60	178	28 58	60	178	27 58	60	178	26 58	60	178	25 58	60	178
1	30 00	60	179	29 00	60	179	28 00	60	179	27 00	60	179	26 00	60	179
0	30 00	−60	180	29 00	−60	180	28 00	−60	180	27 00	−60	180	26 00	−60	180

LHA	25° Hc	d	Z	26° Hc	d	Z	27° Hc	d	Z	28° Hc	d	Z	29° Hc	d	Z	LHA
	° ′	′	°	° ′	′	°	° ′	′	°	° ′	′	°	° ′	′	°	
69	−1 19	41	122	−2 00	42	123	−2 42	42	124	−3 24	42	124	−4 06	42	125	291
68	−0 40	42	123	−1 22	42	124	−2 04	42	124	−2 46	43	125	−3 29	42	126	292
67	−0 01	43	123	−0 44	42	124	−1 26	43	125	−2 09	42	126	−2 51	43	126	293
66	00 37	43	124	−0 06	43	125	−0 49	43	126	−1 32	43	126	−2 15	42	127	294
65	01 15	−43	125	00 32	−44	125	−0 12	−43	126	−0 55	−43	127	−1 38	−43	127	295
64	01 52	43	125	01 09	44	126	00 25	43	127	−0 18	44	127	−1 02	43	128	296
63	02 30	44	126	01 46	44	127	01 02	44	127	00 18	44	128	−0 26	44	129	297
62	03 07	44	127	02 23	45	127	01 38	44	128	00 54	44	129	00 10	44	129	298
61	03 43	44	127	02 59	45	128	02 14	44	129	01 30	45	129	00 45	44	130	299
60	04 20	−45	128	03 35	−45	129	02 50	−45	129	02 05	−45	130	01 20	−45	131	300
59	04 56	45	129	04 11	46	129	03 25	45	130	02 40	45	131	01 55	45	131	301
58	05 31	45	129	04 46	46	130	04 00	45	131	03 15	46	131	02 29	45	132	302
57	06 07	46	130	05 21	46	131	04 35	46	131	03 49	46	132	03 03	46	133	303
56	06 42	46	131	05 56	47	132	05 09	46	132	04 23	46	133	03 37	47	133	304
55	07 16	−46	132	06 30	−47	132	05 43	−46	133	04 57	−47	133	04 10	−47	134	305
54	07 51	47	132	07 04	47	133	06 17	47	134	05 30	47	134	04 43	47	135	306
53	08 24	47	133	07 37	47	134	06 50	47	134	06 03	48	135	05 15	47	136	307
52	08 58	48	134	08 10	47	134	07 23	48	135	06 35	48	136	05 47	47	136	308
51	09 31	48	134	08 43	48	135	07 55	48	136	07 07	48	136	06 19	48	137	309
50	10 03	−48	135	09 15	−48	136	08 27	−48	136	07 39	−49	137	06 50	−48	138	310
49	10 36	49	136	09 47	49	137	08 58	48	137	08 10	49	138	07 21	49	138	311
48	11 07	49	137	10 18	48	137	09 30	50	138	08 40	49	138	07 51	49	139	312
47	11 39	50	137	10 49	49	138	10 00	49	139	09 11	50	139	08 21	49	140	313
46	12 10	50	138	11 20	50	139	10 30	49	139	09 41	50	140	08 51	50	141	314
45	12 40	−50	139	11 50	−50	140	11 00	−50	140	10 10	−50	141	09 20	−50	141	315
44	13 10	50	140	12 20	51	140	11 29	50	141	10 39	51	141	09 48	50	142	316
43	13 39	50	141	12 49	51	141	11 58	51	142	11 07	50	142	10 17	51	143	317
42	14 08	51	141	13 17	51	142	12 26	51	142	11 35	51	143	10 44	51	143	318
41	14 37	51	142	13 46	52	143	12 54	51	143	12 03	52	144	11 11	51	144	319
40	15 05	−52	143	14 13	−51	143	13 22	−52	144	12 30	−52	145	11 38	−52	145	320
39	15 32	52	144	14 40	52	144	13 48	52	145	12 56	52	145	12 04	52	146	321
38	15 59	52	145	15 07	52	145	14 15	53	146	13 22	52	146	12 30	53	147	322
37	16 26	53	145	15 33	53	146	14 40	52	146	13 48	53	147	12 55	53	147	323
36	16 52	53	146	15 59	53	147	15 06	54	147	14 12	53	148	13 19	53	148	324
35	17 17	−53	147	16 24	−54	148	15 30	−53	148	14 37	−54	148	13 43	−53	149	325
34	17 42	54	148	16 48	54	148	15 54	53	149	15 01	54	149	14 07	54	150	326
33	18 06	54	149	17 12	54	149	16 18	54	150	15 24	54	150	14 30	54	151	327
32	18 29	54	150	17 35	54	150	16 41	55	151	15 46	54	151	14 52	54	151	328
31	18 52	54	150	17 58	55	151	17 03	55	151	16 08	54	152	15 14	55	152	329
30	19 15	−55	151	18 20	−55	152	17 25	−55	152	16 30	−55	153	15 35	−55	153	330
29	19 36	55	152	18 41	55	153	17 46	55	153	16 51	56	153	15 55	55	154	331
28	19 58	56	153	19 02	55	154	18 07	56	154	17 11	56	154	16 15	55	155	332
27	20 18	56	154	19 22	55	154	18 27	56	155	17 31	56	155	16 35	56	156	333
26	20 38	56	155	19 42	56	155	18 46	56	156	17 50	56	156	16 54	57	156	334
25	20 57	−56	156	20 01	−57	156	19 04	−56	157	18 08	−56	157	17 12	−57	157	335
24	21 16	57	157	20 19	57	157	19 22	56	157	18 26	57	158	17 29	57	158	336
23	21 33	56	158	20 37	57	158	19 40	57	158	18 43	57	159	17 46	57	159	337
22	21 51	57	159	20 54	58	159	19 56	57	159	18 59	57	160	18 02	57	160	338
21	22 07	57	160	21 10	58	160	20 12	57	160	19 15	57	160	18 18	58	161	339
20	22 23	−58	160	21 25	−57	161	20 28	−58	161	19 30	−58	161	18 32	−57	162	340
19	22 38	58	161	21 40	58	162	20 42	58	162	19 44	57	162	18 47	58	163	341
18	22 52	58	162	21 54	58	163	20 56	58	163	19 58	58	163	19 00	58	163	342
17	23 06	58	163	22 08	59	164	21 09	58	164	20 11	58	164	19 13	58	164	343
16	23 19	59	164	22 20	58	165	21 22	59	165	20 23	58	165	19 25	59	165	344
15	23 31	−59	165	22 32	−58	165	21 34	−59	166	20 35	−59	166	19 36	−58	166	345
14	23 42	59	166	22 43	58	166	21 45	59	167	20 46	59	167	19 47	59	167	346
13	23 53	59	167	22 54	59	167	21 55	59	168	20 56	59	168	19 57	59	168	347
12	24 03	59	168	23 04	59	168	22 05	60	169	21 05	59	169	20 06	59	169	348
11	24 12	59	169	23 13	60	169	22 13	59	169	21 14	59	170	20 15	59	170	349
10	24 20	−59	170	23 21	−60	170	22 21	−59	170	21 22	−59	171	20 23	−60	171	350
9	24 28	60	171	23 28	59	171	22 29	60	171	21 29	59	172	20 30	60	172	351
8	24 34	59	172	23 35	60	172	22 35	59	172	21 36	60	172	20 36	60	173	352
7	24 40	59	173	23 41	60	173	22 41	60	173	21 41	59	173	20 42	60	174	353
6	24 46	60	174	23 46	60	174	22 46	60	174	21 46	59	174	20 47	60	174	354
5	24 50	−60	175	23 50	−60	175	22 50	−59	175	21 51	−60	175	20 51	−60	175	355
4	24 54	60	176	23 54	60	176	22 54	60	176	21 54	60	176	20 54	60	176	356
3	24 56	59	177	23 57	60	177	22 57	60	177	21 57	60	177	20 57	60	177	357
2	24 58	60	178	23 58	60	178	22 58	59	178	21 59	60	178	20 59	60	178	358
1	25 00	60	179	24 00	60	179	23 00	60	179	22 00	60	179	21 00	60	179	359
0	25 00	−60	180	24 00	−60	180	23 00	−60	180	22 00	−60	180	21 00	−60	180	360

S. Lat. {LHA greater than 180°....... Zn=180−Z; LHA less than 180°........... Zn=180+Z}

TABLE 5.—Correction to Tabulated Altitude for Minutes of Declination

d ′	1	2	3	4	5	6	7	8	9	10	11	12	13	14	15	16	17	18	19	20	21	22	23	24	25	26	27	28	29	30	31	32	33	34	35	36	37	38	39	40	41	42	43	44	45	46	47	48	49	50	51	52	53	54	55	56	57	58	59	60	d ′
0	0	0	0	0	0	0	0	0	0	0	0	0	0	0	0	0	0	0	0	0	0	0	0	0	0	0	0	0	0	0	0	0	0	0	0	0	0	0	0	0	0	0	0	0	0	0	0	0	0	0	0	0	0	0	0	0	0	0	0	0	0
1	0	0	0	0	0	0	0	0	0	0	0	0	0	0	0	0	0	0	0	0	0	0	0	0	0	0	0	0	0	0	1	1	1	1	1	1	1	1	1	1	1	1	1	1	1	1	1	1	1	1	1	1	1	1	1	1	1	1	1	1	1
2	0	0	0	0	0	0	0	0	0	0	0	0	0	0	0	1	1	1	1	1	1	1	1	1	1	1	1	1	1	1	1	1	1	1	1	1	1	1	1	1	1	1	1	1	2	2	2	2	2	2	2	2	2	2	2	2	2	2	2	2	2
3	0	0	0	0	0	0	0	0	0	0	1	1	1	1	1	1	1	1	1	1	1	1	1	1	1	1	1	1	1	2	2	2	2	2	2	2	2	2	2	2	2	2	2	2	2	2	2	2	2	2	3	3	3	3	3	3	3	3	3	3	3
4	0	0	0	0	0	0	0	1	1	1	1	1	1	1	1	1	1	1	1	1	1	1	2	2	2	2	2	2	2	2	2	2	2	2	2	2	2	3	3	3	3	3	3	3	3	3	3	3	3	3	3	3	4	4	4	4	4	4	4	4	4
5	0	0	0	0	0	0	1	1	1	1	1	1	1	1	1	1	1	2	2	2	2	2	2	2	2	2	2	2	2	2	3	3	3	3	3	3	3	3	3	3	3	4	4	4	4	4	4	4	4	4	4	4	4	4	5	5	5	5	5	5	5
6	0	0	0	0	0	1	1	1	1	1	1	1	1	1	2	2	2	2	2	2	2	2	2	2	2	3	3	3	3	3	3	3	3	3	4	4	4	4	4	4	4	4	4	4	4	5	5	5	5	5	5	5	5	5	6	6	6	6	6	6	6
7	0	0	0	0	1	1	1	1	1	1	1	1	2	2	2	2	2	2	2	2	2	3	3	3	3	3	3	3	3	4	4	4	4	4	4	4	4	4	5	5	5	5	5	5	5	5	5	6	6	6	6	6	6	6	6	7	7	7	7	7	7
8	0	0	0	1	1	1	1	1	1	1	1	2	2	2	2	2	2	2	3	3	3	3	3	3	3	3	4	4	4	4	4	4	4	5	5	5	5	5	5	5	5	6	6	6	6	6	6	6	7	7	7	7	7	7	7	7	8	8	8	8	8
9	0	0	0	1	1	1	1	1	1	2	2	2	2	2	2	2	3	3	3	3	3	3	3	4	4	4	4	4	4	4	5	5	5	5	5	5	6	6	6	6	6	6	6	7	7	7	7	7	7	8	8	8	8	8	8	8	9	9	9	9	9
10	0	0	0	1	1	1	1	1	2	2	2	2	2	2	2	3	3	3	3	3	4	4	4	4	4	4	4	5	5	5	5	5	6	6	6	6	6	6	6	7	7	7	7	7	8	8	8	8	8	8	8	9	9	9	9	9	10	10	10	10	10
11	0	0	1	1	1	1	1	1	2	2	2	2	2	3	3	3	3	3	3	4	4	4	4	4	5	5	5	5	5	6	6	6	6	6	6	7	7	7	7	7	8	8	8	8	8	8	9	9	9	9	9	10	10	10	10	10	10	11	11	11	11
12	0	0	1	1	1	1	1	2	2	2	2	2	3	3	3	3	3	4	4	4	4	4	5	5	5	5	5	6	6	6	6	6	7	7	7	7	7	8	8	8	8	8	9	9	9	9	9	10	10	10	10	10	11	11	11	11	11	12	12	12	12
13	0	0	1	1	1	1	2	2	2	2	2	3	3	3	3	3	4	4	4	4	5	5	5	5	5	6	6	6	6	6	7	7	7	7	8	8	8	8	8	9	9	9	9	10	10	10	10	10	11	11	11	11	11	12	12	12	12	13	13	13	13
14	0	0	1	1	1	1	2	2	2	2	3	3	3	3	4	4	4	4	4	5	5	5	5	6	6	6	6	7	7	7	7	7	8	8	8	8	9	9	9	9	10	10	10	10	10	11	11	11	11	12	12	12	12	13	13	13	13	14	14	14	14
15	0	0	1	1	1	2	2	2	2	2	3	3	3	4	4	4	4	4	5	5	5	6	6	6	6	6	7	7	7	8	8	8	8	8	9	9	9	10	10	10	10	10	11	11	11	12	12	12	12	12	13	13	13	14	14	14	14	14	15	15	15
16	0	1	1	1	1	2	2	2	2	3	3	3	3	4	4	4	5	5	5	5	6	6	6	6	7	7	7	7	8	8	8	9	9	9	9	10	10	10	10	11	11	11	11	12	12	12	13	13	13	13	14	14	14	14	15	15	15	15	16	16	16
17	0	1	1	1	1	2	2	2	3	3	3	3	4	4	4	5	5	5	5	6	6	6	7	7	7	7	8	8	8	8	9	9	9	10	10	10	10	11	11	11	12	12	12	12	13	13	13	14	14	14	14	15	15	15	16	16	16	16	17	17	17
18	0	1	1	1	2	2	2	2	3	3	3	4	4	4	4	5	5	5	6	6	6	7	7	7	8	8	8	8	9	9	9	10	10	10	10	11	11	11	12	12	12	13	13	13	14	14	14	14	15	15	15	16	16	16	16	17	17	17	18	18	18
19	0	1	1	1	2	2	2	3	3	3	3	4	4	4	5	5	5	6	6	6	7	7	7	8	8	8	9	9	9	10	10	10	10	11	11	11	12	12	12	13	13	13	14	14	14	15	15	15	16	16	16	16	17	17	17	18	18	18	19	19	19
20	0	1	1	1	2	2	2	3	3	3	4	4	4	5	5	5	6	6	6	7	7	7	8	8	8	9	9	9	10	10	10	11	11	11	12	12	12	13	13	13	14	14	14	15	15	15	16	16	16	17	17	17	18	18	18	19	19	19	20	20	20
21	0	1	1	1	2	2	2	3	3	4	4	4	5	5	5	6	6	6	7	7	7	8	8	8	9	9	9	10	10	10	11	11	12	12	12	13	13	13	14	14	14	15	15	15	16	16	16	17	17	18	18	18	19	19	19	20	20	20	21	21	21
22	0	1	1	1	2	2	3	3	3	4	4	4	5	5	6	6	6	7	7	7	8	8	8	9	9	10	10	10	11	11	11	12	12	12	13	13	14	14	14	15	15	15	16	16	16	17	17	18	18	18	19	19	19	20	20	21	21	21	22	22	22
23	0	1	1	2	2	2	3	3	3	4	4	5	5	5	6	6	7	7	7	8	8	8	9	9	10	10	10	11	11	12	12	12	13	13	13	14	14	15	15	15	16	16	16	17	17	18	18	18	19	19	20	20	20	21	21	21	22	22	23	23	23
24	0	1	1	2	2	2	3	3	4	4	4	5	5	6	6	6	7	7	8	8	8	9	9	10	10	10	11	11	12	12	12	13	13	14	14	14	15	15	16	16	16	17	17	18	18	18	19	19	20	20	20	21	21	22	22	22	23	23	24	24	24
25	0	1	1	2	2	2	3	3	4	4	5	5	5	6	6	7	7	8	8	8	9	9	10	10	10	11	11	12	12	12	13	13	14	14	15	15	15	16	16	17	17	18	18	18	19	19	20	20	20	21	21	22	22	22	23	23	24	24	25	25	25
26	0	1	1	2	2	3	3	3	4	4	5	5	6	6	6	7	7	8	8	9	9	10	10	10	11	11	12	12	13	13	13	14	14	15	15	16	16	16	17	17	18	18	19	19	20	20	20	21	21	22	22	23	23	23	24	24	25	25	26	26	26
27	0	1	1	2	2	3	3	4	4	4	5	5	6	6	7	7	8	8	9	9	9	10	10	11	11	12	12	13	13	14	14	14	15	15	16	16	17	17	18	18	18	19	19	20	20	21	21	22	22	22	23	23	24	24	25	25	26	26	27	27	27
28	0	1	1	2	2	3	3	4	4	5	5	6	6	7	7	7	8	8	9	9	10	10	11	11	12	12	13	13	14	14	14	15	15	16	16	17	17	18	18	19	19	20	20	21	21	21	22	22	23	23	24	24	25	25	26	26	27	27	28	28	28
29	0	1	1	2	2	3	3	4	4	5	5	6	6	7	7	8	8	9	9	10	10	11	11	12	12	13	13	14	14	14	15	15	16	16	17	17	18	18	19	19	20	20	21	21	22	22	23	23	24	24	25	25	26	26	27	27	28	28	29	29	29
30	0	1	2	2	2	3	4	4	4	5	6	6	6	7	8	8	8	9	10	10	10	11	12	12	12	13	14	14	14	15	16	16	16	17	18	18	18	19	20	20	20	21	22	22	22	23	24	24	24	25	26	26	26	27	28	28	28	29	30	30	30
31	1	1	2	2	3	3	4	4	5	5	6	6	7	7	8	8	9	9	10	10	11	11	12	12	13	13	14	14	15	16	16	17	17	18	18	19	19	20	20	21	21	22	22	23	23	24	24	25	25	26	26	27	27	28	28	29	29	30	30	31	31
32	1	1	2	2	3	3	4	4	5	5	6	6	7	7	8	9	9	10	10	11	11	12	12	13	13	14	14	15	15	16	17	17	18	18	19	19	20	20	21	21	22	22	23	23	24	25	25	26	26	27	27	28	28	29	29	30	30	31	31	32	32
33	1	1	2	2	3	3	4	4	5	6	6	7	7	8	8	9	9	10	10	11	12	12	13	13	14	14	15	15	16	16	17	18	18	19	19	20	20	21	21	22	23	23	24	24	25	25	26	26	27	28	28	29	29	30	30	31	31	32	32	33	33
34	1	1	2	2	3	3	4	5	5	6	6	7	7	8	8	9	10	10	11	11	12	12	13	14	14	15	15	16	16	17	18	18	19	19	20	20	21	22	22	23	23	24	24	25	26	26	27	27	28	28	29	29	30	31	31	32	32	33	33	34	34
35	1	1	2	2	3	4	4	5	5	6	6	7	8	8	9	9	10	10	11	12	12	13	13	14	15	15	16	16	17	18	18	19	19	20	20	21	22	22	23	23	24	24	25	26	26	27	27	28	29	29	30	30	31	32	32	33	33	34	34	35	35
36	1	1	2	2	3	4	4	5	5	6	7	7	8	8	9	10	10	11	11	12	13	13	14	14	15	16	16	17	17	18	19	19	20	20	21	22	22	23	23	24	25	25	26	26	27	28	28	29	29	30	31	31	32	32	33	34	34	35	35	36	36
37	1	1	2	2	3	4	4	5	6	6	7	7	8	9	9	10	10	11	12	12	13	14	14	15	15	16	17	17	18	18	19	20	20	21	22	22	23	23	24	25	25	26	27	27	28	28	29	30	30	31	31	32	33	33	34	35	35	36	36	37	37
38	1	1	2	3	3	4	4	5	6	6	7	8	8	9	10	10	11	11	12	13	13	14	15	15	16	16	17	18	18	19	20	20	21	22	22	23	23	24	25	25	26	27	27	28	28	29	30	30	31	32	32	33	34	34	35	35	36	37	37	38	38
39	1	1	2	3	3	4	5	5	6	6	7	8	8	9	10	10	11	12	12	13	14	14	15	16	16	17	18	18	19	20	20	21	21	22	23	23	24	25	25	26	27	27	28	29	29	30	31	31	32	32	33	34	34	35	36	36	37	38	38	39	39
40	1	1	2	3	3	4	5	5	6	7	7	8	9	9	10	11	11	12	13	13	14	15	15	16	17	17	18	19	19	20	21	21	22	23	23	24	25	25	26	27	27	28	29	29	30	31	31	32	33	33	34	35	35	36	37	37	38	39	39	40	40
41	1	1	2	3	3	4	5	5	6	7	8	8	9	10	10	11	12	12	13	14	14	15	16	16	17	18	18	19	20	20	21	22	23	23	24	25	25	26	27	27	28	29	29	30	31	31	32	33	33	34	35	36	36	37	38	38	39	40	40	41	41
42	1	1	2	3	4	4	5	6	6	7	8	8	9	10	10	11	12	13	13	14	15	15	16	17	18	18	19	20	20	21	22	22	23	24	24	25	26	27	27	28	29	29	30	31	32	32	33	34	34	35	36	36	37	38	38	39	40	41	41	42	42
43	1	1	2	3	4	4	5	6	6	7	8	9	9	10	11	11	12	13	14	14	15	16	16	17	18	19	19	20	21	22	22	23	24	24	25	26	27	27	28	29	29	30	31	32	32	33	34	34	35	36	37	37	38	39	39	40	41	42	42	43	43
44	1	1	2	3	4	4	5	6	7	7	8	9	10	10	11	12	12	13	14	15	15	16	17	18	18	19	20	21	21	22	23	23	24	25	26	26	27	28	29	29	30	31	32	32	33	34	34	35	36	37	37	38	39	40	40	41	42	43	43	44	44
45	1	2	2	3	4	4	5	6	7	8	8	9	10	10	11	12	13	14	14	15	16	16	17	18	19	20	20	21	22	22	23	24	25	26	26	27	28	28	29	30	31	32	32	33	34	34	35	36	37	38	38	39	40	40	41	42	43	44	44	45	45
46	1	2	2	3	4	5	5	6	7	8	8	9	10	11	12	12	13	14	15	15	16	17	18	18	19	20	21	21	22	23	24	25	25	26	27	28	28	29	30	31	31	32	33	34	34	35	36	37	38	38	39	40	41	41	42	43	44	44	45	46	46
47	1	2	2	3	4	5	5	6	7	8	9	9	10	11	12	13	13	14	15	16	16	17	18	19	20	20	21	22	23	24	24	25	26	27	27	28	29	30	31	31	32	33	34	34	35	36	37	38	38	39	40	41	42	42	43	44	45	45	46	47	47
48	1	2	2	3	4	5	6	6	7	8	9	10	10	11	12	13	14	14	15	16	17	18	18	19	20	21	22	22	23	24	25	26	26	27	28	29	30	30	31	32	33	34	34	35	36	37	38	38	39	40	41	42	42	43	44	45	46	46	47	48	48
49	1	2	2	3	4	5	6	7	7	8	9	10	11	11	12	13	14	15	16	16	17	18	19	20	20	21	22	23	24	24	25	26	27	28	29	29	30	31	32	33	33	34	35	36	37	38	38	39	40	41	42	42	43	44	45	46	47	47	48	49	49
50	1	2	2	3	4	5	6	7	8	8	9	10	11	12	12	13	14	15	16	17	18	18	19	20	21	22	22	23	24	25	26	27	28	28	29	30	31	32	32	33	34	35	36	37	38	38	39	40	41	42	42	43	44	45	46	47	48	48	49	50	50
51	1	2	3	3	4	5	6	7	8	8	9	10	11	12	13	14	14	15	16	17	18	19	20	20	21	22	23	24	25	26	26	27	28	29	30	31	31	32	33	34	35	36	37	37	38	39	40	41	42	42	43	44	45	46	47	48	48	49	50	51	51
52	1	2	3	3	4	5	6	7	8	9	10	10	11	12	13	14	15	16	16	17	18	19	20	21	22	23	23	24	25	26	27	28	29	29	30	31	32	33	34	35	36	36	37	38	39	40	41	42	42	43	44	45	46	47	48	49	49	50	51	52	52
53	1	2	3	4	4	5	6	7	8	9	10	11	11	12	13	14	15	16	17	18	19	19	20	21	22	23	24	25	26	26	27	28	29	30	31	32	33	34	34	35	36	37	38	39	40	41	42	42	43	44	45	46	47	48	49	49	50	51	52	53	53
54	1	2	3	4	4	5	6	7	8	9	10	11	12	13	14	14	15	16	17	18	19	20	21	22	22	23	24	25	26	27	28	29	30	31	32	32	33	34	35	36	37	38	39	40	40	41	42	43	44	45	46	47	48	49	50	50	51	52	53	54	54
55	1	2	3	4	5	6	6	7	8	9	10	11	12	13	14	15	16	16	17	18	19	20	21	22	23	24	25	26	27	28	28	29	30	31	32	33	34	35	36	37	38	38	39	40	41	42	43	44	45	46	47	48	49	50	50	51	52	53	54	55	55
56	1	2	3	4	5	6	7	7	8	9	10	11	12	13	14	15	16	17	18	19	20	21	21	22	23	24	25	26	27	28	29	30	31	32	33	34	35	35	36	37	38	39	40	41	42	43	44	45	46	47	48	49	49	50	51	52	53	54	55	56	56
57	1	2	3	4	5	6	7	8	9	10	10	11	12	13	14	15	16	17	18	19	20	21	22	23	24	25	26	27	28	28	29	30	31	32	33	34	35	36	37	38	39	40	41	42	43	44	45	46	47	48	48	49	50	51	52	53	54	55	56	57	57
58	1	2	3	4	5	6	7	8	9	10	11	12	13	14	14	15	16	17	18	19	20	21	22	23	24	25	26	27	28	29	30	31	32	33	34	35	36	37	38	39	40	41	42	43	44	44	45	46	47	48	49	50	51	52	53	54	55	56	57	58	58
59	1	2	3	4	5	6	7	8	9	10	11	12	13	14	15	16	17	18	19	20	21	22	23	24	25	26	27	28	29	30	30	31	32	33	34	35	36	37	38	39	40	41	42	43	44	45	46	47	48	49	50	51	52	53	54	55	56	57	58	59	59

Appendix F

Sight-Reduction Form

Sight Reduction by H.O. 249 or H.O. 229

Course ______ Date ____________________ Assumed Lat. °N S

Speed ______ DR pos. ____________________________ at _______ GMT

LMT GMT

Civil twil.

Height of eye ___ ft. LHAγ ____° at ______ LMT

___ m.

Body						
GMT						
Hs						
I.C (+or-)						
Dip (-)						
R LL UL						
Ho						
GHA (hr.)/v						
SHA ★						
GHA (m & s)						
v corr (moon,plan.)						
GHA (total)						
Long. -W +E						
LHA						
Dec. /d						
d corr						
Dec. (tot.)						
	↕	↕	↕	↕	↕	↕
Alt. /d						
d corr.						
Hc						
Ho						
Intercept						
Z						
Zn						

Advance AP (mi.) ______ ______ ______ ______ ______ toward ___°

H.O. 249 Vol. 1: Move fix ___ mi. toward ____°.